Information Systems Innovation and Diffusion: Issues and Directions

Tor J. Larsen
The Norwegian School of Management

Eugene McGuire
American University

IDEA GROUP PUBLISHING
Hershey, USA • London, UK

Senior Editor:	Mehdi Khosrowpour
Managing Editor:	Jan Travers
Copy Editor:	Terry Heffelfinger
Printed at:	BookCrafters

Published in the United States of America by
Idea Group Publishing
1331 E. Chocolate Avenue
Hershey PA 17033-1117
Tel: 717-533-8845
Fax: 717-533-8661
E-mail: jtravers@idea-group.com
Website: http://www.idea-group.com

and in the United Kingdom by
Idea Group Publishing
3 Henrietta Street
Covent Garden
London WC2E 8LU
Tel: 171-240 0856
Fax: 171-379 0609

Library of Congress Cataloging-in-Publication Data

Information systems innovation and diffusion: issues and directions/
[edited by], Tor J. Larsen, Eugene McGuire.
p. cm. -- (Series in information technology management)
Includes bibliographical references and index.
ISBN 1-878289-43-8 (hc). -- ISBN 1-878289-46-2 (soft)
1. Information technology. 2. Management information systems.
3. Diffusion of innovations. I. Larsen, Tor Jermud. II. McGuire, Eugene, 1953- .
III. Series.
T58.5.I533 1998
658.4'06--dc21 97-45778
CIP

British Cataloguing in Publication Data
A Cataloguing in Publication record for this book is available from the British Library.

IDEA GROUP PUBLISHING

Hershey, PA, USA • London, UK

Series in Information Technology Management

The surge in information technology during the latter part of the 20th century has forced organizations to meet its challenges with an increased use and management of information resources. This series takes an in-depth look at trends, current practices, and problem resolution in information technology management and offers you a first-class source for expanding the reader's knowledge in this ever-growing field.

Books in this series:

- **Information Systems Innovation and Diffusion: Issues and Directions (Tor J. Larsen and Eugene McGuire)**
- **The Virtual Workplace (Magid Igbaria and Margaret Tan)**
- **Information Systems Success Measurement (Edward Garrity and G. Lawrence Sanders)**
- **Cases on Information Technology Management in Modern Organizations (Jay Liebowitz and Mehdi Khosrowpour)**
- **Collaborative Technologies and Organizational Learning (R. Neilson)**
- **Information Systems Outsourcing Decision Making: A Managerial Approach (L.A. de Looff)**
- **Information Technology and Organizations: Challenges of New Technologies (Mehdi Khosrowpour)**
- **Management Impacts of Information Technology (Edward Szewczak)**
- **Managing Information Technology Investments with Outsourcing (Mehdi Khosrowpour)**
- **Business Process Change: Reengineering Concepts, Methods and Technologies (Varun Grover and William Kettinger)**
- **Reengineering MIS: Aligning Information Technology and Business Operations (Kevin Coleman, Jim Ettwein, Clelland Johnson, Dick Pigman and Deborah Pulak)**

For more information, or to submit a proposal for a book in this series, please contact:

Idea Group Publishing
1331 E. Chocolate Avenue
Hershey, PA 17033-1117
Tel: 1/800-345-4332 or 717/533-8845
Fax: 717/533-8661
E-mail: jtravers@idea-group.com
Website: http://www.idea-group.com

Author guidelines also available at http://www.idea-group.com

Information Systems Innovation and Diffusion: Issues and Directions

Table of Contents

Preface

It is more true than ever that information systems (IS) and information technology (IT) are prerequisites for business success. Business organizations are reported to be investing a considerable proportion of their free capital in IS/IT. However, it is also reported that as much as 50 per cent of these undertakings lead to IS/IT solutions judged as outright failures or deemed highly unsatisfactory.

As the impressive amount of published research on IS/IT innovation and diffusion has already shown, the development and use of IS/IT requires continuous innovation and diffusion processes among an organization's line employees, internal IT experts, and external support mechanisms (e.g, research centers, consultants, software houses, and vendors). Because of the alarmingly high IS/IT investment failure rate, however, one can easily assume that this published research on IS/IT innovation and diffusion is thus far fragmented, incomplete, and has not provided clear direction to effectively understanding technology innovation and diffusion issues. Of course, practitioners may also not be applying the lessons already learned from existing research, but that is another matter.

Part of the reason for this seeming disconnect between research and practice is simply that IS/IT is continuing to develop at such a rapid pace. More and more information and complex relationships are being captured, stored, processed, and transmitted to ever larger numbers of knowledge workers within and between organizations. For many industry segments, the very nature and definition of their business is undergoing continual and radical change. The way work is done, decisions are made, products are tailored, production is automated, material is distributed, and customers are satisfied is all changing on a global stage. IT stands squarely in the middle of this transformation and looks to a near future where business trends will require the use of information in ways not yet even imagined.

Readers of this book should not therefore expect to find an integrated and comprehensive approach to IT innovation and diffusion issues. The chapters in this book represent the diverse and unique views of authors who approach the technology innovation and diffusion process from a wide variety of perspectives. Each, in their respective chapter, offers a piece of the whole picture. Taken together, these chapters present the careful and insightful exploration of many variables affecting the innovation and diffusion process. As such, this book, by bringing together these chapters, makes a valuable contribution to the IT innovation and diffusion literature. The continuing challenge in this field of inquiry is to persist in exploring developing trends and to assist researchers and practitioners in better grasping future challenges. The contributions in this book help address that challenge.

The book concentrates on four broad innovation and diffusion areas. First, single aspects that may influence innovation and diffusion outcome are examined. These aspects are human oriented (e.g., power distribution) or technology related (e.g., a graphical simulator.) Second, diffusion theory as applied to the introduction and spreading of software application packages that are commonly found in organizations is scrutinized. Third, positive and negative aspects related to the use of formally organized diffusion facilitation mechanisms are explored. Fourth, integrated views and models describing innovation and diffusion processes and future potential are forwarded.

Last, but not least, we take the opportunity to thank our contributors for their insightful thoughts and their patience with us during the process of completing the book. We extend our gratitude to the people who helped us in the reviewing process. This book would not have been completed without the professional and thorough support of Jan Travers and Mehdi Koshrowpour. As is always the case, we extend our gratitude to our wives and children who continue to support and love us despite the fact that we spend so much time at work.

Tor J. Larsen
Oslo, Norway

Eugene G. McGuire
Washington, D.C., U.S.A.

Section I
Influences on the Innovation and Diffusion Process

This section contains five chapters that examine various influences on the technology innovation and diffusion process within organizations. Power distribution, accelerated and continuous change, mental workload, professional resistance, and hardware evolution are all issues that directly affect innovation and diffusion of information technology in today's organizations. These issues are developed by the authors in this section through new models and conceptual frameworks that are interdisciplinary in nature, expansive in scope, and thought provoking in application. All the chapters in this section, each in their own way, suggest that traditional approaches to understanding the innovation and diffusion process need to be expanded when addressing the use of information technology in today's organizations.

Butler and Gibbons examine the relationship between power distribution within organizations and the diffusion of technology throughout those organizations. Their central thesis is that power of all types: formal, social, resource-based, and personal, strongly influences the diffusion of technology throughout an organization. As organizational power becomes more decentralized as evidenced by the growth of distributed and participative decision making among employees, the influence from individuals and groups concerning innovative uses of information technology is likely to have more impact than the direct actions of management. As new technologies emerge and diffuse, however, they can change both the existing formal and informal power structures in the organization. Butler and Gibbons have developed an interactive model of power distribution and technology diffusion to explain the relationship between the two.

Allen develops a new conceptual framework for explaining organizational use of information systems in an environment characterized by continual, rather than occasional, innovation. The basis for his view is that organizations are responding to rapid changes in their business environment such as the globalization of markets and flexible production techniques by increasing their own internal and external rates of innovation. According to Allen's framework, higher rates of innovation create more frequent changes in work situations, and work problems. The paradigm for information systems use therefore changes from one of applying information systems to relatively stable and well-known situations to one of attempting to integrate information systems capabilities with continuously changing situations. The conceptual framework discussed in this chapter clarifies some of the implications of this paradigm shift.

Evaristo and Karahanna-Evaristo propose a model to examine the linkages among individual mental workload, innovation attributes, and the evaluation and acceptance of innovations. Specifically, they propose that new technologies which impose high levels of mental workload (or information load) may lead an individual to mental (or information) overload and result in that individual expressing higher resistance to the adoption of this technology. In their model, the level of mental workload is hypothesized to have a direct effect on the quality of the evaluation of the innovation and the subsequent adoption or rejection decision. The area of mental workload has previously been primarily researched in psychology-related fields and has only recently been addressed in the business literature. The model developed for this chapter is the first that links mental workload research to the innovation/adoption decisionmaking process.

By applying the innovation and diffusion framework to a technical process as opposed to a specific technology, Gibson explores the reluctance of information systems professionals to adopt software process improvement models. This chapter is based on the proposition that the quality of a software product is highly dependent on the quality of the process used to develop it. As such, three categories of related innovations are seen as necessary for change: new software development process models, new process modelling techniques, and the acceptance of software process maturity assessment. Gibson examines these issues and notes that a major obstacle to software process improvement is the attitude and culture of the people involved in the software development process. Strategies for increasing the rate of innovation adoption in this environment must directly address these concerns.

Warren, van Dijk, Jobing, Seeley and Marci present a case study of how a simulation software tool is being used for decision support by managers and front line employees in an organization. The authors note that historically, simulation has been a technology controlled by technicians and consultants. In this case study, however, the use of this tool has been successfully and effectively diffused to those employees who have direct knowledge of the organization's productive processes. The authors show that the use of simulation tools in organizations has followed the innovation and diffusion patterns in computer hardware from mainframes to PCs, into graphical and direct manipulation environments, and onto the desktops of a wide variety of workers who are not technicians or engineers.

CHAPTER

1

Power Distribution as a Catalyst and Consequence of Decentralized Technology Diffusion

Brian Butler
Carnegie Mellon University

Deborah E. Gibbons
Georgia State University

Power, whether formal, social, resource-based, or personal, influences the diffusion of technology among members of an organization. Technology diffusion, in turn, can alter the bases and distribution of organizational power. Yet the tendency of power structures to shape individuals' willingness and constrain their ability to participate in technology diffusion has often been overlooked in prior work, which has dealt primarily with technology distribution mandated by a central source. This chapter applies theories about power from fields such as sociology, organization studies, and social psychology to social systems in which autonomous professionals actively participate in decisions regarding new technology adoption and diffusion. The result is a socio-political model of decentralized technology diffusion.

Three fundamental principles are proposed to determine the dynamic relation between power and technology diffusion. These principles are applied separately to diffusion, adoption, and indirect interventions by individuals operating within a decentralized diffusion system. As the socio-political model is developed, the role of power distribution as catalyst and consequence of technology diffusion becomes clear. Finally, practical and theoretical implications of the model are discussed.

Technology management has been modeled as a centralized process in which a change agent discovers or creates a new technology and directs its diffusion within a particular social system (Kwon & Zmud, 1987). In this model, information systems (IS) personnel play the traditional role of change agents managing top-down diffusion of technology (Fougere, 1991; Cooper & Zmud, 1990). However, organizations have been changing in ways that can undermine or preclude centralized diffusion. Many organizations are becoming less centralized, enabling professional individuals and teams to make more decisions locally.

Information technology is also becoming less centralized. Once there were strong economic and technical reasons to develop and manage information technology centrally (Grosch, 1953; Knight, 1968), but development of microcomputers and cost-effective telecommunications has increased the viability of smaller scale or widely distributed applications (Mendelson, 1987; Kang, 1990). Improved access to a wide range of specialized and often competing suppliers of system components enables individuals or departments to develop their own information systems. These technological changes, alongside decentralization in organizations, have shifted information technology from a concentrated, centrally managed tool to one that is widely used and often locally applied.

In this environment, top-down efforts to direct and manage the adoption of new technologies are likely to be less effective. Increasing diversity of applications reduces the viability of many centrally developed systems. Distributed decision-making and multiple sources of new technologies reduce the viability of centrally mandated diffusion. Instead, new technologies are often identified and applied by multiple groups and individuals throughout the organization (McLean, Kappelman, and Thompson, 1993). As employees become more involved in decisionmaking, innovative uses of information technology are likely to spread through peer networks. In these decentralized diffusion systems, influence from individuals and groups within the organization is likely to have more impact than the direct actions of management. As a result, the study of decentralized diffusion systems becomes increasingly important for understanding technology distribution and use within modern organizations.

Decentralized diffusion has been observed in many cases, yet principles governing such diffusion remain poorly understood. Rogers (1995), in his comprehensive review of diffusion research, states that user-dominated diffusion has been largely overlooked, and its effects have been underestimated. Yet theories and research from various disciplines provide clues about factors such as power structures, which are likely to influence the diffusion of new technologies from one professional to another. By integrating existing works on power, diffusion, and technology adoption, we build a theory to explain the interactions between power and decentralized diffusion processes in organizations.

If we approach diffusion as a centralized process, power issues reduce to the differential between central change agents and potential adopters. This view recommends strategies for recruiting opinion leaders or overcoming resistance to technology that management has determined to be beneficial for the group (Leonard-Barton & Kraus, 1985). Yet decentralized diffusion involves more complex interdependencies and broader sources of relevant power. In this context, both formal and informal power structures influence dissemination of new technology because each individual's position within those structures affects personal adoption and diffusion behaviors. In turn, emergence and diffusion of new technologies can change both formal and informal power structures in an organization. To explain this process, we build an interactive model of power distribution and technology diffusion within professional settings.

Socio-Political Model of Technology Diffusion

Diffusion of technology in organizations has been attempted by various means and has been studied through many lenses. However, the results of diffusion attempts vary widely, and the reasons often elude researchers. Some find that managerial support and the ability to "sell" the innovation to workers contributes to success (e.g., Kanter, 1983). Others insist that local opinion leaders should be recruited to support acceptance of new technologies (e.g., Leonard-Barton & Kraus, 1985). Despite ongoing attempts to diffuse innovations, however, many efforts are stifled by constraints on the ability or willingness of individuals to support change (Kanter, Stein, & Jick, 1992; Joshi, 1991). Even studies that assume a centralized diffusion model often acknowledge the importance of informal influence and personal willingness to accept an innovation. By doing so, they indirectly consider the roles of informal power structures. Yet investigation of non-centralized actions tends to be secondary—undertaken primarily to inform the study of management actions. In contrast, decentralized models focus on emergent behaviors, and study of management or other central direction is secondary. Decentralized diffusion is characterized by local changes within a social system, while centralized diffusion is characterized by global action. For this reason, models of centralized diffusion frequently assume that social structures are stable and not significantly affected by diffusion of the innovation. In contrast, we argue that decentralized diffusion systems are characterized by constant change, as structures and innovation diffusion interact within the social system.

This has been observed in several studies that describe the interaction of organizational structures and technology use following small local changes (e.g. Barley, 1986; Orlikowski, 1996; DeSanctis & Poole, 1994). For example, Markus (1983) reports how individuals' power concerns played a role in the implementation of an organizational accounting system. Barley (1990) demonstrates a three way interaction among individual roles, social/organizational structure, and information technology use. In his detailed ethnographic studies, Barley highlights the link between new technology and changes in the power structure of the organization. Burkhardt and Brass (1990) also found that new technology changed power structures within a government agency. Results of these studies demonstrate links between changing technology and power, but work at the juncture of power and technology remains sparse and disconnected. However, considerable research has examined sources, uses and effects of power, and this significant theoretical literature may be applied to technology diffusion.

Development of coherent theory regarding technology diffusion in organizations requires a more complete treatment of power. Without considering this critical factor, we cannot understand the processes surrounding decentralized diffusion in organizations. We know that the tension between power to promote innovation and motivation to do so has stalled many innovation attempts (Kanter, 1983). Individuals may suppress or facilitate diffusion based on their organizational positions and potential for change related to the technology (Gibbons & Butler, 1995). Although issues of personal or group-based power have been considered as potential blocks to diffusion (Joshi, 1991), they may also become facilitators of technology diffusion (Butler & Gibbons, 1996). Finally, the impact of new information technology on power structures in organizations must be included in a complete model of decentralized diffusion within organizations. Toward that goal we will discuss organizational power structures, individuals' behavior in decentralized diffusion systems, and the consequences of information technology dissemination for subsequent power distribution. This framework is presented in Figure 1.

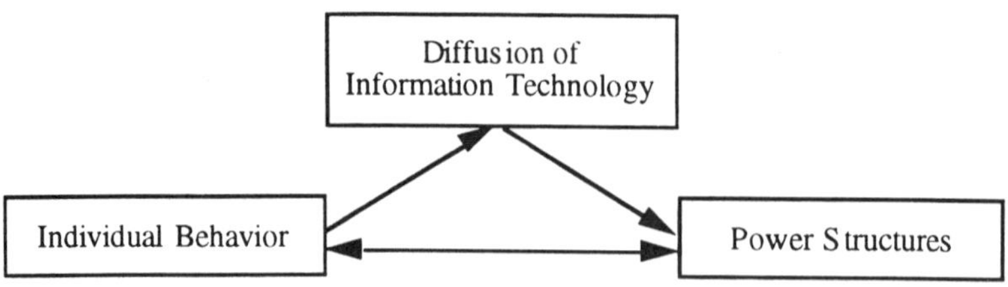

Figure 1: Socio-Political Model of Decentralized Technology Diffusion

Organizational Power Structures

Power is a very intuitive concept. Everyone is familiar with power and has witnessed its effects. At the same time, defining power in a way that is not tautological can sometimes be difficult (Pfeffer, 1981). Though there are many descriptions of organizational power, one definition that captures many of the common features describes power as "the ability to get things done the way one wants them done; it is latent ability to influence people." (Shafritz & Ott, 1992). This definition highlights two key aspects of organizational power. First, as Pfeffer (1981) states, power is tied to particular social contexts. Individuals are not powerful or powerless in general, but only relative to other parties in a specific social setting. Second, power is not concerned only with outcomes; it can also be used to control processes. Therefore, sources of power may enhance one's ability to influence individuals, processes, and outcomes in the surrounding organization.

Many potential sources of power have been identified, but we will group them within the following categories: formal, resource-based, interpersonal, and personal. By exploring each source of power we lay foundations for understanding the interaction of power structures, diffusion processes, and the dissemination of information technology within organizations.

The most visible source of power in any organization is the formal structure. Formal power, or authority, is based on an individual's position in the organizational structure that exists for purposes of coordination and control. Formal power is prearranged. Whether implicitly or explicitly, it is agreed that individuals who occupy high-ranking positions in the structure have the right, within certain bounds, to influence activities of lower-ranked participants (Mechanic, 1962). Norms and expectations regarding formal relationships are important contributors to the existence of formal power. Thus, a person's formal power arises directly from the way an organization is structured and from the individual's role in that structure. For example, a supervisor wields formal power over subordinates who, if they choose to participate in the organization, are expected to meet his requests. Similarly, a department head may possess formal power over systems developers assigned to her if it is their job to meet the needs of her department.

Formal power reflects the authority that is built into organizational roles. Therefore, formal power does not depend on the individual; rather, it depends on the position that he or she holds (Kanter, 1979). As a result, it makes no difference who occupies the position, which resources are controlled, or how the individual's talents and skills fit into the social context. Formal power is based strictly on one's position within the organization's structure. However, alternative sources of power, though distinct, are rarely independent. For example, those who fill high-ranking roles in organizations often control resources and maintain many social contacts. Yet these power sources can shift independently. As Mechanic (1962) explains, it is possible for a low-ranked individual, who has little or no formal power, to derive substantial power from other sources. Similarly, while increased access to assets and information may accompany acquisition of formal power, use of formal power does not rely on these resources. Thus, formal power can be examined separately from other sources of organizational power.

Resource-based power derives from the direct control of valuable resources (Pettigrew, 1972). Exchange theory (Blau, 1964) and resource dependence theory (Pfeffer & Salancik, 1978) both posit that control of resources is the primary source of organizational power. In models of exchange and dependence, resources generally refer to information and assets that are needed by individuals within the organizational context. Those who possess or can obtain critical resources have the potential to influence people who need or want them. Through strategic allocation or exchange of resources, one obtains power over others. Consequently, exchange and dependence may be described as foundations of power.

Resource dependence theory argues that all entities, whether organizations or individuals, seek to reduce uncertainty around their critical needs. Those who control resources that can meet such needs hold power over all who depend on them to do so (Kotter, 1977; Emerson, 1962). Scarcity of a resource or limited access to alternative sources increases dependence and hence increases the resource holder's power. Therefore, an important source of power within organizations arises from control over scarce, valuable resources that reduce critical uncertainties (Pfeffer & Salancik, 1978). As with formal power, resource-based power is relative to a particular context. Power increases with the scarcity and criticality of the resource, so an asset that brings power in one organization may be ignored in a setting where is it common or unimportant.

For example, consider a plant manager who is intimately familiar with operations at her facility. Because she is in close contact with the production process and may be involved in operations, she has a great deal of information about the current status of production. In contrast, the VP of operations, who is rarely involved in production, lacks this information which he may need to plan and evaluate the company's operations. In this case, the VP of operations is dependent on the plant manager for current information. This dependence increases the plant manager's resource-based power relative to the VP.

Now consider the plant manager's relationship with other individuals in the firm. The plant manager's knowledge about the production facilities is of little value to the corporate legal staff. In this relationship, the information that the plant manager possesses provides no power because the corporate lawyers do not value it. Neither does the plant manager's knowledge of current production increase her power with employees in her own department. In this case, the crucial issue is not value, but scarcity. Because local staff are also involved with ongoing production activities, they have access to other sources of information about the current state of production. As a result, the plant manager may possess a valuable resource, but it affords no power because the existence of multiple sources precludes the formation of dependencies. In all of these cases, the plant manager's power varies as others depend more or less on her to meet their critical needs.

A person's power increases with the number of others who depend on him (Emerson, 1962), and with the extent to which they are dependent. Critical resources may be directly allocated to dependents in order to maintain power over them, or such resources may be exchanged for other necessities (Blau, 1964).

As previously discussed, formal and resource-based power structures are related. On one hand, individuals who occupy high-ranking positions in organizations often have access to more critical resources. At the same time, the nature of managerial work is such that individuals holding high rank may depend on many others for information (Kotter, 1977; Mechanic, 1962; Scheff, 1961). Further, possession of critical resources may allow an individual to change her official position or alter organizational structures to enhance her formal power. Once a higher position in the formal hierarchy is obtained, it provides power with or without the resources that enabled the original power shift (Pfeffer, 1981). As a result, it is useful to consider resource control and formal structure as distinct, but mutually-enhancing, power bases within an organization.

A third kind of power arises directly from participation in the informal social networks within an organization. This interpersonal power differs in subtle ways from resource-based power. Models of resource-based power predict that people will form relations to obtain access to particular critical resources; these models focus on the resources. Models of interpersonal power argue that social relations enable people to exert influence or obtain resources, but these models focus on the relations (Brass, 1984).

Interpersonal power comes from one's position within a social structure. Someone who maintains positive social ties with many others possesses interpersonal power (Brass & Burkhardt, 1992); if the others are likewise well-situated in the social structure, this may further enhance his power (Bonacich, 1987). Interpersonal power provides social influence and relation-based access to other sources of power. An example of interpersonal power involves those who perform gatekeeping functions, providing a social connection between two otherwise disconnected segments of the organization. These individuals have interpersonal power because of their ability to control the flow of other people's knowledge and information within the social system (Burt, 1992). An individual who maintains contact with many different parts of an organization may act as a conduit for information flow within the organization. This ability to provide and control information results from her position within a complex set of social connections, and it serves as a basis for power.

As with formal and resource-based power, interpersonal power is structural (Brass, 1984). That is, this type of power does not depend on attributes of the person; rather, it depends on the position that she occupies in a social system. Individual attributes contribute to interpersonal power only to the degree that they support the development and maintenance of social relations. Unlike formal and resource-based power, interpersonal power does not rely on organizational authority or specific resources. Instead, it is based on general access to others. As a result, an individual may have a great deal of interpersonal power in the absence of resource-based or formal power.

Yet interpersonal power is related to formal and resource-based power. Those whose formal roles place them at the center of critical processes are more likely to build informal ties with coworkers. Those who possess more informal ties to coworkers are better able to obtain new formal roles (Granovetter, 1973). Similarly, possession of needed resources attracts new relationships that may prove useful for general purposes, and participation in work relationships facilitates procurement of new resources. Because formal, resource-based, and interpersonal power depend on the embedding social structure, interactions among these three power sources also depend heavily on the context.

It has been argued that the opportunities and constraints arising from a person's position in various social structures are the dominant forces shaping power in organizations (Brass, 1984). Yet within each social context, individuals' attributes also contribute to their personal power (French & Raven, 1968), which we define as power derived from an individual's skills, expertise, or innate characteristics. The key concept is that an attribute of the individual creates power. Personal power can result from the influence that one individual has over another who admires or identifies positively with him. French and Raven (1968) called this referent power. Personal power also results from the influence of an expert on those who value her expertise. In both of these cases, an individual possesses the source of the power. Referent power accrues as people develop cognitive bonds with someone in whom they perceive desirable characteristics. These bonds make them more susceptible to influence from the admired person. Expert power accrues as people recognize and value the expertise someone possesses. This recognition of competency makes them more likely to seek and follow her advice even when they are not necessarily dependent on her knowledge.

Referent and expert power may coexist. For example, the senior person in an engineering group may be able to influence attitudes and behavior of colleagues through a combination of referent and expert power. Some people may listen to him because they see him as a role model and aspire to be like him, in which case his referent power sways them. Others may listen to him because they believe he possesses the expertise to make good decisions. The desirable characteristics that underlie referent power and the knowledge that supports expert power belong to the individual and may be taken with him from one social context to another. Further, they need not be specific to a particular field of endeavor; attributes such as general intelligence or pleasant personality may contribute to personal power across organizational boundaries. Because personal power derives from attributes that are not context-dependent, it can be readily distinguished from formal, resource-based, and interpersonal sources of power.

Everyone in an organization is affected by all four sources of power. Formal power belongs to those whom the organization authorizes to control and coordinate actions of persons in subordinate roles. Resource-based power accrues to those who hold the assets and information that can reduce critical uncertainties for others. Interpersonal power emerges from social structures that place some people in key positions while others fall to the periphery. Finally, personal power results from individual characteristics that are noticed and valued by others. All four types of power influence decisions and behavior that can impact dissemination of technology.

Individual Behavior in Decentralized Diffusion Systems

We have discussed types of power structures within organizations and have suggested that each source of power can potentially influence

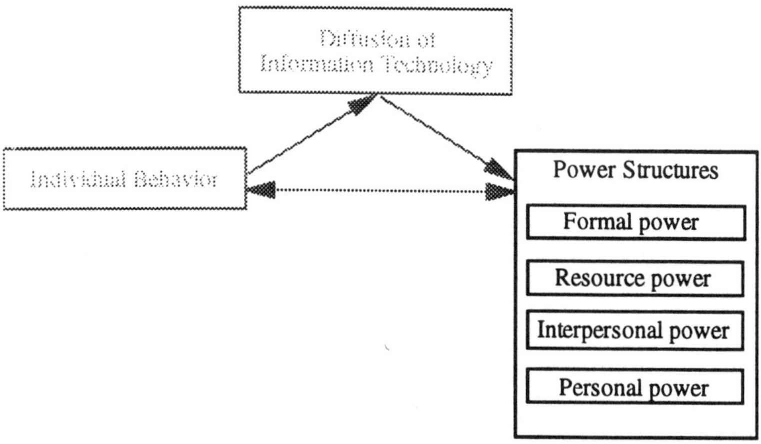

Figure 2: Power Structures in the Socio-Political Diffusion Model

individual behaviors related to dissemination of technology. Before we can understand the linkage between power and technology diffusion, then, we must understand intermediary behaviors that emerge to play significant roles in the functioning of a decentralized diffusion system.

What individual behaviors are relevant to diffusion within a decentralized system? How do they affect the final distribution of technology? By answering these questions, we begin to understand the process of diffusion. This, in turn, enables us to analyze potential interactions between power and technology diffusion in decentralized systems. The behavior categories that we will consider are innovating, adopting an innovation, and diffusing an innovation.

The act of innovating, of creating or discovering a new technique or tool, has received considerable attention in studies of diffusion. In centralized diffusion models, innovators are synonymous with the earliest adopters (Rogers, 1995). These models often assume that innovations develop outside the immediate social context, then knowledgeable gatekeepers evaluate and introduce them to the organization as stable programs or products. Thus, characterization of the earliest adopters as innovators implies a relative definition of innovation based on the identification and acceptance of technologies developed elsewhere. Despite the popularity of this approach within the diffusion literature, several studies indicate that later adopters often generate new innovations (e.g., Von Hippel, 1988). Even in centralized diffusion systems, innovations are re-invented by individual adopters (Danzinger, 1977).

The separation of innovation and adoption is even less clear in decentralized diffusion systems (Rogers, 1993) where individuals make decisions regarding use of the technology, then pass modified information to other persons. This ongoing process suggests that innovation and adoption occur simultaneously during decentralized diffusion. Such adaptive processes in technology dissemination deserve careful attention in future research. Nevertheless, we exclude innovation from further explicit discussion in this chapter because it is assumed to accompany adoption.

Adoption behavior dominates current research on decentralized diffusion. We define adoption as the initial use of a technology by an individual who had not previously used it, whether or not innovations were incorporated at the time of adoption. A great deal of work has gone into characterizing individuals who adopt innovations at different stages of diffusion. In addition to personality attributes, socio-economic status, and age, research shows that mass media exposure and personal networks (communication and relational) affect individuals' adoption behaviors (see Rogers, 1995, for review).

At the system level, the classic S-curve is used to describe cumulative adoption behavior. Social contagion, in its simplest form, is modeled using a stable probability of adoption given contact with someone who has already adopted (Rogers and Kincaid, 1981). This type of model produces the familiar S-curve in which adoption rate increases as the proportion of adopters increases, until half of the potential adopters have accepted the technology. At this point, the rate of adoption decreases until infusion of the technology throughout the system approaches its maximum density. This simple view of social contagion, however, overlooks important individual and social factors.

Slightly more complex models incorporate individual attributes that interact with social context to influence likelihood of adoption at any given point in the diffusion process. For example, some people tend to adopt an innovation if they observe a few adopters. Others require evidence of many adoptions within their social world before they will try something new (Valente, 1991, 1993). The density of adoption within the surrounding community that is required before an individual will likewise adopt is called an adoption threshold (Granovetter, 1978). Thresholds for adoption differ across persons and innovation types, and models have been developed to examine the effects of varying individual thresholds on patterns of diffusion (Valente, 1995). In general, a higher threshold for adoption across a population results in less diffusion. If thresholds are too high within a system, the innovation will not diffuse.

Other models of adoption behavior assign values to the innovation that indicate the ratio of objective versus social influence that is likely to affect adoption behavior (e.g. Bass, 1969). These models can only be formulated after diffusion has occurred, but they may then be helpful in predicting subsequent diffusion of similar innovations (Mahajan, Muller & Shrivastava, 1990). Like contagion models that involve static probabilities or individual thresholds for adoption, these models focus on adoption as a response to social information. Social context serves as a source of information that does not otherwise impact the adoption decision.

In contrast, critical mass models propose that an individual's social context can directly influence the value of the innovation to an adopter. These models describe innovations, such as telephones or electronic mail, for which the value of the innovation is a function of the number of adopters (Markus, 1990). Social context plays a dual role as an information source and an objective factor for persons evaluating the innovation. As with other models, the focus remains on adopters, not diffusers. Yet another way that social context can directly influence technology dissemination is through concerns about power and control that also affect potential diffusers.

Ironically, individual diffusion behavior is the most overlooked, yet possibly the most critical, aspect of technology diffusion. Most diffusion models assume that each adopter immediately becomes a source for further dissemination of the innovation. In models that consider personal networks, this means that all individuals who have a social relation with an adopter will be exposed to positive influence toward adoption. Likewise, models that do not explicitly consider personal networks make the tacit assumption that adopters at time *t* become information sources at time *t+1*. At best, this is an unvalidated assumption; at worst, it misrepresents many participants in any system as advertisers of innovations that they may not actually have the desire to diffuse. Individuals may promote an innovation by providing positive information, but they may also impede dissemination by withholding support or providing negative information. Therefore, a complete examination of diffusion processes must recognize the possibility that individuals will exercise conscious control over their diffusion activities. By eliminating the assumption of passive dissemination, we can explore the options available to potential diffusers alongside the options available to potential adopters. Studies of opinion leaders' roles in diffusion provide a good starting point.

Substantial research has examined the positive or negative influences of opinion leaders on adoption of innovations. An opinion leader is someone who, by virtue of his or her social position or behavior, serves as a role model or source of information for others in the social system (Lazarsfeld & Menzel, 1963). Opinion leaders' support (Rogers, 1995) or resistance (Leonard-Barton, 1985) significantly impacts the dissemination of innovations in social systems. As a result, change agents are advised to target opinion leaders first so that their adoption will facilitate dissemination among others (Leonard-Barton & Kraus, 1985). However, opinion leaders are typically characterized as passive participants whose importance in the diffusion process derives primarily from the large number of individuals who come to them for information or look to them as examples.

Similar work in the implementation literature has emphasized the importance of champions, or individuals who will support the innovation in the organization. Schön (1963) claims that "[t]he new idea either finds a champion or dies." In contrast to the more passive opinion leaders, champions are characterized as active supporters and promoters of ideas (Howell & Higgins, 1990b). These streams of research primarily document the importance of opinion leaders or champions and describe the individuals who fill these roles (e.g. Howell & Higgins, 1990a).

Research that has examined how and why opinion leaders and champions support or hinder the diffusion of innovations may provide insight into diffusion behavior in decentralized systems. Beath (1991) and Howell & Higgins (1990b) found that power concerns played a significant role in determining the support of technology champions. Rogers (1995) argues that opinion leaders carefully evaluate their behaviors to ensure that they do not lose their status by becoming too innovative. Therefore, they tend to support innovations that extend the boundaries of current practices without violating strong norms

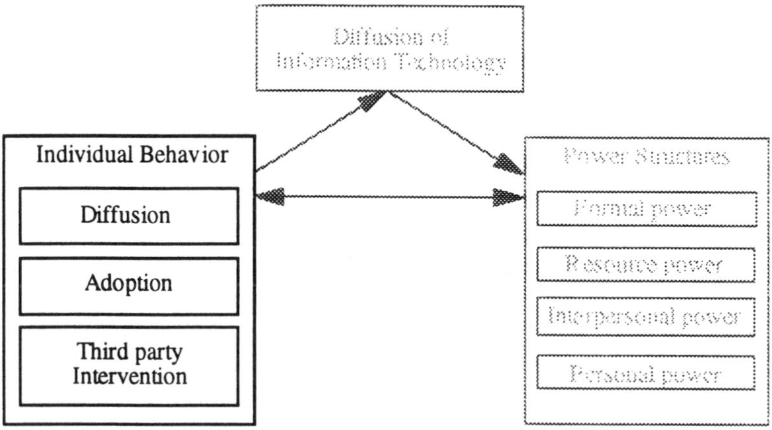

Figure 3: Individual Behaviors in the Socio-Political Diffusion Model

within the community. We will return to this in our discussion of power and diffusion behaviors.

Since most research ignores the factors driving individuals' diffusion choices, and research on opinion leaders applies primarily to centralized diffusion, possible effects in decentralized diffusion systems are unclear. Motivations for positive or negative attitudes toward diffusion remain poorly understood. Yet we know that people will act to maintain or enhance their organizational power (Salancik & Pfeffer, 1977; Mintzberg, 1983) and will do so in the context of innovation (Kanter, Stein & Jick, 1992). Therefore, the potential of technology diffusion to alter power structures should influence adoption and diffusion behaviors. It may also influence third-party interventions in the diffusion of technology among colleagues or competitors.

Linking Power, Behavior and Technology

We have discussed four general sources of power: formal, resource-based, interpersonal, and personal. We have also discussed three potential behaviors relevant to dissemination of technology: diffusion, adoption, and third-party intervention in transfers between other persons. We now combine what we know about power structures with our understanding of dissemination behaviors to build a predictive model of decentralized diffusion.

How Power Shapes Adoption and Diffusion Behaviors

Decentralized diffusion involves local innovation, evaluation and transfer from one person to another through a web of interconnected relations. In this context, individual interests and social structure combine to determine whether and to whom an innovation spreads. Possession of power from any source enhances one's ability to adopt or reject, hoard or diffuse a new technology. Lack of power, on the other hand, limits one's ability to make choices but may increase the motivation to make changes. Finally, power held by other members of the organization constrains each person's role in dissemination of technology.

As new technology diffuses, it can redistribute power within the organization. This may occur because of changes in the formal hierarchy, through redefinition of critical resources, or through new patterns of personal interaction. We begin this section by considering, in detail, the interaction of organizational power and the individual-level diffusion behaviors discussed above. We will then examine how the dissemination of information and communication technology can affect, on a large scale, the power structures within an organization.

Power plays a role in individuals' adoption and diffusion behaviors by impacting their *willingness* and *ability* to act. Specifically, each type of power will influence adoption and diffusion behaviors as follows:

Principle 1: Effects of power on willingness to adopt or diffuse new technology:

(a) If adoption is likely to increase an individual's power bases, then the individual will be more willing to adopt.

(b) If diffusion is likely to increase an individual's power bases, then the individual will be more willing to engage in diffusion behaviors.

Principle 2: Effects of power on general support for technology diffusion:

(a) Individuals will be more likely to actively support diffusion of new technologies that increase their organizational power.

(b) Individuals will be more likely to actively hinder diffusion of new technologies that decrease their organizational power.

Principle 3: Effects of power on ability to adopt or diffuse new technologies: Individuals who possess greater organizational power will be more capable of adopting, diffusing, or impacting the diffusion of new technologies than will less powerful others.

As noted above, the most visible form of power in most organizations is formal. In many information systems contexts, there is concern about changes in tasks and responsibilities that often accompany the adoption of a new information system. Jobs may be deskilled and routinized, significantly altering the roles and responsibilities of individuals who adopt the technology. In other cases, technologies provide individuals with the opportunity to restructure their tasks in ways that improve their role in the organization. For example, office support technologies, such as word processing software, typically reduce the amount of time spent on purely clerical tasks and enable office staff to take on more complex (and important) administrative and organizational jobs. For these employees, adoption of office automation and support tools provides an opportunity to increase their formal power by modifying their responsibilities within the organization.

Resource-based power may also be positively or negatively affected by adoption of technology. Information technology can be an asset (i.e., a resource), a tool for generating information resources, and a means for accessing others' sources of information. It can enhance adopters' power in at least two ways. First, adoption reduces dependence on others, thereby increasing personal autonomy. Second, new technology may allow an individual to become a source of information that others in the organization depend on, thereby increasing personal control. On the other hand, use of information systems may reduce resource-power by requiring people to relinquish privately-held information (Connolly & Thorn, 1990). In these cases, individuals' willingness to adopt the information system will be affected by the expected shift in relative power and control.

In some ways, resource-based power maintenance and technology diffusion are at crossed purposes. If the technology provides a valuable resource, individuals will act to obtain or preserve control over it. Given this situation, diffusion behavior may reduce the diffuser's resource-based power by increasing availability of the information and decreasing the scarcity of the resource. In other cases, diffusion of innovations creates new dependencies that increase resource-based power. This is a common strategy employed by manufacturers of consumer products ranging from shaving razors to cellular phones. The producer actively diffuses one part of a system, such as the cellular phone handset, because adoption of that technology creates dependence on the diffuser for complementary services, goods, or information. Within organizations, similar strategies may foster diffusion of new technologies by those whose specialized skills, data, or resources will be valued by new adopters. In general, persons whose resources become more critical or valuable in the presence of new technology are likely to support its dissemination.

Because telecommunications technology provides access to resources, relevant adoption and diffusion behaviors should be particularly affected by power considerations. For example, external communication technologies such as electronic mail and the World Wide Web (WWW) provide access to many information sources outside the organization. Those who adopt the technology strictly for personal use may be unconcerned about its diffusion within the organization. However, some adopters use information technology to support internal gatekeeping positions (Gibbons & Butler, 1995) which are sources of interpersonal power. For these people, enhancement of resource-based power encourages adoption of a technology while protection of interpersonal power discourages diffusion behavior. A gatekeeper whose position is supported by dissemination of acquired information could lose power following diffusion of communication technology through which the information is obtained. Therefore, individuals in these positions may be expected to adopt the technology but avoid propagation of the tool.

Competition to retain one's position within the surrounding social structure may also lead to adoption of innovations. Burt (1992) argues that people who hold equivalent positions within the social structures of an organization will observe, imitate, and compete with each other. This is proposed to occur because they are structurally equivalent, which means they maintain the same types of relations with the same others in their shared social setting (Burt, 1987). Because they occupy the same social position, they also play the same roles for their contacts. Perceiving their similarity, each may regard the other as a standard for imitation and a target for competition. In the context of our power discussion, this competition translates to a desire to maintain or enhance one's interpersonal power relative to the most similar and potentially threatening others. If one person adopts a new information technology, this theory predicts that structurally equivalent competitors will also adopt.

Besides the use of technology to support existing bases of interpersonal power, early adoption of new information technologies can also improve adopters' social positions (Burkhardt & Brass, 1990). Similarly, technology diffusion can serve as the basis for developing relationships or forming coalitions that enhance interpersonal power (Butler & Gibbons, 1996). In both of these studies, power increases accrued to those who acted early in the diffusion process, which suggests that timing matters when building interpersonal power through strategic technology dissemination.

Personal power also plays a role in determining willingness to adopt or diffuse technology. Expert power may be enhanced by adopting or diffusing innovations, and referent power will be affected if the use of technology creates social identification between individuals and a group. Rogers (1995) argues that the influence of opinion leaders depends on their appropriateness as role models and their similarity with others in the social context. However, the same people are often sought for advice because they have knowledge and experience with new technologies. This creates the following dilemma: if opinion leadership can be gained or lost, then opinion leaders must adopt and diffuse some innovations to support their expert power, while avoiding innovations that would excessively reduce their similarity with the larger community.

Individual *ability* to adopt new technology may also depend on power in the organization. Lacking formal authority, adopters may be unable to modify their work routines or their surroundings to obtain the benefits of new technologies. Information systems typically require maintenance, a task which takes a great deal of time and effort. High-ranking individuals can integrate support tasks into the organizational structure to ensure that their needs are met. However, lower-ranking persons may rely on other power bases to obtain needed support. For example, individuals who control important resources can implicitly exchange them for informal information systems support. As a result, they will be more able to adopt and use an innovative information technology.

Further, adoption of new information technologies in an organizational setting may depend on coordination with others. Interrelated tasks, shared resources, and common goals all contribute to the many interdependencies that complicate individuals' decisions to adopt new technologies. The need for altered technical interfaces, communication conventions, and coordination mechanisms requires implicit acceptance, though not necessarily adoption, of the new technology by others in the immediate organizational context. Formal authority, resource exchanges, and personal persuasion are all mechanisms by which powerful people can acquire cooperation.

An individual's formal position within the organization can also impact his ability to diffuse an innovative technology. Project champions who hold high rank are in good positions to support innovations by using their formal authority (Howell & Higgins, 1990b). Although mandating technology use often has unexpected negative consequences, top-management support can facilitate the diffusion of new technologies within an organization (Leonard-Barton & Deschamps, 1988).

When a technology's value depends on a critical mass of adopters, individuals' ability to support diffusion influences their willingness to adopt and continue using the innovation. People who are not powerful must rely on passive diffusion to increase the number of adopters and the overall value of the innovation. In contrast, powerful individuals may actively raise the value of the innovation by persuading or coercing others to adopt the innovation. As a result, powerful individuals should be more likely to adopt early and continue using new communication technologies within the organization.

In summary, concerns over individual power can influence both adoption and diffusion behaviors, sometimes in opposite directions. Models of rational adoption or probabilistic social contagion that neglect context overlook these critical factors in the decision process. Power concerns that have a direct influence on individual behaviors aggregate and interact across a social system. They influence coalition formation and group action (Mintzberg, 1983), with the potential for broad effects on dissemination. As a result, existing power structures within an organization significantly influence the overall diffusion of technology.

Redistribution of Power as a Consequence of Technology Diffusion

Diffusion of technology within an organization can also affect wholesale changes in power structures. Information often undergirds organizational power. To the extent that diffusion of new information technology alters access and control over information, it has the potential to redistribute power. Technology diffusion is particularly likely to influence formal, resource-based, and interpersonal power structures, but it may also have contingent effects on personal power.

Formal structure exists to coordinate tasks that are crucial for the organization's continued existence. Distribution of new information technologies may impact the organization's ability to change its formal structure. On one hand, widespread diffusion of new information and communication technologies throughout a firm may support a wider range of coordination structures (Applegate, 1995). As a result, options for restructuring increase, and dramatic reorganizations that were not previously possible may become feasible. On the other hand, adoption of particular systems within organizations can significantly increase the cost of changing organizational structures. Therefore, technology diffusion may be used to solidify existing power structures. One characteristic of decentralized diffusion systems is the ubiquitous presence of local adaptation (Rogers, 1995). For new information technology, this often takes the form of customization to fit the needs of the immediate organizational context (Danzinger, 1977). Once distributed and customized, some technologies can be difficult to standardize or otherwise modify on a large scale. Yet changes in formal structure may depend on modification of existing systems. Therefore, diffusion of locally-adapted technologies within the organization may hinder subsequent modification of formal power structures.

For example, banks have traditionally been organized around products, such as savings accounts, loans, and mortgages. Information technologies adopted by units within these organizations to support day-to-day operations were adapted to align with this product orientation. In recent years, changing regulations and increased competition have driven many banks to reorganize their formal structures to focus on categories of customers. However, the current distribution of product-oriented systems supports prior structures and creates a barrier to this type of reorganization. Thus, the diffusion of IT hinders large scale change in formal power structures.

Another consequence of increased technology use is that the technology itself becomes a critical resource. The discussion of centralized versus decentralized IT spending is at one level a technical issue of standardization and cost savings, and at another a battle to control a resource that is critical throughout the organization. Dissemination of technology may initially increase dependence on central sources of technical support. As technology use is incorporated into work processes, the demand for local control and customization increases. This fosters adoption of technologies that can be adapted and managed locally, which in turn reduces dependence on central sources of support. Even when central sources provide service for local systems, they may lack the power to control adoption and adaptations that affect their ability to provide the expected services. This lack of needed power, in turn, leads to increased efforts to create formal policies and diffuse technologies that reestablish centralized control.

Competition for control over information technologies often causes organizations to oscillate between centralized and decentralized systems. Information systems managers press for standardization which will make their jobs easier and their control over resources tighter. Department managers and end-users, in turn, demand customization and local control over the technologies on which they depend.

This issue has driven diffusion of new information technologies such as PC's and local area networks. Individuals' demands to control information technologies applicable to their work promote widespread diffusion of end-user computing in organizations. At the same time, information systems managers may attempt to retain control over the newly distributed systems. Particularly if formal responsibility does not correspond with actual control of the resources, political conflict is likely to result and intensify as individuals and departments vie for control. The cycle continues as technology management moves toward centralization, then toward decentralization, in a dance between organizational power and technological control.

Diffusion of technology can create new sources of information for some people and negate the value of prior sources used by others. Therefore, technologies that provide access to information have the potential to influence social structures built around organizational communication. Interpersonal power often depends on control of information flow. Technology that reroutes information or changes its availability to the general population alters the landscape of information control and thereby alters the structure of interpersonal power. For example, a person's interpersonal power may increase following admittance to a centralized information repository, but her power may decrease if access provided to others enables them to circumvent her in future transfers of information. Similarly, new communications technologies may dramatically alter the interpersonal structure by reducing the cost of maintaining contacts. As a result, new technologies may reduce the powerlessness of geographically or temporally isolated individuals who are otherwise unable to maintain relations with colleagues (Sproull & Kiesler, 1991). This, in turn, reduces the ability of more central players to maintain interpersonal control.

The distribution of new information technologies can also modify personal power. Widespread use of alternative communication technologies can reduce opportunities to exercise referent or expert power simply by reducing a person's relative visibility. Conversely, technologies may extend the reach of previously unnoticed individuals by connecting them with a larger proportion of the organization. These changes may have a systematic tendency to increase personal power among peripheral members of the organization and to decrease personal power among central members. Nevertheless, personal attributes limit the potential to develop personal power, so the level of change will always be idiosyncratic to the individual.

Power Distribution as Catalyst and Consequence of Decentralized Diffusion

Formal, resource-based, interpersonal, and personal power influence individual adoption and diffusion behaviors. These individual behaviors combine and interact in predictable ways to shape technology diffusion within a decentralized system. The dissemination of new technology, in turn, alters structures that support individuals' and groups' power within a social system.

The complete Socio-Political Diffusion Model, presented in Figure 4, integrates the seemingly disparate worlds of organizational power and technology diffusion. It provides a foundation for understanding processes and impacts of technology diffusion in complex organizational settings. Several practical and theoretical implications arise from understanding these dynamic links between power and dissemination of technology.

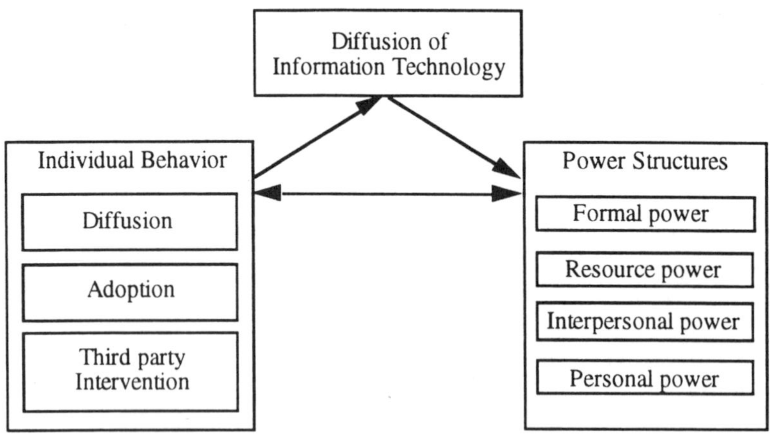

Figure 4: The Socio-Political Diffusion Model

Practical Applications and Implications for Future Research

Rogers (1995) notes that diffusion research has typically focused on successful innovations—those that diffuse to a large portion of the social setting. Relatively little empirical work has examined reasons for nondiffusion or for selective diffusion of one technology rather than another. Two related questions, seldom asked, may guide the unraveling of these mysteries. How does the organizational context influence people's decisions to adopt a new technology? And how does the same context affect people's decisions to disseminate what they have adopted? Our model provides a foundation to address these issues. Power influences the adoption (or non-adoption) of information technology by individuals and across organizations. Power concerns have the potential to serve as a hindrance, slowing or limiting the diffusion of new technologies, or to serve as a catalyst, speeding and expanding dissemination. Finally, power issues influence the decision to support or suppress the diffusion of new technology.

Recognition that active diffusion cannot be taken for granted in a decentralized diffusion system suggests rich and novel directions for future research. The Socio-Political Diffusion Model provides a theoretical foundation for understanding the relations among various kinds of power and diffusion behaviors. Further work on decision processes guiding opinion leaders' and project champions' participation in technology diffusion would be helpful. While prior work indicates that political support plays an important role in IT diffusion, questions remain about conditions under which formal versus informal sources of power dominate the process. This also provides fruitful ground for future research.

In the discussion of the model, we noted that power bases may change over time. Although diffusion studies often track the progress of system level diffusion, dynamics of power and of related diffusion behaviors have received little or no attention. However, it is reasonable to expect that individuals' power, positions, and concerns will change over time. Further, some behaviors that support a power base early in the diffusion process may become dysfunctional later. Concurrent examination of changing power structures and dissemination behaviors over time would advance research in both fields and provide insights into the broader issues surrounding technology dissemination.

To understand effects of organizational context on decentralized diffusion processes, we must understand how and when power issues impact individuals' adoption and diffusion decisions. The issues are too big to be addressed in a single study or series of studies; rather, adequate exploration of power and diffusion will require integration of work conducted in various types of organizations under more or less volatile social conditions. In addition, simulation models, such as those currently used to study network dynamics (e.g., Kaufer & Carley, 1993), could provide a better understanding of reciprocal effects between power and diffusion at the system level.

For practitioners, the Socio-Political Diffusion Model suggests answers to the question: How can we manage the diffusion of new technologies within our organization? Proactive consideration of power concerns and political contexts surrounding organizational interventions can improve the outcomes of change efforts (Pettigrew, 1975). Toward that end, we offer three general suggestions for enhancement of a diffusion effort. First, the timing of changes in formal structure and in related technology may be crucial to influence people's willingness to participate in diffusion. If new technology is introduced to a setting wherein it threatens the power bases of key players, power concerns are likely to hinder diffusion (Joshi, 1991). In contrast, if appropriate structural changes can be arranged before the diffusion effort begins, power concerns may be harnessed to support diffusion. Placement of key people in positions which will be served by diffusion may garner support that might otherwise be lost.

The second area of practical interest is ability to participate in diffusion. Individuals who delay adoption, often called laggards in the diffusion literature, may avoid a new technology for various reasons. If lack of power constrains their ability to cooperate with a diffusion effort, the reasons should be analyzed and neutralized. Without explicit examination of differential power bases among key players in an organization, power-related limitations on ability to adopt or diffuse may go unnoticed. A person whose more powerful contacts refuse to adopt may be unable to forge ahead without them. Likewise, someone who cannot obtain adequate support for the transition to a new system needs assistance or incentives to switch. Providing formal power, resources, or interpersonal contacts to individuals who lack the ability to support their own adoption can be a useful way of turning potential laggards into adopters.

Finally, change agents should pay separate attention to adoption and diffusion behaviors and the role that power issues may play in each. In contexts where motivations for adoption and diffusion differ, early targeting of likely adopters may be a mistake or a waste of resources. Instead, one should target for early adoption those who are most likely to promote diffusion of the technology. Current opinion leaders may fit this description. Yet because the diffusion of IT can influence or undermine existing power bases, individuals who are currently opinion leaders may actually be poor candidates for promoting diffusion of technology. In every case of planned technology dissemination, consideration of potential participants' power bases will smooth the dissemination process.

Conclusion

Technology diffusion and the impact of information technology in organizations are phenomena with important practical and research implications. Managers need to understand how their organizations cope with an increasingly dynamic and complex technological environment. These phenomena also provide opportunities for researchers to develop greater understanding of how complex social systems change and evolve. Similarly, organizational power and politics are familiar to all managers, and they provide researchers with a valuable theoretical lens for investigating individual behaviors. The Socio-Political Diffusion Model merges these to further the development of social theories of technology by conceptualizing power as both a catalyst and consequence of diffusion in organizations.

References

Applegate, L. (1995). *In search of a new organization: Lessons from the field.* Paper presented at the Academy of Management Annual Meeting, Vancouver, BC.

Barley, S. R. (1986). Technology as an occasion for structuring: Evidence from observations of CT scanners and the social order of radiology departments. *Administrative Science Quarterly*, 31, 78-108.

Barley, S. R. (1990). The alignment of technology and structure through roles and networks. *Administrative Science Quarterly*, 35, 61-103.

Bass, F. M. (1969). A new product growth model for consumer durables. *Management Science*, 13, 215-227.

Beath. C. M. (1991). Supporting the information technology champion. *MIS Quarterly*, 9, 115-123.

Blau, P. M. (1964). *Exchange and power in social life.* New York: John Wiley & Sons.

Bonacich, P. (1987). Power and centrality: A family of measures. *American Journal of Sociology*, 92, 1170-1182.

Brass, D. J. (1984). Being in the right place: A structural analysis of individual influence in an organization. *Administrative Science Quarterly*, 29, 518-539.

Brass, D. J. & Burkhardt, M. E. (1992). Centrality and power in organizations. In N. Nohria & R. G. Eccles (Eds.), *Networks and organizations: Structure, form, and action* (pp. 191-215). Boston: Harvard Business School Press.

Burkhardt, M. E. & Brass, D. J. (1990). Changing patterns or patterns of change: The effect of change in technology on social structure and power. *Administrative Science Quarterly*, 35, 104-127.

Burt, R. S. (1987). Social contagion and innovation: Cohesion versus structural equivalence. *American Journal of Sociology*, 92, 1287-1335.

Burt, R. (1992). *Structural holes: The social structure of competition.* Cambridge, MA: Harvard University Press.

Butler, B. S. & Gibbons, D. E. (1996). *Suppression or diffusion? Using technology as a vehicle for power and control.* Paper presented at the Eastern Academy of Management Annual Meeting.

Connolly, T. & Thorn, B. K. (1990). Discretionary databases: Theory, data, and implications. In J. Fulk & C. Steinfield (Eds.), *Organizations and communication technology*. (pp. 219-233). Newbury Park, CA: Sage.

Cooper, R. B. & Zmud, R. W. (1990). Information technology implementation research: A technological diffusion approach. *Management Science*, 36, 123-139.

Danzinger, J. N. (1977). Computers, local governments and the litany to EDP. *Public Administration Review*, 37, 28-37.

DeSanctis, G. & Poole, M. S. (1994) Capturing the complexity in advanced technology use: Adaptive structuration theory. *Organization Science*, 2, 121-147.

Emerson, R. M. (1962). Power-dependence relations. *American Sociological Review*, 27, 31-40.

Fougere, K. T. (1991). The future role of the systems analyst as a change agent. *Journal of Systems Management*, 42(11), 6-9.

French, J. R. P., Jr., & Raven, B. (1968). The bases of social power. In D. Cartwright & A. Zander (Eds.), *Group Dynamics* (3rd ed., pp. 259-269). New York: Harper & Row.

Gibbons, D. E. & Butler B. S. (1995). *Early adopters of electronic communications: Missionaries, archivists, or standard-bearers?* Paper presented at the Academy of Management Annual Meeting, Vancouver, BC.

Granovetter, M. S. (1973). The strength of weak ties. *American Journal of Sociology*, 78, 1360-1380.

Granovetter, M. S. (1978). Threshold models of collective behavior. *American Journal of Sociology*, 83, 1420-1443.

Grosch, H. A. (1953). High speed arithmetic: The digital computer as a research tool. *Journal of Optical Society of America*, 43, 306-310.

Howell, J. M. & Higgins, C. A. (1990). Champions of Technological Innovation. *Administrative Science Quarterly*, 35, 317-341.

Howell, J. M. & Higgins, C. A. (1990). Champions of change: Identifying and supporting champions of technological innovations. *Organizational Dynamics*, Summer, 40-57.

Joshi, K. (1991). A Model of users' perspective on change: The case of information systems technology implementation. *MIS Quarterly*, 9, 229-242.

Kang, Y. M. (1990). Computer hardware performance: Production and cost function analysis. *Communications of the ACM*, 32, 586-591.

Kanter, R. M. (1979). Power failure in management circuits. *Harvard Business Review*, 57, 65-75.

Kanter, R. M. (1983). *The change masters: Innovation for productivity in the American corporation*. New York: Simon & Schuster.

Kanter, R. M., Stein, B. A., & Jick, T. D. (1992). *The challenge of organizational change: How companies experience it and leaders guide it*. New York: The Free Press.

Kaufer, D. & Carley, K. (1993). *Communication at a Distance: Influence of Print on Socio-Cultural Organization and Change*. Hillsdale, N.J.: Laurence Erlbaum.

Knight, K. E. (1968). Changes in computer performance 1963-1967, *Datamation*, 14(1), 31-35.

Kotter, J. P. (1977). Power, dependence, and effective management. *Harvard Business Review*, 55, 125-136.

Kotter, J. P. (1985). *Power and influence: Beyond formal authority*. New

York: Free Press.

Kwon, T. H. & Zmud, R. W. (1987). Unifying fragmented models of information systems implementation. In R. Boland & R. Hirschheim (Eds.), *Critical Issues in Information Systems Research* (pp. 227-252). New York: John Wiley.

Lazarsfeld, P. F. & Menzel, H. (1963). Mass media and personal influence. In W. Schramm (Ed.), *The science of human communication.* New York: Basic Books.

Leonard-Barton, D. (1985). Experts as negative opinion leaders in the diffusion of a technological innovation. *Journal of Consumer Research,* 11, 914-926.

Leonard-Barton, D. & Deschamps, I. (1988). Managerial influences in the implementation of a new technology. *Management Science,* 34, 1252-1265.

Leonard-Barton, D. & Kraus, W. A. (1985). Implementing new technology. *Harvard Business Review,* November-December, 102-110.

Markus, M. L. (1983). Power, politics, and MIS implementation. *Communications of the ACM,* 26, 430-444.

Markus, M. L. (1990). Toward a "critical mass" theory of interactive media. In J. Fulk & C. Steinfield (Eds.), *Organizations and communication technology.* (pp. 194-218). Newbury Park, CA: Sage.

Mahajan, V., Muller, E. & Shrivastava, S. (1990). Determination of adopter categories by using innovation diffusion models. *Journal of Marketing Research,* 27, 37-50.

McLean, E. R., Kappelman, L. A., & Thompson, J. P. (1993). Converging end-user and corporate computing. *Communications of the ACM,* 36(12), 79-92.

Mechanic, D. (1962). Sources of power of lower participants in organizations. *Administrative Science Quarterly,* 7, 349-364.

Mendelson, H. (1987). Economies of scale in computing: Grosch's law revisited. *Communications of the ACM,* 30, 1066-1072.

Mintzberg, H. (1983). *Power in and around organizations.* Englewood Cliffs, NJ: Prentice-Hall.

Orlikowski, W. J. (1996). Improvising organizational transformation over time: A situated change perspective. *Information Systems Research,* 7(1).

Pettigrew, A. M. (1972). Information control as a power resource. *Sociology,* 6, 187-204.

Pettigrew, A. M. (1975). Towards a political theory of organizational intervention. *Human Relations,* 28, 191-208.

Pfeffer, J. (1981). Power in organizations. Marshfield, MA: Pitman.

Pfeffer, J. & Salancik, G. R. (1978). *The external control of organizations: A resource dependence perspective.* New York: Harper & Row.

Rogers, E. M. (1993). Diffusion and the re-invention of project D.A.R.E. In T. Backer & E. Rogers (Eds.), *Organizational aspects of health communications campaigns: What works?* Newbury Park, CA: Sage.

Rogers. E. M. (1995). *Diffusion of Innovations.* New York: Free Press.

Rogers, E. M. & D. L. Kincaid (1981). *Communication networks: Towards a new paradigm for research.* New York: Free Press.

Salancik, G. R. & Pfeffer, J. (1977). Who gets power – and how they hold on to it: a strategic contingency model of power. *Organizational Dynamics,* 5(3), 3-21.

Scheff, T. J. (1961). Control over policy by attendants in a mental hospital. *Journal of Health and Human Behavior,* 2, 93-105.

Schön, D. A. (1963). Champions for radical new innovations. *Harvard Business Review*, 41, 77-86.

Shafritz, J. M. & Ott J. S., (1992). Power and politics in organization theory. In J. M. Shafritz & J. S. Ott (Eds.), *Classics of organization theory* (3rd ed). (pp. 397-403). Belmont, CA: Wadsworth Publishing.

Sproull, L. & Kiesler, S. B. (1991). *Connections: New ways of working in the networked organization.* Cambridge, MA: MIT Press.

Valente, T. W. (1991). *Thresholds and the critical mass in mathematical models of the diffusion of innovation.* Ph.D. Thesis, Los Angeles, University of Southern California.

Valente, T. W. (1993). Diffusion of innovation and policy decision-making. *Journal of Communication*, 43, 30-41.

Valente, T. W. (1995). *Network models of the diffusion of innovations.* Creskill, NJ, Hampton Press.

Von Hippel, E. (1988). *The sources of innovation.* New York, Oxford University Press.

CHAPTER
2

Information Systems Use in Continuously Innovative Organizations

Jonathan P. Allen
University of Cambridge, UK

This chapter develops a new conceptual framework for explaining information systems use in the presence of continuous, rather than occasional, innovation. According to a diverse range of management theorists, higher rates of organizational innovation are changing work into more knowledge-intensive forms; new forms of work are being described as a process of converting some specialized, symbolically-mediated potential into action that satisfies clients. Drawing on these theories, the new framework describes information systems use as a set of activities: access, participation, framing, and action. This new framework allows us to ask empirical questions which differ from those asked by traditional information systems use theory. A review of recent collaborative technology use studies suggests that explanations which focus on continuous innovation are already having a significant impact. The chapter calls upon future studies of information systems use to make their organizational assumptions explicit, consider the rate of innovation in use settings, and to help clarify the process of converting knowledge potential into organizational action.

An important task for any theory of information systems is to explain why information systems are used (or not used) the way they are. From a practical standpoint, organizational leaders do not want their massive investments in information systems to lie unused, or to be used in ways that appear to undermine formal organizational goals. From a theoretical standpoint, the programmability and richly-combinable features of information technology, along with the cognitive flexibility of information itself, make it possible for information systems to be interpreted (Orlikowski and Robey, 1991), first appropriated for use (DeSanctis and Poole, 1994), and used routinely in a wide variety of ways. If the beginning of modern information systems history was design-heavy, with most of the attention of information systems professionals focused on hardware and software building (Friedman, 1989), it is now increasingly use-heavy, with a larger share of the effort going towards extracting the most value from information systems use.

All theories of information systems use make assumptions about the most important features of organizational life (Kling, 1980). The dominant view of information systems use in the research literature (e.g., DeLone and McLean, 1992) makes 'rational systems' assumptions (Scott, 1992) about organizations. According to this view, organizations are most usefully seen as places of rational, goal-driven action where individual decisions about systems quality, information quality, and fitness to task greatly determine how information systems are used. In contrast, a smaller set of information systems use research departs from these 'rational systems' assumptions, focusing on the political and cultural aspects of organizational life (Kling and Allen, 1996; Robey, 1995). Each perspective offers a different insight into information systems use, and encourages us to ask different empirical questions. For example, should an analyst try to explain the use of a system by measuring the quality of the information provided, or should they look instead at the political bargaining process behind a use decision?

In this chapter, the reader will see the development of a conceptual framework for explaining information systems use that critiques the dominant 'rational systems' view for its neglect of high rates of innovation in organizations, rather than its neglect of organizational politics and culture. The dominant view of information systems use tends to assume that organizations are places of relatively stable, repeated action, and that organizational changes (i.e., 'implementation' efforts) are large scale and relatively infrequent. Fixed qualities of the system (such as 'reliability') or the information (such as 'accuracy' and 'relevance') lead to greater, more effective use. A number of management theorists claim; however, that many organizations are experiencing higher rates of innovation than in the past, and that the rate of innovation is changing work in fundamental ways (e.g., Drucker, 1993; Reich, 1991; Leonard-Barton, 1995). If these assumptions are correct, then the work changes described by these theorists—emphasizing knowledge work, symbolic manipulation, experimentation, and collaboration—call into question, for a growing number of organizations, the importance of some traditional information systems use concepts (why not talk about 'experimentability'?) and the meaning of others (what does 'relevance' or 'fitness to task' mean in a world of continuously shifting requirements?).

The purpose of this chapter is to develop a new conceptual framework for explaining information systems use which assumes continuous, rather than occasional, innovation in the workplace. By drawing on new definitions of work as the conversion of a symbolically-mediated potential into action (Stehr, 1994), I describe information systems use as a set of repeated activities: access, participation, framing, and action. According to this framework, higher rates of innovation create more frequent changes in work situations, and in work problems. The process of using information systems is therefore seen less as a decision to apply a 'good' or 'bad' system to a well-known situation, and more as an attempt to fit together information systems capabilities and a changing situation. The conceptual framework clarifies some of the implications of this different view.

Mouzelis (1995) distinguishes between two types of theoretical development in the social sciences: conceptual frameworks (i.e., theory as a resource for asking questions), and substantive theory (i.e., theory as an end product). Conceptual frameworks provide theoretical tools that suggest ways of looking at the world, while substantive theory makes claims that are supported or refuted by empirical evidence. Conceptual frameworks are judged by their "utility...for generating interesting questions and for generally facilitating empirically-oriented research" (Mouzelis, 1995: 1-2), while substantive theory is assessed on the basis of available empirical evidence. The question of what makes a new conceptual framework useful, such as the framework presented in this paper, is a difficult one. By linking the framework to a growing literature on innovation and work change, showing how the framework complements existing information systems use theories, and by using the framework to generate new questions about collaborative technology research, I intend to show some of its conceptual utility. Section two describes the underlying argument that links an increasing rate of innovation in organizations to changes in work. Section three reviews the new management theories of innovative work as converting knowledge potential into action. Section four reviews existing theories of information systems use. Section five defines the new conceptual framework, showing how a focus on the activities required for innovation complements existing theories of information systems use. Finally, section six illustrates the use of the new conceptual framework by applying it to recent research in collaborative technology use.

Continuous Innovation and Changes in Work

Much has been said about how work is changing. Scholars have debated at length how new technologies (e.g., McLoughlin and Clark, 1992; Zuboff, 1988) and new forms of industrial relations between employers and employees (e.g., Wood, 1989) have changed work in the past few decades. In these literatures, changes in work are related to specific changes in technology, or to changes in the larger institutional environment.

A diverse set of management theorists argue, however, that an increasing rate of organizational innovation is changing work in fundamental ways. An organizational innovation is the adoption of an idea or behavior that is new to the organization's general environment (Daft, 1992: 254). The argument is that organizations are responding to more rapid changes in their business environment, due to forces such as globalization (Axford, 1995) and flexible production (Piore and Sabel, 1984), by increasing their own rate of innovation. The organizational innovations are both internal (changes in production technology, administrative systems, and human skills) and external (changes in product and service outputs).

High rates of organizational innovation, according to this same argument, lead to less stability in work situations at all levels of the organization, not simply at the top of the management hierarchy. The definition of the work task to be accomplished is less certain, and the process of bringing resources to bear on the work task is seen as the skillful matching of problem and tools (Starbuck, 1992; Reich, 1991), rather than the execution of a tightly specified sequence of rules. Across a diverse range of management theorizing, work itself is being described as the process of converting some specialized, symbolically-mediated potential into action that satisfies clients. While this conceptualization puts information at the center of the work process, it emphasizes that understanding the process of use—fitting information resources and a changing work situation together in a useful way—is critical (Quinn et al, 1996; Stehr, 1994).

Knowledge Potential and Conversion

Today's management theorists show an interest in both organizational innovation (e.g., Leonard-Barton, 1995; Kay and Willman, 1992), and in knowledge-intensive forms of work (e.g., Starbuck, 1992; Blackler et al, 1993; Drucker, 1993). While each concept has been studied independently, a diverse set of theorists are combining the two ideas by arguing that an increasing rate of organizational innovation is changing work into increasingly knowledge-intensive forms. This link between the rate of innovation and work change cuts across traditional management literatures.

Table 1 lists a set of theoretical arguments that link an increasing rate of organizational innovation to knowledge-intensive forms of work. While the arguments are diverse in their research questions and traditions—discussing theories of innovation, knowledge work, skills, professionalism, and organizational competence—they all share a surprisingly similar view of how to best conceptualize work in the presence of continuous innovation. Work involves *converting* some specialized, symbolically-mediated *potential* into action that satisfies clients.

First, what do these theorists have to say about the potential for action, and why is it so important? These arguments assume that workers, coping with a changing situation, draw upon an intangible resource that acts as a potential for innovative action. The intangible resource is most often conceptualized as knowledge. Having access to knowledge makes a person capable of novel, and perhaps more effective, action because it enables new ways of looking at work situations, and describes new tools and techniques for coping with changing situations (Drucker, 1989: 251). Symbolically mediated knowledge allows workers to distance themselves from their current work situation, and investigate the consequences of different courses of action (Reich, 1991: 178). The higher the rate of organizational innovation, the more important it is to have access to an intangible resource that helps people act differently in a reasoned manner.

Clearly, this concept of knowledge as an intangible resource differs from the traditional definition of knowledge as true, justified beliefs about the external world (Stehr, 1994). The importance of knowledge, for these theorists, lies in its ability to support action. In an unchanging world, 'true' and accurate representations of the external world are an important resource for determining action. In a world of changing situations and capabilities, knowledge acts as a resource for the skillful creation of new ways of understanding, and new programs of action. The benefit of defining knowledge in a more dynamic way, as a condition for the possibility of action, is that "knowledge, as a capacity for action, does not signal that specific knowledge claims always convey or carry a kind of constant and fixed 'value' enabling actors to translate and employ them for identical purposes and for closely similar outcomes" (Stehr, 1994: 96). Seeing knowledge as a potential for action allows the use process to be analyzed.

Emphasizing knowledge as a resource, rather than as true belief, makes access to knowledge a more complicated issue than physical access to knowledge representations (e.g., information). The ability to understand, and make use of, knowledge depends on a "context of specific social and intellectual conditions" (Stehr, 1994: 96). The key concept here is social. To understand what the knowledge potential

Theory	References	Potential	Conversion
Knowledge Work	(Blackler et al, 1993)	Extraordinary changes in contemporary life are making new demands on individual, organizational, and institutional knowledge (p. 857)	Sophisticated know-how is a core requirement for business success, not simply access to codified, commodifiable knowledge (p. 854)
Information-Based Organization	(Drucker, 1989) (Drucker, 1993)	Work now organized around knowledge—information that makes one capable of different or more effective action (1989, p. 251)	Main function of organizations is to make knowledge potentials productive (1993, p. 44)
Organizational Competencies	(Hirschhorn and Mokray, 1992)	Skill is a necessary, but not sufficient, concept to explain work effectiveness (p. 19)	A worker's ability to apply their potential skillfully is shaped by the roles they occupy (p. 18)
Managing Innovation	(Kay and Willman, 1992)	Innovation and adaptation requires building upon an existing knowledge base (p. 23)	Conversion of individual and organizational expertise into business specific know-how is key to effective technology innovation (p. 25)
Sources of Innovation	(Leonard-Barton, 1995)	Key innovation activities: importing knowledge and knowledge building through experimentation (p. 8-13)	Key innovation activities: shared problem solving and implementing/integrating new technologies (p. 8-13)
Managing Intellect	(Quinn et al, 1996)	Professionals distinguished by their command of a body of knowledge: know-what, know-how, know-why, and care-why (p. 72)	Converting intellect into useful products and services is the key to success in a changing world (p. 71)
Symbolic Work	(Reich, 1991)	Analysts manipulate, experiment with, and communicate abstract symbols representing a changing reality (p. 178)	Value comes from the capacity to effectively and creatively use knowledge potential (p. 182)
Knowledge-Intensive Firms	(Starbuck, 1992)	In changing, learning-dependent firms, value comes from access to exceptional, esoteric expertise (p. 716)	Expert work: creating, applying, and preserving knowledge that must be framed within current issues (p. 722)
Knowledge-Based Occupations	(Stehr, 1994)	Growing demand for novel and more specialized knowledge, defined as a capacity for action (p. 95)	In expert work, knowledge is contested, and subject to interpretation—application is rarely straightforward (p. 166)

Table 1: Management Theories Linking Innovation and Work Change

means, a worker must take participate in a community of experts (Stehr, 1994), colleagues (Reich, 1991), or professionals (Quinn et al, 1996). This is because knowledge is no longer conceptualized as timeless, objective fact. Knowledge is seen instead as more fragile, tentative claims and techniques with inherent points of view (Blackler et al, 1993). True access to knowledge potential, according to this view, requires on-going interaction with a community. By participating in the work of a community, the shifting meanings of knowledge claims, and the assumptions behind them, are clarified enough for knowledge to serve as the basis of action.

These observations on knowledge potential lead to the first summary claim from these theories of innovation and work: *work in continuously innovative organizations requires access to, and participation in, a body of specialized knowledge that serves as a capability for new action.*

Second, the theorists that link higher innovation rates and work change describe work as a process of converting knowledge potential into action. In a changing work situation, according to these arguments, work activity is no longer dominated by the application of pre-selected resources to a pre-defined problem. The task to be accomplished, or the problem to be solved, must be defined by the worker. The worker must ask which tools and techniques can be usefully applied. Because the objectives of work are not pre-defined, the criteria for successful work are conceptualized more as client satisfaction (Reich, 1991) and less as achieving pre-defined goals.

For some management theorists, such as Drucker, the process of conversion is so important that they claim the "primary function" of today's organizations is to turn knowledge potentials into productive activity (Drucker, 1993: 44). Other analysts of innovation and work change also see the conversion of knowledge potential into activities that satisfy clients as a major source of added value for organizations (Quinn et al., 1996: 71; Kay and Willman, 1992: 25; Blacker et al., 1993: 854; Leonard-Barton, 1995).

While important, these same theorists do not see the process of converting knowledge potential into action as straightforward, or easily imitable. Access to, and a deep understanding of, specialized knowledge is necessary, but not enough. Under conditions of rapid innovation, the application of knowledge to a particular problem is rarely straightforward (Stehr, 1994: 166) and requires considerable skill (Starbuck, 1992). The difficulty of the conversion process, however, is a source of competitive business advantage in an era when codified knowledge, in the form of information, is more widely available than ever (Reich, 1991).

The theories outlined in Table 1 do a better job of raising the conversion process as an important issue than of describing the conversion process itself. To the extent that they all agree on the importance of continuous innovation in work change, they agree that knowledge-intensive work must deal with at least two challenges. The first is to create a useful definition of what the problem is that must be solved (Starbuck, 1992: 722; Reich, 1991)—this can no longer be taken for granted. Following Fairhurst and Sarr (1996), I call the process of clarifying the assumptions that will underlie and direct a particular work activity framing.

Once a changing work situation has been framed, the second challenge for knowledge-intensive work is to take action in ways that are valued by clients. According to Stehr (1994), the ability to create valued action requires access to knowledge potential, and the ability to interpret the applicability of knowledge potential in specific situations. But creating valued action also requires something more— the resources needed to actually perform the action must be under the control of the worker. "Knowledge which pertains to moving a heavy object from one place to another is insufficient to accomplish the movement...the value which resides in knowledge is linked to its capacity to set something into motion. Yet, knowledge always requires some kind of attendant interpretive skills and a command of the situational circumstances." (Stehr, 1994; 120). This final process of combining knowledge potential, and suitable interpretations, with command over the situation to produce valued activity is what I define as *action* in this framework.

From these observation, we can make a second summary claim from these theories of innovation and work: *work in continuously innovative organizations involves a conversion of knowledge potential into action that satisfies clients by creating a useful problem framing, and by combining knowledge potential, a problem framing, and control over resources to produce valued action.*

While these two summary claims leave out a great deal of what has been said about work under the demands of continuous innovation, it compactly summarizes two claims made across a diverse range of management theorizing. These descriptions of work, and the assumptions they make about organizational action, form a useful contrast with the pictures of work assumed by traditional theories of information systems use.

Theories of Information Systems Use

Kling (1980) published the first comprehensive review of information systems use theories. He classified theories into two types, depending on the assumptions made about how organizations work. The first type of theory (which Kling calls sys*tems rationalist*) assumes there is consensus about the goals of information systems use, and that organizational efficiency is the predominant shared value. The second type of theory (which Kling calls *segmented institutionalist*) does not assume consensus about appropriate goals, but does assume that the main value held by groups is to maximize control over their own destiny.

In later work, Kling and associates (Kling and Jewett, 1994) adopted Scott's (1992) framework for classifying theoretical models of organizations. Scott's framework distinguishes between rational systems and natural systems models. According to Scott, rational systems models of organizational action (corresponding to Kling's systems rationalist perspective) stress goal consensus, formal structure, and efficiency. Natural systems models of organizational action (corresponding to Kling's segmented institutionalist) do not assume consensus about organizational goals, and emphasize survival as the main value held by groups.

The prevailing view of information systems use found in the information systems literature assumes a rational systems model of organizations. These theories assume that organizations are predominantly places of rational, consensus, goal-driven action where systems quality, information quality, and fitness to task greatly determine how information systems are used. DeLone and McLean (1992), in their review of information systems success measures, create a model which breaks down the factors leading to successful information systems use into two types: systems quality factors, and information quality factors. Systems quality factors imply that attributes such as equipment reliability, ease of use, availability of equipment, and response time determine why information systems are used (or not used) the way they are. Information quality factors imply that attributes of the information supplied by the system, including information accuracy, relevance, timeliness, and completeness, determine why information systems use takes place. Theories that explain information systems use in terms of systems quality and information quality assume a rational systems outlook on organizational action: people use information systems that are dependable, and that maximize organizational efficiency.

In addition to systems quality and information quality, two other streams of rational systems theorizing deserve mention: implementation, and perceived benefits. The implementation literature explains information systems use in terms of the quality of the delivery process for new information systems. The implementation literature relates individual skill and motivational factors, and task-technology fit, with successful information systems use or user satisfaction (Kwon and Zmud, 1987). The implementation literature explains successful information systems use as skill-building, creating motivation and ownership in the user population, and ensuring that the information system fits with what the business is trying to achieve (Walton, 1989).

The literature on the perceived benefits of information systems use draws upon psychological studies of the links between attitudes and behaviors, and studies of the diffusion of innovation which focus on the perceptions of the potential adopter. Perceived benefits studies argue that information systems will be used if individual users perceive that the expected future benefits of using an information system will outweigh expected future costs (Davis, 1989; Adams et al., 1992).

Rational systems theories of information systems use have been critiqued by an alternative view, based on the assumptions of a natural systems model of organizations. Natural systems theories are open to understanding organizational action as political and cultural processes involving subgroups, with the potential for different goals. While Kling (1980) found mostly political theories in his early review of this alternative perspective—explaining information systems use as power politics, coalition building, and forwarding agendas—the cognitive and linguistic turns in recent organizational research have brought cultural concerns to the fore (Walsham, 1993, for example, includes both cultural and political processes in his information systems and organizational change framework). Cultural theories explain how information systems use is shaped by shared meanings and assumptions within a subculture, and by clashes of meaning between subcultures. Recent examples of research on how cultural interpretations and political resources shape information systems use are discussed in Orlikowski and Robey (1991).

Figure 1 summarizes this broad overview of theories explaining why information systems use takes place the way it does. Rational systems explanations focus on how the quality of the information technology, the quality of information, the quality of the implementation process, and perceived benefits lead to greater, and more organizationally effective, information systems use. Natural systems explanations focus on how information systems are used for political gain by groups within organizations, and used according to the cultural assumptions of different subcultures. How these explanations relate to, and differ from, the arguments about the nature of work in continuously innovative organizations is the topic of the next section.

Information Systems Use and Innovation: A Framework

Given the wide range of theories that already exist to explain information systems use, do we need new explanations of information systems use in continuously innovative work environments? If the

Rational Systems Explanations	SYSTEMS QUALITY (SQ) - greater reliability, ease of use and availability lead to greater and more effective use.
	INFORMATION QUALITY (IQ) - greater accuracy, relevance, timeliness, and completeness of information lead to greater and more effective use.
	IMPLEMENTATION (IM) - building appropriate skills, motivation, and fit with business/task needs during the delivery process leads to greater and more effective use.
	PERCEIVED BENEFIT (PB) - people's expectations that future benefits will outweigh future costs lead to greater and more effective use.
Natural Systems Explanations	ORGANIZATIONAL POLITICS (OP) - information systems are used to forward political agendas and to increase control over people's own destiny.
	ORGANIZATIONAL CULTURE (OC) - information systems are used according to cultural interpretations and schemes of appropriate behavior.

Figure 1: Theories of Information Systems Use

demands of continuous innovation do, as the theorists in section three argue, move work more towards a process of converting some knowledge potential into a problem framing or action that meets the demands of a new situation, then there is value in new conceptual frameworks that pay attention to how workers maintain their mastery of a knowledge potential, and go about converting that potential into action. In addition, new conceptual frameworks can stimulate fresh thinking about how information systems should be used to support innovation. How would a framework based on theories of continuous innovation and work compare to traditional theories of information systems use?

The traditional, rational systems explanations of information systems use emphasize serving fixed, or slowly-changing, organizational goals. According to these explanations, systems are either poorly constructed or well constructed, and provide either good information or bad information. A good system providing good information will be widely used, and will be used to serve stable, widely-known organizational goals; bad systems will be ignored. What this account does not capture very well is information systems use which explores the meaning of the information (i.e., any sort of participation in the knowledge potential), and which is used to identify and create a new set of assumptions appropriate for a new, uncertain problem. Using information systems for innovative solutions to new problems seems to require an active, skillful search for ways to understand and apply information. The theories of implementation and perceived benefit do incorporate a limited account of the conversion process, but assume it is relatively infrequent and straightforward. Looking at work as continuously innovative requires a shift from providing good systems or good information to supporting the process through which people skillfully fit together information systems capabilities and a changing situation.

Natural systems explanations of information systems use see the 'natural' aspects of social interaction—particularly political behavior and cultural values—as the most important facets of organizational life. Information systems are built, and used, if different parties can negotiate an acceptable solution, or one party can impose its will on others. Information systems are used, according to these explanations, if the systems are compatible with the cultural expectations of important groups. Natural systems explanations share with theories of continuous innovation a concern with how the desires of important groups (i.e., clients) are satisfied. However, theories of continuous innovation understand information systems use as an active and deliberate search for the 'appropriate meanings' that will lead to action that satisfies a particular client. Theories of continuous innovation therefore provide a more 'rational systems' view of how the search for understanding and agreement shapes the process of information systems use.

Following the claims made in section three, derived from theories of innovation and work, I present a new framework for thinking about information systems use. In continuously innovative work situations, where a body of information must often be applied to novel situations, we can think of information systems use as a set of four repeating activities: access, participation, framing, and action. I define each of these processes below.

Access is the process of obtaining information reliably, when desired, and without excessive effort. As in rational systems explanations of information systems use, the technical problem of creating reliable, easy to use hardware and software is the first challenge faced by those wanting to support continuously innovative work. Access is the largely technical act of making the codified knowledge potential easily available.

For the information to make sense in novel situations, however, access is not enough. A potential user must participate in a community which helps them interpret the meaning of the information, and thus its applicability to novel situations. The participation can occur through conversation, instruction, and general exploration not intended to solve any particular problem. While rational systems theorists seek to measure the accuracy of information provided by a system, a continuous innovation view would ask how people are using the information system to explore the meaning and quality of the information it provides.

In continuously innovative work, the knowledge potential codified in information must somehow be converted into a valued product or service. Because of the novelty of problem situations, and the indeterminate meanings of the knowledge potential, a problem must first be appropriately framed. Framing is the process of choosing one particular meaning, or set of meanings, over others in order to interpret a situation (Fairhurst and Sarr, 1996: 3). According to this framework, workers in continuously innovative organizations use information systems to help identify and create appropriate sets of assumptions for a novel problem.

Once a changing work situation has been framed, then the worker must skillfully combine this interpretation of the problem, together with the knowledge potential and the resources at their disposal, to create useful activity that satisfies clients. More routine work situations presumably have clearer problem framings, resources with more clearly defined capabilities, and a more clearly defined authority to act in a pre-defined manner. With more continuous innovation, the gap between problem framing and action that satisfies clients requires more cognitive and interpretive skill. Producing valued action is both an issue of individual skill, and organizationally-approved support (i.e., authority) for using resources in a creative, uncertain manner.

ACCESS (ACC) - Information systems are used to reliably make information available on demand.

PARTICIPATION (PART) - Information systems are used to explore the quality and meaning of information in general.

FRAMING (FRAM) - Information systems are used to identify and create an appropriate set of assumptions for a particular problem.

ACTION (ACT) - Information systems are used to support action that satisfies clients by combining information and control over resources with a problem definition.

Figure 2: Information Systems Use Processes in Continuously Innovative Organizations

Figure 2 summarizes the four activities of access, participation, framing, and action. The main benefit of this framework is that it emphasizes use processes (such as exploring meaning, creating assumptions, and converting into action) that are considered to be important by a group of management theorists, but are not highlighted by existing theories. Because of its assumptions about innovation and work, I would predict that this framework would be more appropriate for explaining information systems use in more innovative situations, where work situations and available resources are less clearly defined. Rather than asking how reliable or accurate information systems are, it may help to ask questions such as: Are people able to explore the meaning of the information they have access to?; Are they able to create appropriate sets of assumptions?; or Are information systems enabling people to take specific actions that satisfy clients?

This framework is meant to complement, not replace, other explanations of why information systems are used the way they are. To the extent that organizational action consists of turning action potentials into valued problem framings and/or resolutions, concepts of innovation and work can be a useful way of explaining information systems use. Because work is never completely static nor completely changing in the ends it tries to achieve, this framework will always complement, and not replace, traditional explanations of information systems use.

Applying the Framework to Collaborative Technology Use

The processes identified in section five should be particularly appropriate for explaining information systems use in dynamic, innovative work. New collaborative technologies, such as bulletin board systems, shared project management systems, and shared editors, have been designed in the past decade with an eye towards supporting work more flexibly across space and time (Baecker, 1993). Though collaborative technologies do not have to be used to support innovation, they are often intended to, and many of the uses of collaborative technology studied so far have been for relatively innovative activities such as design and consulting. Collaborative technology use would therefore seem to be a good testing ground for the access/participation/framing/action framework—would the explanations of why information systems are used, and the use issues raised, reflect continuous innovation as well as rational and natural systems concerns?

As a preliminary test, I reviewed 23 articles selected from the proceedings of Computer-Supported Cooperative Work (CSCW) conferences in the 1990's (Tatar, 1990; Turner and Kraut, 1992; Futura and Neuwirth, 1994). The 23 articles (out of 117 total) were chosen because they described, in some depth, the use of a collaborative technology outside of a controlled experiment. Articles that described the use of a predominantly synchronous technology (i.e., a technology meant to support real-time interaction, such as desktop video) were excluded, because the use issues mentioned concentrated on attributes of the communication medium, rather than the information itself.

The CSCW articles were coded based on the concepts included in their explanations of how and why information systems were being used. When systems quality, information quality, implementation process, and perceived benefits were invoked as explanations, these were coded as rational systems explanations. Explanations using organizational politics and culture were coded under the appropriate natural systems explanations. When information systems use was explained in terms of participation in establishing the meaning of information, framing the definition of a specific work problem, or combining information, framing, and resources to take valued action, then the articles were coded as using explanations consistent with the new framework presented in this chapter. (Explanations in terms of access were coded as a traditional, rational systems explanation.)

Table 2 lists the information systems use theories that were referred to in articles on seven different types of collaborative technologies. The coding scheme, which differentiates between the traditional use theories described in section four, and the processes of use under continuous innovation described in section five, is a crude indicator of what CSCW researchers feel are the most important explanations of CSCW technology use.

From a continuous innovation standpoint, what is striking about these CSCW articles is the extent to which the search for 'useful' meanings explains how the systems are used. In design rationale systems, for example, the ability to create a shared context and explore different possible solutions is given more attention than the accuracy or availability of information. The use of bulletin board systems is explained in rational, political, and cultural terms, but the ability (or inability) to collaborate in the search for what the information in the repository means plays an important role in explaining use. The ability to create meaningful categories for the different discussion groups is one example of how continuous innovation processes are given more

Information System Used	Rational/Natural Explanations of Use	Potential/Conversion Explanations of Use
Rationale Documentation (2 articles)	• Used to obtain quality information (IQ)	• Used to refine terminology over time (PART) • Used to discuss assumptions behind a debate (FRAM) • Used to focus agenda for a debate (FRAM) • Used to define different possible solutions to a problem (FRAM) • Used to take different perspectives (PART)
Bulletin Board/ Repository (5 articles)	• Used to obtain quality information (IQ) • Used if incentives were appropriate (PB) • Used if cultural interpretations of technology appropriate (OC) • Used if willing to share information (OP) • Used to maintain status (OP)	• Used to define role of system (PART) • Used to define new information categories (PART) • Used to filter information collaboratively (PART) • Used to define general relevance collaboratively (PART) • Used to resolve ambiguity in the meanings of categories (PART)
Office Information Systems (8 articles)	• Used if network is reliable (SQ) • Used if support is adequate (SQ) • Use shaped by conflict over legitimate work practices (OP) • Use shaped by conflict over ownership of shared data (OP) • Use shaped by conflict over visibility of work process (OP)	• Used to communicate meanings across reports (PART) • Used to assess the trustworthiness of information (PART) • Used to assess the formality of information (PART) • Used to share domain knowledge (PART)
Scheduling (2 articles)	• Used if training adequate (SQ) • Use shaped by conflict over scheduling (OP)	• Used to provide additional contextual information (PART)
Shared Editor (2 articles)	• Used to support culturally defined social roles (OC)	• Used to provide additional contextual information (PART)
Control Room (2 articles)	• Used to obtain quality information (IQ)	• Used to define a common work situation (FRAM) • Used to define relevance to a specific task collaboratively (FRAM)
Structured Email (2 articles)	• Used to obtain quality information (IQ) • Used shaped by conflict over ownership of shared data (OP)	

Table 2: Explanations of Collaborative Technology Use in CSCW Articles

importance than accuracy and implementation concerns in these early studies.

While the access/participation/framing/action framework does highlight many of the special concerns of CSCW use, CSCW researchers still employ more traditional theories to explain how and why CSCW systems are used. Technical problems with systems are discussed as an explanation of generic office information systems use (i.e., the use of local area networks in offices). Political processes are identified to explain the use of bulletin board, scheduling, structured email, and office information systems. The cultural interpretation of a bullctin board/rcpository system—whether it was interpreted as a collaborative technology or a traditional database—was used in one case to explain a use failure. I hope that the reader will be convinced, however, that an exclusive focus on traditional explanations of information systems use offers an incomplete account, which can be usefully supplemented by explanations that assume the presence of more continuous innovation.

Analyzing articles from CSCW proceedings has limitations. One limitation has already been mentioned: there is no guarantee that all of the systems studied looked at continuously innovative work situations, or indeed at particularly innovative technologies. Some studies looked at fairly traditional office information systems, that happened to be used over a network. Another more important limitation is that the interests of the CSCW research community bias the findings towards particular kinds of theorizing, regardless of what is 'actually happening'. For example, there is little or no mention of how the ability (or inability) to take action that satisfies clients affects collaborative technology use. That final delivery of service is of comparatively little interest to CSCW researchers. With these limitations in mind, we can still say that the processes associated with work in continuously innovative organizations—particularly participation and framing—have played a significant role in explaining collaborative technology use in the CSCW literature so far.

Conclusion

In this chapter, I have argued that concepts explaining how work has changed in continuously innovative organizations can be a useful addition to our explanations of information systems use. Our existing concepts of information systems use assume models of organizational action that were in place before the recent emphasis on innovation. The framework presented in this chapter complements, but does not replace, traditional models of information systems use in rapidly innovating organizations.

Besides the contributions of the framework itself, the arguments made in this chapter have implications for future attempts to explain why information systems are used or not used, whether they be in academic studies or practical projects:

Make assumptions about organizations clear. The assumptions made about the most important features of organizational life (Rules and regulations? Change? Politics?) shape the kinds of explanations that can be made, and determine which questions can even be asked. Information systems use is a special kind of organizational action. Organizational action is complex, and can be usefully understood from many different perspectives (e.g., Scott, 1992). The inevitable choice to focus only on certain aspects of organizational action, including information systems use, should be made explicitly.

State which explanations of information systems use will or will not be considered. Instead of viewing information systems use as a phenomena that many different people and studies are trying to explain, different research traditions make little reference to each other. Within the rational systems traditions, improved models of the psychology of people's use expectations make little use of implementation factor studies, for example, and even less use of political and cultural studies. By viewing the different traditions as attempts to explain the same basic phenomena, it is possible to ask why an issue receiving attention in the management literature (the rate of innovation and work change) was not being addressed in studies of information systems use generally. Placing an explanation of information systems use in the context of all attempts to explain information systems use forces attention on other, very different, explanations, and contributes to a common project of understanding use.

Consider the rate of innovation when explaining information systems use. This chapter has reviewed the arguments for why and how the rate of innovation plays an important role in work change. The use of information, according to these arguments, has very different dynamics depending on whether innovation is more continuous or occasional.

Help clarify the process of converting knowledge potential into organizational action. The process of converting knowledge potential into action, which has information systems use at its center, is seen as a major and growing source of added value for organizations. Yet little is known about this process. The access/participation/framing/action framework, like the theories of innovation and work it draws upon, does more to clarify the issue than to describe how this key activity actually takes place. Clarifying this process should place the issue of information systems use—usually seen as a specialized research area—closer to the heart of general management thinking.

I present the access/participation/framing/action framework as a set of sensitizing concepts to be explored, and contested, in future work. Understanding why a technology as flexible as an information system comes to be used (or not used) in a particular set of ways is a difficult challenge. Hopefully, this framework will allow people to systematically ask questions that they might not have asked before.

Acknowledgments

I would like to thank Matthew Jones and Séamas Kelly for a thoughtful reading of an earlier draft.

References

Adams, D. A., Nelson, R. R., and Todd, P. A. "Perceived Usefulness, Ease of Use, and Usage of Information Technology Applications." *MIS Quarterly*, June (1992), 227-247.

Axford, B. *The Global System: Economics, Politics and Culture.* Cambridge, U.K.: Polity, 1995.

Baecker, R. M. *Readings in Groupware and Computer-Supported Cooperative Work: Assisting Human-Human Collaboration.* San Mateo, California: Morgan Kaufmann, 1993.

Blackler, F., Reed, M., and Whitaker, A. "Knowledge Workers and Contemporary Organizations." *Journal of Management Studies*, 30 (1992), 851-862.

Daft, R. L. *Organizational Theory and Design.* Saint Paul, Minnesota: West Publishing, 1992.

Davis, F. D. "Perceived Usefulness, Perceived Ease of Use, and User Acceptance of Information Technology. *MIS Quarterly*, September (1989), 319-340.

DeLone, W. H., and McLean, E. R. "Information Systems Success: The Quest for the Dependent Variable. *Information Systems Research*, 3 (1992), 60-95.

DeSanctis, G., and Poole, M. S. "Capturing the Complexity in Advanced Technology Use: Adaptive Structuration Theory." *Organization Science*, 5 (1994), 121-147.

Drucker, P.F. *Post-Capitalist Society.* Oxford, U.K.: Butterworth-Heinemann, 1993.

Drucker, P. F. *The New Realities.* New York: Harper & Row, 1989.

Drucker, P. F. "The Coming of the New Organization." *Harvard Business Review*, January-February (1988), 45-53.

Fairhurst, G. T., and Sarr, R. A. *The Art of Framing: Managing the Language of Leadership.* San Francisco: Jossey-Bass, 1996.

Friedman, A. L. *Computer Systems Development: History, Organization, and Implementation.* Chichester, U.K.: John Wiley & Sons, 1989.

Futura, R., and Neuwirth, C., eds. *CSCW '94: Proceedings of the Conference on Computer-Supported Cooperative Work.* New York: ACM Press, 1994.

Hirschhorn, L., and Mokray, J. "Automation and Competency Requirements in Manufacturing: A Case Study. In P. S. Adler, ed., *Technology and the Future of Work.* New York: Oxford University Press, 1992, 15-45.

Kay, J. and Willman, P. "Managing Technological Innovation: Architecture, Trust and Organizational Relationships in the Firm. In P. Swann, ed., *New Technologies and the Firm: Innovation and Competition.* London: Routledge, 1993, 19-35.

Kling, R. "Social Analyses of Computing: Theoretical Orientations in

Recent Empirical Research." Computing Surveys 12 (1980), 61-110.

Kling, R., and Allen, J. P. "Can Computer Science Solve Organizational Problems? The Case for Organizational Informatics. In R. Kling, ed., *Computerization and Controversy: Value Conflicts and Social Choices*, 2nd ed. San Diego: Academic Press, 1996, 261-276.

Kling, R., and Jewett, T. "The Social Design of Worklife with Computers and Networks: An Open Natural Systems Perspective. In M. C. Yovitz, ed., *Advances in Computers*, vol. 39. San Diego: Academic Press, 1994.

Kwon, T. H., and Zmud, R. W. "Unifying the Fragmented Models of Information Systems Implementation." In R. J. Boland, and R. A. Hirschheim, eds., *Critical Issues in Information Systems Research.* Chichester, U.K.: John Wiley & Sons, 1987, 227-251.

Leonard-Barton, D. *Wellsprings of Knowledge: Building and Sustaining the Sources of Innovation.* Boston: Harvard Business School Press, 1995.

McLoughlin, I. and Clark, J. *Technological Change at Work.* Milton Keynes, U.K.: Open University Press, 1992.

Mouzelis, N. *Sociological Theory: What Went Wrong?* London: Routledge, 1995.

Orlikowski, W. J. and Robey, D. "Information Technology and the Structuring of Organizations." *Information Systems Research*, 2 (1991), 143-169.

Piore, M. J., and Sabel, C. F. *The Second Industrial Divide.* New York: Basic Books, 1984.

Quinn, J. B., Anderson, P., and Finkelstein, S. "Managing Professional Intellect: Making the Most of the Best." *Harvard Business Review*, March-April (1996), 71-80.

Reich, R. *The Work of Nations: Preparing Ourselves for 21st-Century Capitalism.* New York: Vintage, 1991.

Robey, D. "Theories that Explain Contradiction: Accounting for the Contradictory Organizational Consequences of Information Technology." In *Proceedings of the Sixteenth International Conference on Information Systems*, December 1995, 55-64.

Scott, W.R. *Organizations: Rational, Natural, and Open Systems*, 3rd ed. Englewood Cliffs, New Jersey: Prentice-Hall, 1992.

Starbuck, W. H. "Learning by Knowledge-Intensive Firms." *Journal of Management Studies*, 29 (1992), 713-740.

Stehr, N. *Knowledge Societies.* London: Sage, 1994.

Tatar, D., ed. *CSCW '90: Proceedings of the Conference on Computer-Supported Work.* New York: ACM Press, 1990.

Turner, J. and Kraut, R., eds. *CSCW '92: Proceedings of the Conference on Computer-Supported Cooperative Work.* New York: ACM Press, 1992.

Walsham, G. *Interpreting Information Systems in Organizations.* Chichester, UK: John Wiley & Sons, 1993.

Walton, R . *Up and Running: Integrating Information Technology and the Organization.* Boston: Harvard Business School Press, 1989.

Wood, S., ed. *The Transformation of Work: Skill, Flexibility, and the Labour Process.* London: Unwin Hyman, 1989.

Zuboff, S. *In the Age of the Smart Machine: The Future of Work and Power.* New York: Basic Books, 1988.

CHAPTER

3

The Impact of Mental Workload in the Evaluation of Innovations

J. Roberto Evaristo
University of Denver

Elena Karahanna
University of Cyprus

Several models have been developed to describe the process of new technology introduction. Most models, whether at the individual or organizational level, describe a sequence of stages through which organizations or individuals pass from initial knowledge of the new technology to a decision to adopt or reject to an implementation stage. All these models involve at some phase the evaluation of the innovation and the adoption or rejection of the innovation by the organization or individual.

At the micro level, most research available on how individuals evaluate new technologies addresses rational issues such as cost/benefit analysis and evaluation of certain attributes of the technology such as relative advantage, compatibility, complexity, trialability, observability, and visibility. What has not been studied is the contribution of a new technology to overall individual mental workload (or information load) and its subsequent effect on the adoption decision. The level of the ensuing mental workload is proposed to have a direct effect on the quality of the evaluation of the innovation and the subsequent adoption or rejection decision. This study proposes a model of the relationship among individual mental workload, innovation attributes, and the evaluation and acceptance of innovations.

A number of models have been developed to describe the process of new technology introduction (e.g., Brown and Chervany 1994; Cash, McFarlan, McKenney and Applegate, 1992; Cooper and Zmud, 1990; Kwon and Zmud, 1987; Rogers, 1983; Zmud and Apple, 1989). Most models, whether at the individual or organizational level, describe a sequence of stages through which organizations or individuals pass from initial knowledge of the new technology to a decision to adopt or reject to an implementation stage. All these models involve at some phase the evaluation of the innovation and the adoption or rejection of the innovation by the organization or individual.

At the micro level, most research available on how individuals evaluate new technologies addresses rational issues such as cost/benefit analysis and evaluation of certain attributes of the technology such as relative advantage, compatibility, complexity, trialability, observability, and visibility (Rogers, 1983). What has not been studied is the contribution of a new technology to overall individual mental workload (or information load). Technologies which might otherwise have many advantages but which induce high levels of mental workload, leading an individual closer to the point of overload, might imply lower productivity and therefore higher resistance to adoption of this technology. This study proposes a model of the relationship among individual mental workload, innovation attributes, and the evaluation and acceptance of innovations.

Initially, a theoretical background will be offered. It includes a description of the evaluation process of Information Technology innovation as well as recent advances in the mental workload literature. Based on this framework, a model implementing the linkage between these two areas is developed, and key research questions and hypotheses representing our current understanding of the topic are proposed.

Theoretical Background

This project lies at the intersection of two areas. The first is the innovation adoption decision making process. The second is mental workload, previously researched in psychology-related fields, particularly in the human factors and ergonomics areas, and only recently introduced to the business literature (Evaristo, 1993).

The Innovation Decision Process

The innovation-decision process, the five-stage process through which an individual in a social system passes from first knowledge of the innovation to rejection or acceptance and implementation of the innovation, is an information intensive activity aimed at reducing uncertainty about the advantages and disadvantages of the innovation (Rogers, 1983). During this information-seeking and information processing activity, individuals obtain two kinds of information in order to reduce uncertainty about the innovation: software information and innovation-evaluation information. Initially, at the knowledge stage, individuals search for software information to determine what the innovation is, what it does, and why it works (principles knowledge). In this manner, uncertainty about the cause-effect relationship regarding the innovation's ability to achieve a desired outcome is reduced.

As individuals move to the persuasion stage of the innovation-decision process, they search for innovation-evaluation information about the innovation's advantages and disadvantages to reduce uncertainty about the consequences of their adoption and use of the innovation. At this stage, Innovation Diffusion suggests that certain attributes of innovations, as perceived by individuals in a social system, may affect the rate at which innovations diffuse and are adopted. These include perceptions of relative advantage (or perceived usefulness), computer avoidance, image, compatibility, complexity (or ease of use), trialability, visibility, and result demonstrability (Rogers, 1983; Moore and Benbasat 1991). Based on these important attributes, defined in table 1 and Appendix A, individuals form a favorable or unfavorable attitude toward the innovation. Attitude along with normative pressures from the social environment are precursors of adoption or rejection of the innovation (Moore, 1989; Karahanna, 1993).

Mental workload may also be an important factor in the innovation evaluation process and the final adoption or rejection decision. The innovation evaluation process may induce mental workload through several distinct processes. For instance, since innovation evaluation is an information intensive activity, it imposes certain demands for information processing to determine what the innovation

Innovation Attribute	**Definition**
Relative Advantage (Perceived Usefulness)[i]	the degree to which using an innovation is perceived as being better than using the practice it supersedes.
Computer Avoidance	beliefs that may affect attitude toward adopting an innovation negatively and may result in individuals avoiding use of the innovation
Image	degree to which adoption of an innovation is perceived to enhance one's image or status in one's social system
Compatibility	degree to which using an innovation is perceived as consistent with the existing sociocultural values and beliefs and past and present experiences of potential adopters
Complexity (Ease of Use)	the degree to which the innovation is perceived as difficult to understand and use
Trialability	degree to which one can experiment with an innovation on a limited basis before making an adoption or rejection decision
Visibility	degree to which the innovation is apparent to the sense of sight (visible)
Result Demonstrability	degree to which the results of using the innovation are observable and communicable to others

Table 1: Innovation Attributes

does and what are its advantages and disadvantages, creating a certain level of mental workload. Another possibility is the use of the innovation itself, which may further increase the user's mental workload. The extent of this increase depends partly on the complexity of the user interface and the user's familiarity and experience with similar types of innovations.

Mental Workload

It is well known that very high or very low mental workload levels have dysfunctional effects on individual performance (Miller, 1978). For instance, Quastler and Wulff (1955) investigated typing performance for different levels of input (measured in symbols per second), and Vince (1949) measured subjects' responses to input information in the form of dots and numbers. Both experiments showed remarkably similar results: the graph of input versus output was smooth, with performance first increasing and then decreasing gradually at the overload point. In the innovation evaluation process, high levels of mental workload may lead to poor evaluation of the innovation, or to a poor innovation decision (both type I and type II errors), perhaps because relevant information was disregarded, resulting in premature rejection of the innovation.

Mental workload is defined as "that portion of the operator's (or user's) limited capacity actually required to perform a particular task" (O'Donnell and Eggemeier, 1986, p. 42-2). The implication of this definition is that an individual may decide, based on external and internal factors such as motivation, to allocate a portion of his/her information processing capacity to perform a given task. This partially allocated information processing capacity is called *available information processing resources* (AIPR). The mental workload level that the individual experiences in performing a given task (see Figure 1) is then dependent on how much of the available information processing resources is being "used" by the demands posed by the task.

Figure 1 presents the two antecedents of mental workload: (a) available information processing resources (AIPR), and (b) demand for available information processing resources. The amount of AIPR depends on two main determinants of resource allocation. The first is attentional, encompassing certain psychological states that an individual experiences, such as happiness, anxiety, fear, anger, and intrinsic and extrinsic motivation. The second determinant of the processing resource allocation is physiological; it includes fatigue, drugs, and stress.

Demand for available information processing resources, on the other hand, can be influenced by several task and information characteristics (Evaristo, 1993). By its turn, both task characteristics and information characteristics can be influenced by innovation attributes and subjective norms, which is the crucial connection

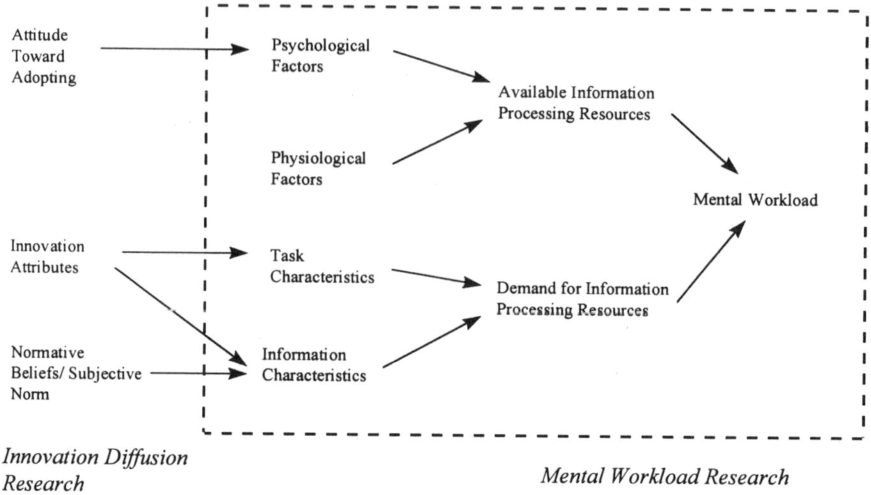

Figure 1: Antecedents of Mental Workload

between mental workload and the innovation adoption decision process. Three task characteristics, time pressure, task formalization, and task complexity are significant determinants of mental workload through their effect on demand for AIPRs. Time pressure, the amount of time available to perform a given task, is one of the most important determinants of demand for AIPRs and therefore of individual mental workload. Researchers such as Svenson and Edland (1987) and Wright (1974) have provided experimental evidence showing that choices made under time pressure differ significantly from those made under conditions of no time pressure. In fact, this effect can be so strong that preferences were reversed under higher time pressure. These preference reversals were only explainable by subjects over-evaluating negative factors on the most important attributes. Moreover, under increased time pressure, people adapt by filtering and omitting additional information (Huntley, 1972), by shedding or lowering performance of certain tasks (Hart, 1986), by being less risky (Ben and Bresnitz, 1981) and by using fewer cues (Wright, 1975). However, individuals might be able to develop improved ways of thinking over time, therefore enabling them to process greater information loads (Miller, 1978; Schroder, Driver and Streufert, 1967). In the innovation evaluation process, if individuals are under time pressure when evaluating whether or not to adopt the innovation, overemphasis may be placed on negative aspects of the innovation attributes and important positive elements may be ignored. Interviews with would-be adopters of Microsoft Windows in a large banking institution showed that time to evaluate and learn the software was a concern for the majority of the respondents (Karahanna, 1993).

Task formalization, the level of specification and structure afforded by the task, is another task characteristic likely to influence mental workload. Jobs being performed by individuals are sometimes defined in a structured way; in others, the understanding of the activities that are required to perform the task, or even of the specific goals of the task, is left open to judgment. The latter includes ambiguous and unstructured tasks. Individuals have to exert less effort in work management when the task is structured[2]; in fact, most of the demand for information processing resources in structured tasks originates from task execution. The structured task affords what has been described by cognitive psychologists as automated responses requiring little overhead cognitive effort. In contrast, individuals might not know completely how to execute unstructured tasks, or tasks in which the level of autonomy is high, therefore they must first plan the execution of the task. The net result is likely to be higher mental workload. In the evaluation of innovations, the more radical the innovation to be evaluated, the less likely that individuals will have mental models and automated responses developed, and therefore, the higher the level of mental workload experienced while using the innovation. As users gain experience with the innovation and climb the learning curve, the level of mental workload experienced while using the innovation decreases.

The last task characteristic is task complexity. Wood (1986) defined three types of task complexity: component, coordinative and dynamic. Each of these types of task complexity affects demand for information processing resources in a slightly different way. For instance, coordinative complexity refers to the nature of relationships between task inputs and task products. It includes aspects such as sequencing or timing of specific acts. Another way of looking at coordinative complexity is examining the form of the relationship: When the task product is a simple linear function of task inputs, complexity is low. The more difficult it is to coordinate relationships between task inputs (e.g., information) and task outcomes (e.g., decision making), the higher is the level of demand for information processing resources. Most of the effect of coordinative complexity on demand for information processing resources stems from the increased work management necessary to coordinate or plan the sequencing and timing of specific acts (this is in excess of the effect of component complexity, which includes only how to manage a large number of acts not necessarily related to each other). The interested reader is referred to Evaristo (1993) for a complete discussion of the other types of task complexity. In the evaluation of innovations, the complexity of the user interface may lead to higher mental workload during use of the innovation.

In addition to task characteristics, information characteristics are important antecedents of mental workload (Evaristo, 1993). Information characteristics, such as information volume, uncertainty, ambiguity, complexity and turbulence, refer to characteristics of the information, or inputs, used in performing a task. The emphasis of the discussion will be on *information uncertainty* because it provides the link to the innovation evaluation process. Uncertainty is usually defined as a function of knowledge, typically as a knowledge inadequacy (Montagna, 1980). Uncertainty may arise from several sources. Among them, (a) unavailability of or incompleteness of information, (b) low information reliability, or (c) information novelty (Smith, Benson and Curley, 1991). A result of uncertainty is an individual's perceived inability to predict something accurately.

There is both a direct and an indirect relationship between uncertainty and mental workload (Evaristo, 1993). The direct effect is based on the fact that an individual's AIPRs are limited, and if one is concerned that one does not have all the information or knowledge necessary to make a decision, part of one's AIPRs is spent managing the planning efforts of how to increase the completeness or reliability of the information. In other words, part of the AIPRs are being allocated to work management. Therefore, fewer AIPRs will be available for task execution and, unless the task is executed at a lower performance standard, mental workload will be higher than if uncertainty was absent.

In addition, there are also indirect effects of uncertainty on mental workload. These effects are "indirect" because there is an intermediate step before the change in uncertainty is reflected in the level of mental workload. For instance, a decision made during the work management process might be to engage in further search for information, adding these efforts to the demands previously identified in the task execution process. Even though this extra information has the potential to mitigate one's uncertainty about task outcomes (decreasing mental workload), it increases future volume of information to be processed (increasing mental workload). The net result is likely to be a lagged increase in mental workload. Moreover, higher uncertainty might force individuals to make explicit trade-offs among alternatives that otherwise might not have been important. This extra processing caused by higher uncertainty also has the potential to increase one's mental workload.

On the other hand, there is a second possible indirect effect of uncertainty on mental workload. Higher uncertainty might cause individuals to completely ignore certain alternatives, in effect reducing the number of possible choices and therefore, decreasing the volume of information to be processed as well as the processing necessary. The final result might be lower mental workload.

Connection Between Mental Workload and the Innovation Adoption Process

As discussed above, innovation evaluation and adoption is conceptualized in terms of a framework based on uncertainty and information (Rogers, 1983). An innovation creates uncertainty about its expected consequences for potential adopters. At the same time, it represents an opportunity to reduce uncertainty. According to Rogers (1983), "the latter type of potential uncertainty reduction (the information embodied in the technological innovation itself) represents the possible efficacy of the innovation in solving an individual's felt need or perceived problem; this advantage provides the motivation that impels an individual to exert effort in order to learn about the innovation. Once such information-seeking activities have reduced the uncertainty about the innovation's expected consequences to a tolerable level for the individual, a decision concerning adoption or rejection will be made. If the new idea is used by the individual, further evaluative information about the technological innovation is thus obtained and uncertainty about its effects is further reduced. Thus the innovation-decision process is essentially an information-seeking and information-processing activity in which the individual is motivated to reduce uncertainty about the advantages and disadvantages of the innovation" (p. 13).

Given that uncertainty—as an information characteristic—is positively related to mental workload (Evaristo, 1993), innovation attributes may potentially affect one's mental workload level. Therefore, an individual experiences a certain level of mental workload while engaging in the process of evaluating the innovation and reducing the uncertainty about its consequences. This mental workload, further increased by the mental workload experienced due to use of the innovation, can be reduced by manipulating some of the task or information characteristics. For instance, alternatives could be to reduce the ambiguity of the information received about the innovation, to provide more information on the innovation's relative advantage with respect to the individual's job and existing technology, to reduce the number of innovation attributes considered during evaluation, to reduce the complexity of evaluation information received, and to decrease time pressure for the innovation decision. In addition, normative beliefs and subjective norm may reduce the level of mental workload through informational and normative influences by reducing uncertainty and the volume of information required for the evaluation decision. Further, positive attitude toward adopting the innovation may increase one's motivation to evaluate the innovation (a psychological antecedent of available information processing resources) and thus may decrease the level of mental workload experienced.

In summary, as shown in Figure 1, the current research proposes that (1) normative beliefs and subjective norm affect information characteristics, and therefore the level of mental workload experienced, (2) attitude toward adopting affects the psychological factors that influence the level of available information processing resources and therefore, the perceived level of mental workload and (3) innovation characteristics influence both task and information characteristics and, therefore, the level of mental workload experienced by individuals. A detailed explanation of these linkages is presented in the following section. The level of mental workload is then hypothesized to have a direct effect on the quality of the evaluation of the innovation and the subsequent adoption or rejection decision. This linkage had not been proposed before in the literature. In the context of innovation evaluation, several specific research questions and hypotheses can be proposed linking innovation attributes, social norms, mental workload, and the final adoption or rejection decision.

Theoretical Model

Figure 1 presented the antecedents of mental workload. Figure 2[3] presents the research model proposed here, used to guide research questions raised in this chapter as well as in future empirical studies. Prior research on the evaluation of innovations, based on the Theory of Reasoned Action (TRA) (Fishbein and Ajzen 1975), has shown that the adoption or rejection decision is based on one's subjective norm and attitude toward adopting the innovation (Moore, 1989, Karahanna, 1993). Attitude toward adopting the innovation reflects the user's favorable or unfavorable evaluation with respect to adopting the innovation. It is determined by the user's beliefs about the consequences of adopting the innovation (i.e., the innovation attributes presented in Table 1). Subjective norm refers to the user's perceptions of social pressures put on her to adopt or reject the innovation. It is in turn determined by the user's normative beliefs that she attributes

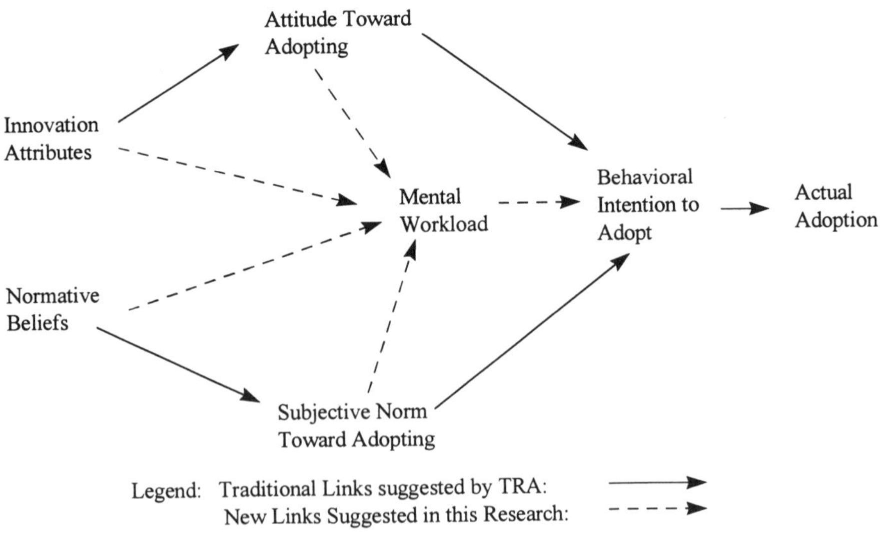

Figure 2: Research Model

to what relevant others want her to do with respect to adopting the innovation. We propose mental workload as an antecedent of the intention to adopt in addition to attitude and subjective norm. For the sake of drawing clarity, certain links are not represented, such as the possible connection between mental workload and actual behavior.

The overall research question that guides this research is then "How does mental workload affect the innovation decision process?" In order to be able to address this problem, one has to first understand the different ways under which mental workload may manifest itself in each of the phases of the innovation adoption process. At any given point in time during the innovation decision process, there is a level of mental workload (MWL) that an individual is experiencing. The underlying assumption is that MWL affects all phases of the innovation decision process. The equations on Table 2 try to give a better explanation of what this level may be at key phases of the process: (1) during the knowledge state; (2) during the persuasion stage; and (3) during the decision making stage.

Individuals experience a certain level of mental workload, called baseline mental workload, irrespective of activities involved in the innovation decision process. Baseline mental workload is a result of physical and cognitive activities, other than the innovation evaluation decision, that an individual engages in at the time of innovation evaluation. As such, baseline mental workload is a component of an individual's total mental workload at every stage of the innovation decision process. Since it is due to factors exogenous to the innovation decision process, it is only shown on Table 2 for the sake of completeness and can be ignored in empirical studies.

During the knowledge stage, individuals search for information to determine what the innovation is, what it does and how it works. In addition to the baseline mental workload, individuals experience a heightened level of mental workload induced by searching and assimilating this type of information ($MWL_{information}$). Innovation characteristics affect the level of mental workload experienced through their effect on information characteristics (see Figure 1 for link). For instance, simpler innovations decrease the amount of uncertainty about the cause effect relationship regarding the innovation's ability to achieve a desired outcome, potentially leading one to experience lower levels of $MWL_{information}$ during this stage.

For the Persuasion stage, MWL has different determinants depending on whether the individual does or does not have the opportunity to try the innovation before an adoption decision is made. When such trial is possible, we will refer to it as experiential or hands-on evaluation, whereas we will refer to cases where an individual evaluates an innovation purely based on readings, brochures or other written or graphical materials as non-experiential or hands-off evaluation.

During the Persuasion stage, individuals search for innovation-evaluation information about the innovation's advantages and disadvantages to reduce uncertainty about the consequences of their adoption and use of the innovation. In addition to $MWL_{baseline}$, individuals experience MWL due to evaluation of innovation. Obviously, innovation attributes are important at this stage. They affect MWL through their ability to reduce uncertainty (an information characteristic) about the consequences of adopting the innovation. Normative beliefs and subjective norm also affect MWL through their effect on information characteristics (uncertainty and volume) as suggested in the relationships presented in Figure 1 and discussed in detail under research question 2. Furthermore, attitude toward adopting may also have an effect on MWL by increasing or decreasing the amount of available information processing resources (AIPR - an antecedent of MWL, as presented on Figure 1). A positive attitude toward adopting an innovation increases one's motivation with respect to evaluating the innovation. All else being equal, increased motivation increases the available information processing resources, thus reducing the level of MWL experienced.

The above description applies to both hands-on and hands-off evaluation. An additional component of MWL during hands-off evaluation is $MWL_{information}$ as discussed above. During hands-on evaluation, there exist two additional sources of mental workload induced by the use of the innovation: $MWL_{learning}$, which is caused by the process of learning the innovation and differs from $MWL_{information}$ in the fact that $MWL_{information}$ concentrates in the "what," instead of in the "how." When an individual moves to trial use, $MWL_{information}$ becomes less relevant, and $MWL_{learning}$ takes priority. Finally, MWL_{usage} refers to MWL induced by the use of innovation. Innovation attributes affect these two components of MWL through their effect on task characteristics (as seen on linkages suggested in Figure 1). For instance, the more user-friendly the interface of an innovation is, or the more compatible an innovation is with one's past experience, the less complex is the task of learning and using the innovation. Therefore, all else being equal, the less the mental workload experienced.

Table 2: MWL Sources for each Innovation Decision Stage

Stage	MWL that an individual is under
Knowledge:	$MWL_{ks} = MWL_{baseline} + MWL_{information}$
$Persuasion_{hands\text{-}off}$:	$MWL_{Phoff} = MWL_{baseline} + MWL_{hands\text{-}off\ evaluation} + MWL_{information}$
$Persuasion_{hands\text{-}on}$:	$MWL_{Phon} = MWL_{baseline} + MWL_{hands\text{-}on\ evaluation} + MWL_{learning} + MWL_{usage}$

Notes

MWL_{Total} = the sum of all different types of mental workload that one is experiencing at a particular stage in the innovation decision process.

$MWL_{baseline}$ = this is the baseline level of mental workload that one is experiencing and is due to all other activities one may be engaged in.

$MWL_{evaluation}$ = mental workload caused by the evaluation of a possible innovation. It may be either hands-on ($MWL_{hands\text{-}on\ evaluation}$) when the individual can actually try the innovation or hands-off ($MWL_{hands\text{-}off\ evaluation}$), when such trial is not possible.

MWL_{usage} = mental workload caused by the usage of an innovation. It occurs only during the hands-on trial and eventual usage of the innovation.

$MWL_{learning}$ = mental workload caused by the process of learning how to use an innovation.

$MWL_{information}$ = mental workload induced by searching and assimilating software information (i.e., information about what an innovation does and why it works).

The more specific research questions that can be gleaned from this theoretical model are the following:

RQ 1. How does mental workload affect the innovation adoption decision in
a. an experiential or hands-on situation?
b. a non-experiential or hands-off situation?

This is arguably the most important question presented here. It addresses the fundamental point of the importance of mental workload as an intervening variable in the innovation decision process. This question is also interesting because one can make a plausible argument in either way. On the one hand, one could argue that the more difficult an innovation is to learn — and therefore the higher the mental workload generated — the least likely it is that such an innovation would be adopted. On the other hand, the same individual may have invested so much effort in working with the innovation that by the principle of escalating commitment she or he may be more likely to actually adopt the innovation.

On balance, however, we would expect that an innovation which poses fewer demands for information processing resources and thus creates lower levels of mental workload to be more likely to be adopted than one that is more demanding.

We also believe that the nature of the evaluation (hands-on versus hands-off) may make a difference in this result, since one knows that the evaluation of the innovation is less complete when based on hands-off experience and, therefore, may compensate for this by becoming more risk averse and therefore more likely to reject innovations, particularly under situations of heightened mental workload. Further, we would expect MWL due to evaluation to be lower during hands-on evaluation than during hands-off evaluation. This is because information derived through direct experience with an innovation is likely to result in the individual being more able to evaluate the object "in a clear, confident, and meaningful way" (Fazio and Zanna 1981, p. 186). This reduces the uncertainty involved in innovation evaluation and thus, $MWL_{evaluation}$. Of course, whether or not MWL_{Total} will be lower will depend on the relative magnitude of the other MWL components.

RQ 2. Is there an interaction effect between levels of MWL (low, high) and subjective norm (low, high) concerning the innovation adoption decision?

Subjective norm, which can be exerted through both information influence and normative pressure, is proposed to reduce uncertainty with respect to the innovation decision. Uncertainty is reduced through informational influence when near-peers of the potential adopter (i.e., individuals who are similar to the potential adopter) inform the potential adopter of their own personal experience and evaluation of the innovation or when the potential adopter can observe peers use the innovation. Use by peers, termed psychological or vicarious trial, can be a very effective source of evaluative information (Bandura, 1977). In addition to informational influence, normative pressure from supervisors and peers to adopt the innovation reduces the risk of adoption and uncertainty since it provides strong evidence indicating that adoption is appropriate in this particular setting. Due to the proposed uncertainty reduction property of MWL, we hypothesize that subjective norm will have a more important role in the adoption decision in cases of high MWL than in cases of low MWL.

For the sake of argument, let us assume that in situations of higher mental workload one is less likely to adopt an innovation. If this is indeed the case, institutions may be able to decrease this effect by manipulating the level of subjective norm. Therefore, such a finding would affect the way marketing of internal systems, for instance, could be done. Of course, if an empirical study based on these ideas were to prove that in situations of lower mental workload one is less likely to adopt an innovation, then we could make the opposite argument about subjective norm.

H2.[4] *Mental workload will be lower in the presence of strong informational and normative influences with respect to adoption of the specific innovation.*

RQ 3. Is there an interaction effect between levels of MWL (low, high) and evaluation (experiential versus non-experiential) concerning the innovation adoption decision?

The response to this question would suggest that in certain situations (say, higher mental workload) one may be better off in reinforcing the adoption likelihood by providing an experiential or hands-on evaluation.

RQ 4. At the persuasion stage, will innovation attributes and normative beliefs significantly influence $MWL_{evaluation}$?

As already discussed, during the persuasion stage innovation attributes and normative beliefs reduce uncertainty about the consequences of adopting the innovation and, thus, mental workload. Here, we propose differences between the types of innovation attributes and normative pressures that influence hands-on and hands-off evaluation. During hands-on evaluation, only some innovation characteristics are salient and contribute to mental workload. The usefulness of an innovation with respect to one's job, how easy it is to use the innovation, and how compatible the innovation is with the way one likes to work are attributes that the individual can readily and confidently evaluate with a hands-on trial of the innovation. During hands-off trial, however, it is likely that all innovation characteristics will be considered during evaluation as well as subjective norm. Both the consideration of more characteristics, as well as normative beliefs, may have the effect of offsetting the higher uncertainty involved in the hands-off evaluation decision.

H4a. For Non-experiential evaluation (hands-off), all innovation attributes and normative beliefs will be important determinants of MWL.

H4b. For Experiential evaluation (hands-on), normative beliefs will not be a significant antecedent of MWL and only some innovation attributes (perceived usefulness, ease of use, and compatibility) will significantly affect MWL.

RQ 5. Through which mechanisms do innovation attributes influence intention to adopt?

A related question to RQ4 examines the mechanisms through which innovation attributes affect the adoption decision during hands-on evaluation. Innovation attributes can affect intention to adopt both through attitude and through mental workload. Which innovation characteristics affect intention to adopt via attitude, which via mental workload, and which via both? For example, we would expect relative advantage to influence intention to adopt through both mental workload and attitude (since it reduces uncertainty about the consequences of adopting), but image to influence intention to adopt only via attitude.

RQ6. Will innovations with a faster rate of decline of MWL be more likely adopted than innovations with a slower rate of decline of MWL?

A component of MWL during hands-on evaluation of the innovation is MWL caused by learning how to use the innovation. It is likely that this component will decline as individuals become more familiar with the innovation during trial use. Moreover, although all innovations are likely to initially induce heightened mental workload in users, the crucial difference is which innovation is likely to cause a lower initial level of mental workload as well as able to decrease this level faster with repeated use. Each innovation will have a declining curve of mental workload against time and repeated use. Some, however, will experience faster decline than others — in other words, the learning effect is enhanced. Innovations that are less likely to be adopted will start from a higher level of mental workload and decline less with repeated use.

Therefore, MWL induced by using the innovation is likely to decline with usage as people performing certain functions will switch from controlled to automatic mode. Controlled processes are described as slow, attention demanding, requiring considerable involvement of short-term memory, exhibiting a large degree of voluntary control by the subject, and requiring little or no training to develop (Shiffrin and Schneider, 1977; Schneider, 1985). Automatic processes, on the other hand, do not require as much attention, are not limited by short-term memory, can be processed in parallel, are not amenable to direct control by the subject, and require extensive training to develop.

The design of the user interface, the ease of use of the innovation and the level of computer experience an individual has with similar types of innovations will all be important in determining the initial level of MWL induced as well as the rate of decline of MWL during the hands-on evaluation process.

H6a. The more experience a user has with similar types of innovations the lower the level of mental workload experienced.

H6b. Mental workload will be lower when the level of innovation complexity (ease of use) is lower.

RQ7. What is the relative importance of attitude toward adopting, subjective norm, and mental workload in determining user intentions to adopt the innovation?

We propose mental workload as an antecedent of the intention to adopt in addition to attitude and subjective norm. Therefore, another important question concerns the relative contribution to the adoption decision of mental workload vis-a-vis attitude and subjective norm. Is an innovation that scores high on many innovation attributes, but induces high mental workload more likely to be rejected?

We would expect that an innovation which poses fewer demands for information processing resources and thus creates lower levels of mental workload to be more likely adopted than one that is more demanding.

H7. The lower the mental workload experienced during use of the innovation, the more likely it is that the innovation will be adopted.

Conclusion

This theoretical work brings together two previously unrelated bodies of research. The intent is to increase our understanding of the process that individuals go through when deciding whether to adopt information technology innovations. Theoretically, the inclusion of mental workload as an important variable in the innovation adoption process has the potential to advance our current level of knowledge and open new directions in the area of new technology introduction and innovation adoption.

The research questions proposed in this chapter investigate important aspects of the theoretical model, but they are not meant to be exhaustive. For instance, a better understanding of the different types of mental workload presented in Table 2 would provide critical insights into this process. Furthermore, the innovation antecedents of the different types of mental workload and their relative contribution to the total mental workload experienced during innovation evaluation remain important, unanswered questions. A number of empirical studies is required to disentangle these effects.

In addition, we have developed a conceptual model and research questions as to the importance of mental workload with respect to only one of many theories of innovation adoption and diffusion at the individual level. The role of mental workload in other innovation theories remains to be investigated. Mental workload at the organizational level and its relationship with theories of organizational and institutional innovation may also be a worthwhile direction for future research.

Practically, innovation is an important source of competitive advantage. Understanding the innovation decision process increases the management effectiveness of the process and the likelihood of deriving the technology's benefits. For instance, results of empirical studies that address the questions raised here will pave the way for further efforts on shedding light on questions such as what kinds of managerial interventions, innovation interface design, marketing efforts, or implementation techniques can be used to reduce (or maybe increase) the level of mental workload involved in evaluating innovations. If indeed mental workload is an important factor in the evaluation and adoption of innovations, reduction in mental workload during innovation evaluation may better facilitate the innovation-decision process.

APPENDIX A: *Innovation Attributes*

Relative Advantage is the degree to which using an innovation is perceived as being better than using the practice it supersedes (Rogers, 1983). This concept is similar to perceived usefulness, "the degree to which a person believes that using a particular system will enhance his or her job performance" (Davis, 1989), but whereas relative advantage deals with the relationship between the new technology and an existing technology, perceived usefulness deals with the relationship between the technology and the job. Within an organizational context, relative advantage focuses primarily on efficiency and effectiveness considerations in performing one's job.

Computer Avoidance refers to those beliefs that may affect attitude toward adopting an innovation negatively and may, therefore, result in individuals avoiding use of the innovation (Moore, 1989). Computer avoidance beliefs include apprehension about potentially adverse effects of computer usage (Susser, 1987 as cited in Moore, 1989), beliefs about work becoming more boring (Sprague and McNurlin, 1986), and work being more monitored and controlled (Keen, 1981).

Image is defined as the degree to which adoption "of an innovation is perceived to enhance one's image or status in one's social system" (Moore and Benbasat 1991, p. 195).

Compatibility is the degree to which using an innovation is perceived as consistent with the existing sociocultural values and beliefs and past and present experiences of potential adopters (Rogers, 1983; Moore and Benbasat, 1991). The definition implies two types of compatibility: normative or cognitive compatibility referring to compatibility with what people feel or think about an innovation, or practical or operational compatibility, referring to compatibility with what people do (Tornatzky and Klein, 1982).

Complexity is defined as "the degree to which an innovation is perceived as difficult to understand and use" (Rogers 1983, p. 230). Conversely, Davis (1989, p. 320) defines ease-of-use as "the degree to which using a particular system would be free of effort." The easier an innovation is to learn and use, the more likely it is to be adopted. This derives from self-efficacy which can be defined as a person's judgment of her "capabilities to organize and execute courses of action required to attain designated types of performance" (Bandura, 1986, p. 391). The more complex an innovation, the less individuals will perceive that they possess the needed capabilities to learn how to use it, and therefore the less likely they will adopt it (Bandura, 1986).

Trialability is the degree to which one can experiment with an innovation on a limited basis before making an adoption or rejection decision (Rogers, 1983). During the persuasion and decision stages, an individual is motivated to seek out innovation-evaluation information (Rogers, 1983). By trying out an innovation, a potential adopter reduces risk and uncertainty about the expected consequences of using the innovation and about the advantages and disadvantages of the innovation. This knowledge may facilitate the adoption decision. In addition to trying the innovation out themselves, individuals may reduce uncertainty by watching others use or try out the innovation. Trial or use by peers, termed psychological or vicarious trial, can be a very effective source of evaluative information (Bandura, 1977).

Visibility is defined as the "degree to which the innovation is apparent to the sense of sight (visible)" (Moore and Benbasat, 1990, p. 9).

Result Demonstrability is defined as the "degree to which the results of using the innovation are observable and communicable to others" (Moore and Benbasat, 1990, p. 9).

APPENDIX B: *Mental Workload Measurement Methods*

a. Secondary Task Method. The secondary task method requires the subject to perform a second task concurrently in order to assess process demands on the primary task (Kerr, 1973). If the subject performs the secondary task well, it is an indication that the primary task is relatively easy, and vice-versa. The use of the secondary task method as a measurement for workload has some drawbacks. The most serious is the intrusive aspect of the secondary task (Meshkati and Loewenthal, 1988b). In fact, the introduction of the secondary task usually degrades performance on the primary task (Williges and Wierwille, 1979). Welford (1973) suggested that a change in strategy in the primary task might be caused by the extra load imposed; therefore, producing misleading assessments of the primary task workload.

b. Performance Measures.This is probably the most obvious method of workload assessment (Meshkati and Loewenthal, 1988b). For instance, if one wanted to know how mental workload during flying is influenced by certain task characteristics, one might want to use flying performance as an indicator of mental workload. However, there is a lack of sensitivity of performance measures to changes in mental workload levels. Meshkati and Loewenthal (1988b) pointed out that the level of mental workload may increase while performance is unchanged, so performance may not be a valid measure of workload.

c. Physiological Measures. Certain physiological functions usually show changes when an individual is under higher mental workload. Several functions have been investigated—among them, sinus arrhythmia (heart beat rate changes) (e.g., Sharit and Salvendy, 1982), the P300 component of the Event Related Brain Potential (ERP) (e.g., Wickens, Heffley, Kramer and Donchin, 1980), mean pulse rate, pulse rate variability, respiration rate, pupil diameter, and voice pattern (Wicrwillc and Conncr, 1983). Although objective and hard to refute (for a selected few functions of the list above), physiological data is difficult to collect in the field or in large quantities in the laboratory. Moreover, special (and expensive) equipment is necessary.

d. Subjective Ratings. A subjective rating of mental workload consists of asking the subject directly and indirectly his opinion of the workload involved in the task. Subjective ratings may come closest to tapping the essence of mental workload and provide the most generally valid and sensitive indicator (Johanssen, Moray, Pew, Rasmussen, Sanders and Wickens, 1979; Sheridan and Stassen, 1979). There are, however, practical problems involved in the translation of an individual's workload experience into a formalized rating. First, even though people might generate evaluations about the difficulty of performing a task, they hardly ever quantify, remember or verbalize these impressions (Hart and Staveland, 1988). Secondly, only the most recent information is directly accessible for verbal reports—which means that a considerable amount of information might be unavailable after the task is performed. An *a posteriori* evaluation also might be affected by several cognitive biases (e.g., recency) that operate to exclude relevant information. Thirdly, there is no objective standard against which people's evaluation of mental workload can be compared. According to Hart and Staveland (1988) the "absence of external validation represents one of the most difficult problems in evaluating a candidate workload assessment technique or the accuracy of a particular rating. There is no objective workload continuum, the 'zero' point and upper limits are unclear, and intervals are often arbitrarily assigned" (p. 143). Finally, in spite of extensive use, the value of ratings has been denied by researchers who consider it just another introspection-based technique. When considering the drawbacks of the other measurement methods, however, the more accepted view is that subjective ratings provide a significant source of information (Hart, 1986). This view becomes even more reasonable when we consider the following example: If an operator considers a task to generate a high workload (and subjectively rate the task as such), she might feel or behave as if she were overloaded, and start employing strategies such as shedding tasks or lowering performance standards without regard for the objectively low task demands.

The different ways to subjectively measure workload include: "(1) unidimensional numerical ratings given with or without behavioral anchors or verbal labels, (2) multidimensional evaluations of task features or psychological variables, (3) rank-ordering of task with respect to workload, and (4) task-specific checklists, questionnaires, and verbal protocols" (Hart, 1986, p. 411). One can rely both on workload ratings given by operators (or subjects) and by observers. Although unidimensional ratings are relatively common and useful for specific tasks which are either simple or when the researcher's interest concentrates on one dimension, the more common technique is the use of multidimensional rating scales based on subjects' ratings. Two techniques in particular have been developed and validated: the subjective workload assessment technique (SWAT), sponsored by the U.S. Air Force and developed by Eggemeier and associates (Eggemeier, 1981), and the NASA-TLX (Task Load Index), developed by Hart and associates (Hart and Staveland, 1988). A more complete review of these methods can be found in Reid and Nygren (1988) and Hart and Staveland (1988).

Endnotes

[1] Relative Advantage is similar to perceived usefulness, "the degree to which a person believes that using a particular system will enhance his or her job performance" (Davis, 1989), but whereas relative advantage deals with the relationship between the new technology and an existing technology, perceived usefulness deals with the relationship between the technology and the job. Within an organizational context, relative advantage focuses primarily on efficiency and effectiveness considerations in performing one's job.

[2] Demand itself is created by two sub-processes: work management and task execution. Work management is the metatask thought, i.e., thinking about the task of managing the task. Therefore, it involves task selection and planning prior to work, and monitoring, evaluation and adjustment of activities during work. Task execution consists of operations performed by the knowledge worker while engaged in a task, contributing to the completion of the task and not classified under work management. In knowledge work task, both work management and task execution place demands simultaneously on AIPRs.

[3] In this figure, the term Mental Workload represents the box labeled "Mental Workload Research" in Figure 1.

[4] The numbering in the hypotheses follows the research question numbering.

References

Achrol, R. (1988). "Measuring Uncertainty in Organizational Analysis," *Social Science Research*, 17, 1988, 66-91.

Achrol, R., & Stern, L (1988). "Environmental Determinants of Decision Making Uncertainty in Marketing Channels," *Journal of Marketing Research*, 25, February, 36-50.

Bandura, A. (1986). *Social Foundations of Thought and Action: A Social*

Cognitive Theory, Prentice Hall, Englewood Cliffs, NJ.

Bandura, A. (1977). *Social Learning Theory*, Prentice Hall, Englewood Cliffs, NJ.

Baron, R.M., & Kenney D.A. (1986). "The Moderator-Mediator Variable Distinction in Social Psychological Research: Conceptual, Strategic, and Statistical Considerations," *Journal of Personality and Social Psychology*, 51, Number 6, 1173-1182.

Ben Zur H., & Breznitz, S. J. (1981). "The Effect of Time Pressure on Risky Choice Behavior," *The Acta Psychologica*, 47, Number 2, February, 89-104.

Brown S. A., & Chervany, N.L. (1994). "An Absorptive Capacity Perspective of the New Information Technology Introduction Process," Working Paper, WP-94-01, University of Minnesota..

Busemeyer, J. R. (1985). "Decision Making Under Uncertainty: A Comparison of Simple Scalability, Fixed-Sample, and Sequential-Sampling Models," *Journal of Experimental Psychology: Learning, Memory, and Cognition*, 11, Number 3, 538-564.

Cash, J. I., McFarlan, F. W., McKenney, J. L., & Applegate, L. M. (1992). *Corporate Information Systems Management*, Third Edition, Irwin, Homewood, IL.

Cooper, R. B., & Zmud, R. W. (1990). "Information Technology Implementation Research: A Technological Diffusion Approach," *Management Science*, 36, Number 2, February, 123-139.

Davis, F. D. (1989). "Perceived Usefulness, Perceived Ease of Use, and User Acceptance of Information Technology," *MIS Quarterly*, 13, Number 3, December, 319-340.

Downs, G. W. Jr., & Mohrs, L. B. (1976). "Conceptual Issues in the Study of Innovation," *Administrative Science Quarterly*, 21, December, 700-714.

Eggemeier, F. T. (1981). "Current Issues in Subjective Assessment of Workload," *Proceedings of the Human Factors Society 25th Annual Meeting, Human Factors Society*, Santa Monica, CA, 513-517.

Evaristo, J. R. (1993). "An Empirical Investigation of the Impact of Information Characteristics and Information Technology on Individual Information Load," Unpublished Doctoral Dissertation, University of Minnesota.

Hart, S.G. (1986). "Theory and Measurement of Human Workload," in J. Zeidner, *Human Productivity Enhancement: Training and Human Factors in Systems Design*, 1, 396-456, Praeger, New York.

Hart, S. G., & Staveland, L. E. (1988). "Development of NASA-TLX (Task Load Index): Results of Empirical and Theoretical Research," in P. Hancock and N. Meshkati, (Eds.) *Advances in Psychology*, 52, North Holland, Amsterdam, 139-184.

Huntley, J. H. (1982). "Task Complexity and Serial Performance under steadily Increasing Input Rates," *Human Factors*, 8, Number 2, 109-113.

Johannsen, G., Moray, N., Pew, R., Rasmussen, J., Sanders, A., & Wickens, C. D. (1979). "Final Report of Experimental Psychology Group," in N. Moray, (Ed.), *Mental Workload: Its Theory and Measurement*, Plenum Press, New York, 101-114.

Karahanna, E. (1993). "Evaluative Criteria and User Acceptance of End-User Information Technology: A Study of End-User Cognitive and Normative Pre-Adoption Beliefs," Unpublished Doctoral Dissertation, University of Minnesota.

Kerr, B. (1973). "Processing Demands During Mental Operations," *Memory and Cognition*, 1, Number 4, 401-412.

Keen, P. G. W. (1981). "Information Systems and Organizational Change," Communications of the ACM, 24, Number 1, January, 24-33.

Kwon, T. H., & Zmud, R. W. (1987). "Unifying the Fragmented Models of Information Systems Implementation," In *Critical Issues in Information Systems Research*, R. J. Boland and R. A. Hirscheim (eds.), John Wiley and Sons Ltd, 227-251.

Miller, J. (1978). *Living Systems*, McGraw-Hill, New York.

Montagna, P. (1980). "Uncertainty as a Scientific Concept and its Application to the Study of Occupations and Organizations," In S. Fiddle (Ed.), *Uncertainty: Behavioral and Social Dimensions*, Praeger, New York, 9-42.

Moore, G. C. (1989). "An Examination of the Implementation of Information Technology by End-Users: A Diffusion of Innovations Perspective," Unpublished Doctoral Dissertation, University of British Columbia.

Moore, G. C., & Benbasat, I. (1990). "An Examination of the Adoption of Information Technology by End-Users: A Diffusion of Innovations Perspective," Working Paper #90-MIS-012, July.

Moore, G. C., & Benbasat, I. (1991). "Development of an Instrument to Measure the Perceptions of Adopting an Information Technology Innovation," *Information Systems Research*, 2, Number 3, September, 192- 222.

Nygren, T. (1991). "Psychometric Properties of Subjective Workload Measurement Techniques: Implications for Their Use in the Assessment of Perceived Mental Workload," *Human Factors*, 33, Number 1, 17-33.

O'Donnel, R., & Eggemeier, T. (1986). "Workload Assessment Methodology," in K. Boff, L. Kaufman, & J. Thomas, (Eds.), *Handbook of Perception and Human Performance*, II, Chapter 42, Wiley, New York.

Perrow, C. (1970). *Organizational Analysis: A Sociological View*, Wadsworth, Belmont, CA.

Quastler, H., & Wulff, V. J. (1953). "Human Performance in Information Transmission," Technical Report R.62. Control Systems Laboratory, University of Illinois, Urbana, Illinois.

Reid, G. B., & Nygren, T. E. (1988). "The Subjective Workload Assessment Technique: A Scaling Procedure for Measuring Mental Workload," (Eds.), *Advances in Psychology* 52, North Holland, Amsterdam, 185-218.

Rogers, E. (1983). *The Diffusion of Innovation*, 3rd edition, Free Press, New York..

Schneider, Walter. (1985). "Toward a Model of Attention and the Development of Automatic Processing," in Michael I. Posner and Oscar S. Marin (Eds.), *Attention and Performance* XI, Lawrence Erlbaum Associates, Publishers, Hillsdale, NJ.

Schroder, H. M., Driver, M.J., & Streufert, S. (1967). *Human Information Processing*, Holt, Rinehart and Winston, New York.

Sharit, J., & Salvendy, G. (1982). "External and Internal Environments II. Reconsideration of the Relationship Between Sinus Arrhythmia and Information Load," *Ergonomics*, 25, 121-132.

Sheridan, T. B., & Stassen, H. (1979). "Definitions, Models and Measures of Human Workload," in N. Moray, (Ed.), *Mental Workload: Its Theory and Measurement*, Plenum Press, New York, 219-234.

Shiffrin, R. M., & Schneider, W. (1977). "Controlled and Automatic Human Information Processing: Perceptual Learning, Automatic Attending, and a General Theory," *Psychological Review*, 84, 119-127.

Smith, G., Benson, P. G., & Curley, S. (1991). "Belief, Knowledge and Uncertainty: A Cognitive Perspective on Subjective Probability," *Organizational Behavior and Human Decision Processes*, 48, 291-321.

Sprague, R., & McNurlin, B. (1986). *Information Systems Management in Practice,* Prentice-Hall Inc., Englewood Cliffs, NJ.

Starbuck, W. H. (1976). "Organizations and their Environments," in M.D. Dunette (Ed.), *Handbook of Industrial and Organizational Psychology,* Rand McNally, Chicago, 1069-1123.

Susser, P. A. (1987). "Modern Office Technology and Employer Relations," *Employee Relations Today,* Winter, 363-371.

Svenson, O., & Edland, A. (1987). "Changes of Preferences Under Time Pressure: Choices and Judgements," *Scandinavian Journal of* Psychology, 28, 322-330.

Vince, M. A. (1949). "Rapid Response Sequences and the Psychological Refractory Period," *British Journal of Psychology,* 40, 23-40.

Welford, A. T. (1973). "Mental Workload as a Function of Demand, Capacity, Strategy and Skill," *Ergonomics,* 21, 157-167.

Wickens, C., Heffley, E., Karmer, A., & Donchin, E. (1980). "The Event-related Brain Potential as an Index of Attention Allocation in Complex Displays," *Proceedings of the Human Factors Society,* 24, 297-301.

Wierville, W. W., & Conner, S. A. (1983). "Evaluation of 20 Workload Measures using a Psychomotor Task in a Moving Base Aircraft Simulator," *Human Factors,* 25, 1-16.

Williges, R.C., & Wierwille, W. W. (1979). "Behavioral Measures of Aircrew Mental Workload," *Human Factors,* 21, 549-574.

Wood, R. E. (1986). "Task Complexity: Definition of the Construct," *Organizational Behavior and Human Decision Processes,* 37, 60-82.

Wright, P. L. (1974). "The Harassed Decision Maker: Time Pressures, Distractions, and the Use of Evidence," *Journal of Applied Psychology,* 59, 555-561.

Wright, P. L. (1975). "Consumer Choice Strategies: Simplifying versus Optimizing," *Journal of Marketing Research,* 11, February, 60-67.

Zakay, D., & Wooler, S. (1984). "Time pressure, training and decision effectiveness," *Ergonomics,* 27, 273-284.

Zmud, R. W., & Apple, L. E. (1989). "Measuring Information Technology Infusion," Unpublished Manuscript.

CHAPTER

4

Software Process Improvement: Innovation and Diffusion

Rick Gibson
American University

Today's information systems (IS) practitioners are, of necessity, innovators. They have been inundated with methodologies, models, techniques, paradigm shifts, tools, and assorted other proposed solutions to software's chronic illnesses. Most recently, a process revolution has substantially altered perceptions within the IS industry. The essence of this revolution is the idea that the quality of a software product is highly dependent on the quality of the process used to develop it. At the forefront of this revolution we can observe three categories of innovations: new process models, new modeling techniques, and the concept of process maturity assessment. Unfortunately, these innovations have yet to be widely diffused—the Software Engineering Institute estimates that over 75% of software development is still done in an immature, ad hoc fashion.

This chapter explores the apparent contradiction between the willingness of IS professionals to embrace technological change, and their corresponding reluctance to adopt process improvement methods. Unlike most other professions, rapid technological development is a defining characteristic of the IS profession. It is easy to forget that in the not-too-distant past, IS departments were tasked only with instituting routine, transaction processing tasks. Wysocki and Young (1990) suggest that this has resulted in a failure to cultivate among IS professionals the skills and processes needed to support innovation. Freedman (1987) suggested that the environments of many IS departments fail to nurture ideas. Later, Couger et al. (1994) suggested that IS managers exhibit different styles of creativity from other professions, which tend to make IS environments inadequate to support innovation.

The most recently proposed solutions to the software crisis—process improvement models—represent a major innovation. The adoption and rate of diffusion of this innovation by the IS community is therefore of interest from both an academic and practitioner point of view. Although there has been much progress in software process improvement, the adoption of software process improvement innovations has been slow. In fact, Balzer (1990) stated that the problems continue to overshadow the progress made. Major obstacles to software process improvement include the attitude and culture of the people involved in the software development process. Couger (1996) observed that terms such as innovation and creativity are used frequently in the IS literature, but found only five relevant papers published over a 35-year period. Moreover, Davis' (1995) *201 Principles of Software Development* omit any discussion of innovation. Many others related works provide only a passing reference to innovation, usually in some prescriptive tone.

Successfully creating software process improvement programs requires top management support, a culture of accepting change as a way of life, and investment for improvements (Curtis & Paulk, 1993). If these elements are missing, the software process improvement program is sure to fail. Software process improvement is an important innovation, which needs some help in increasing its rate of adoption. Given the relative dearth of innovation studies related to the IS field and the economic implications of changes in IS development, it seems worthwhile to examine the coupling of innovation and diffusion theories with process improvement models.

The next section of this chapter presents background information about innovation/diffusion theories relevant to process improvement models. The background information is followed by proposed theory—based guidelines to enable IS managers to better understand, change and control the processes underlying IS development.

Background Preliminaries

The IS literature contains an abundance of publications on innovation as it relates to IS, and the conventional research approach would make use of this material. However, given the rich heritage and diverse origins of innovation/diffusion ideas, it was decided for the purposes of this chapter to return to original source material.

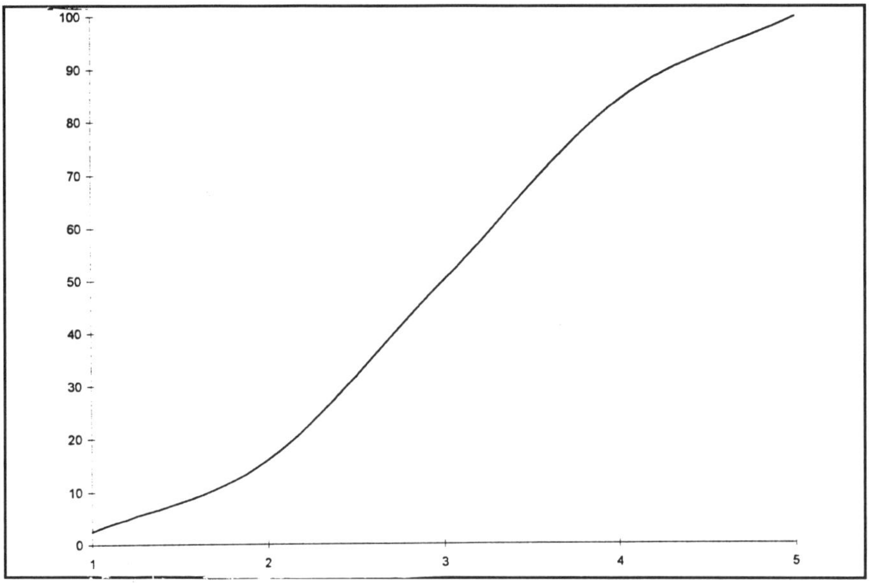

Figure 1: A Model of Diffusion of Innovations

Innovation Adoption

From his substantial work on the subject, Rogers (1983) groups the adopters of innovations into five categories. Innovators form the first 2.5% to adopt an innovation. Early adopters are the next 13.5% to adopt. The early majority are the next 34% to adopt. As are the late majority, who are the next 34% to adopt. Finally, the laggards are the last 16 % to adopt. This division has not been seriously challenged, and was more recently used by Hawkins et al. (1989)

The resulting "S" curve shown in Figure 1 is typical for the rate of adoption for most innovations, although the slope of the curve may vary depending on rapidity of the diffusion. After a first, slow rate when the innovators try out the innovation, the curve moves upward as a result of actions taken by early adopters. Thereafter the curve shows an exponential growth when the early and late majority acquire the innovation, after which the curve again flattens when only the last of the late majority and laggards adopt. SEI data presented in Figure 2 suggest that the adoption of the process maturity assessment innovation follows the same pattern. Note the flattening of the curve for first assessments by 1992.

Rogers describes the innovators, in general, as young, well educated and capable of absorbing the costs of unsuccessful choices. Innovators make use of commercial media and professional sources in learning of new products. In contrast, early majority adopters are cautious with respect to innovation; choosing to wait until the innovation has to be proven successful. These early majority adopters depend

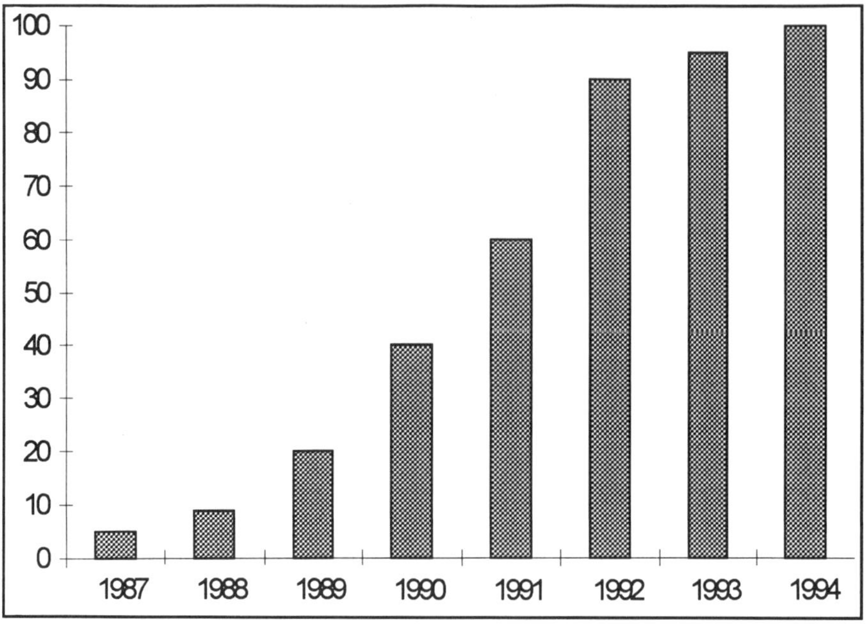

Figure 2: Number of Assessments by Year

heavily on interpersonal sources of information. Late majority adopters are older and skeptical about innovations. They often adopt more in response to decreased availability of the previous product than because of the success of the innovation. Finally, laggards tend to be relatively dogmatic and oriented towards the past; hence, innovations are adopted with reluctance. In summary, the categories of innovators have great relevance for this chapter. IS managers appear subject to the same forces that impede innovation, but must function in an environment that requires rapid adoption and diffusion.

Innovation Diffusion

The subject of diffusion of innovations has been investigated by various authors over the years. In an early, but still undisputed, study of the process of diffusion of innovations, Hagerstrand (1953) asserted that the adoption of an innovation is primarily the outcome of a learning and communication process. Gogan and Cash (1991) provide continued support for this important contention by their model of the information technology innovation process (Figure 3), which begins with information technology literacy.

Hawkins et al. (1989) suggest that one of the questions that should be addressed in connection with adoption of innovations is: What determines how rapidly a particular innovation will spread? In response, they suggest that the diffusion of technological innovations depends to a great degree on the relative advantage (or cost savings) of the innovation. Nabset and Ray (1974) contend that in the early stages the outcome of the innovation is uncertain, and the benefits unclear

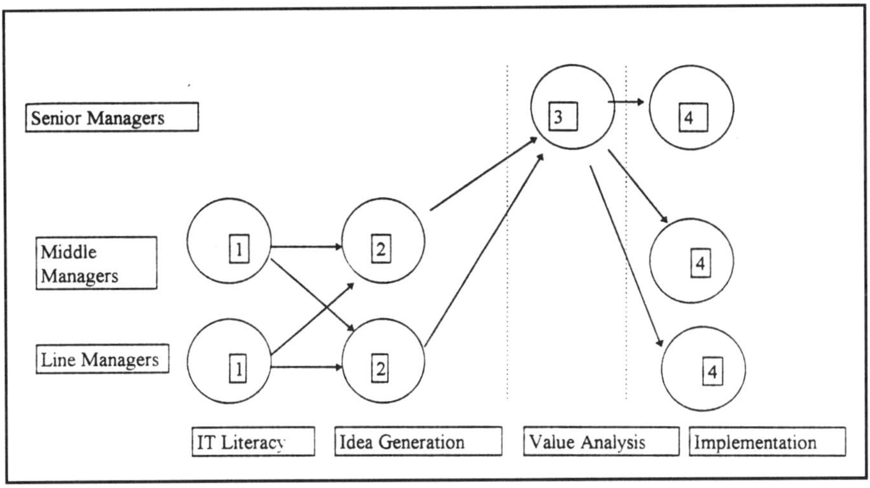

Figure 3: Information Technology Innovation Process

and so diffusion is slow. Furthermore, Brown (1981) points out that the continuous development of the original innovation makes it more desirable over time, but this is difficult to assess properly in the early stages of the diffusion cycle.

A further complication is pointed out by Sahal (1981), who reminds us that the advent of a new technique does not in itself spell an end to the use of the old technique. Rather, the diffusion of an innovation is a process of co-evolution of old and new techniques involving numerous changes in their functional characteristics. Twiss (1986) illustrated this by the example of the material and characteristics of the cord being used in automobile tires. The use of cotton, substituted by steel and later fiberglass have improved the product without changing the appearance or use. (Arguably, software is analogous to these fiber cords—unseen, but critical.)

Mansfield (1968) claims that four principal factors govern how rapidly an innovation will spread: 1) As the number of users increases, the probability of its further adoption increases. Once a critical mass is reached, exponential growth can be expected. 2) The profitability of the innovation is assumed to be directly related to the probability of its adoption. 3) For equally profitable innovations, it is assumed that innovations requiring smaller investments will be adopted faster than innovations requiring larger investments. 4) The more adaptive and risk-prone the industry in which the innovation is introduced, the faster the diffusion will be.

Barczak et al. (1987) identified a number of key elements involved in dissolving existing patterns (i.e., processes) and creating new ones: pattern breaking; experimenting; visioning; and bonding and attunement. Finally, Damanpour (1991) reminds us that radical innovations produce fundamental changes and represent a clear departure from existing practices, whereas incremental changes do not. By reviewing the change management literature, Benjamin and Levison (1993) identified eight principles regarding complex, IS-enabled changes: 1) Develop a systematic process for change; 2) Manage equilibrium and mutual adaptation of the organization, technology and business units; 3) Determine whether there exits enough energy for change; 4) Analyze the size of the change effort; 5) Analyze and manage stakeholder commitment; 6) Identify and understand the champion's role; 7) Prototype organizational response; 8) Build change reviews into the management process.

Armenakis et al. (1993) contend that willingness to change impacts effectiveness and may preempt resistance. More rigorously, Dannemiller and Jacobs (1992) concluded that the product of dissatisfaction with the present situation (D), vision (V) and first process steps toward the vision (F) must be greater than the resistance to change. Concisely, D (V (F > R. It has been shown that process improvement models identify areas of dissatisfaction (D); provide a vision (V) of higher maturity benefits; and specify the necessary first steps (F).

The preceding brief descriptions of the major ideas that are relevant to this chapter are summarized by Table 1. Later in the chapter, these ideas will be coupled with the process improvement concepts described in the next section.

Authors	**Relevant Finding**
Hagerstrand (1953)	Communication and learning processes affect adoption.
Nabset & Ray (1974) Brown (1981	Innovations are difficult to assess properly in the early stages of the diffusion cycle.
Sahal (1981)	Innovative techniques need not supplant old techniques.
Mansfield (1968)	The probability of adoption increases due to: critical mass of users, profitability, smaller investments, the risk-prone industry.
Barczak et al. (1987)	Creation requires: pattern breaking; experimenting; visioning; and bonding and attunement.
Damanpour (1991)	Incremental change is easier than radical change.
Benjamin & Levison (1993)	Principles of change.
Dannemiller & Jacobs (1992) Armenakis et al. (1993)	Resistance to change must be overcome.

Table 1: Major Innovation Concepts

Process Improvement Definitions and Background

In the 1980s, a software process improvement movement emerged because of shortcomings in software productivity and quality. Software development as a process is an old idea, but reemerged because of:

- resources being depleted by economic trends
- offshore development pressures
- outsourcing opportunities
- quality and reliability concerns
- complexity expanding due to marketplace demands

Competitive pressures on the software development industry at home and from abroad led many organizations to examine their software development processes. Heineman et al. (1994) contend that organizations realized that the key to improving the results of their software development operations was to focus on the software process by which the operations were performed. In 1985, several leading U.S. aerospace companies came together as the Software Productivity Consortium. The consortium included companies such as Boeing, General Dynamics, Hughes Aircraft, Martin Marietta, and Rockwell International. The goal of the consortium was to assess and improve the software development process.

Building on the initial work by the consortium, many software process capability assessment and improvement models have been developed. There is a parallel evolution of proprietary and industry models. The proprietary ones (e.g., Andersen's Method/1) are still being refined by the vendors as a method of keeping the present users satisfied and thus avoiding transfer to the less profitable industry model process improvement environment. Some software process capability models in use are: Software Engineering Institute (SEI) Capability Maturity Model (CMM), the International Standards group for Software Engineering's Software Process Improvement and Capability Determination (SPICE) model, and the Quality Improvement Paradigm (QIP). This chapter will focus on the CMM. Detailed discussions of SPICE (e.g., Dorling, 1993) and QIP (e.g., Basili & Green, 1994) are readily available.

The Software Engineering Institute (SEI) was established in 1984 and awarded to Carnegie Mellon University. The SEI's mission is to provide leadership in advancing the state of the practice of software engineering to improve the quality of systems that depend on software. The Capability Maturity Model (CMM) was developed under a federally funded program by the SEI as a common-sense application of process management and quality improvement concepts to software development and maintenance.

The CMM identifies five levels of maturity for measuring the capability of an organization's software development process. It provides a framework for organizations to determine the current level of maturity for their software development process, and the steps to take to increase their maturity level. A more detailed discussion on the CMM can be found in Paulk (1993). The CMM has had a profound impact on the global software industry, but has been criticized by many experts (Bollinger and McGowan, 1991, Jones, 1995), largely due to a lack of emphasis on technology or automation.

Davis and Wallace (1992) assert that the CMM is based on time-proven quality principles in that it applies Total Quality Management to software. Its roots are in statistical quality control, which was developed by Shewart in the 1930s and further developed and successfully demonstrated by Deming and Juran. The CMM casts these principles into a maturity framework, which consists of five maturity levels that describe the progression from a chaotic process to a well-controlled optimizing process. The levels lay successive foundations for continuous process improvement. By determining their position in this framework, organizations can readily identify the most fruitful areas for improvement actions. Each level establishes an intermediate set of goals toward higher levels of process maturity.

Curtis and Paulk (1993) comment on the maturity concept by contrasting mature and immature process impacts on organizations, as summarized in Table 2.

Experience has shown that it typically takes an organization committed to process improvement anywhere from a year and a half to three years to progress by one complete level of maturity; but the progression from Level 1 to Level 2 may require more on the order of three to five years.

Saiedian et al. (1995) suggest that because of its relatively recent

Immature Process	**Mature Process**
Ad hoc; process improvised by practitioners and management	Work gets done in a consistent, defined, documented and continuously improving way.
Procedures not rigorously followed or enforced	Development efforts are supported visibly by management and others.
Highly dependent on current practitioners, key individuals	Individual performance is still important, but institutionalized processes endure after the people who originally defined them are gone.
Product functionality and quality may be compromised in order to meet schedule.	Product and process measurement is used.
Use of new technology is risky.	Disciplined use of technology.
Quality is difficult to predict.	Process fidelity is controlled, audited and enforced.

Table 2: Maturity Concept

arrival on the system development assessment scene, only time will tell if this effort will really work in terms of an overall improvement of software produced. Nevertheless it seems clear that the government (especially the military) can have a major effect on adoption of the CMM by the software contracting industry by requiring Level 3 maturity of all contractors. They discuss the fact that 80% of all companies assessed are currently at Level 1 and why this is the case. A problem with Level 1 is that it is very broad, ranging from completely incapable of developing software to those who are producing excellent software albeit their needed process improvements.

Constance (1996) explains that beginning in the early 1980s, when the command-and-control systems were a focal point, the prime software contractors for the federal government, particularly the Department of Defense, would generally get the hardware delivered on time, but the software was on the average of 18 months late. Looking at software acquisitions across the board began to concern officials because those in place were continually causing trouble. It is then that government leaders began to look for ways to make sure the contractors that were being awarded the contracts were in fact well qualified to do the job timely and accurately. Agencies like the Air Force are using 1998 as a deadline for all contractors doing more than 25% of the coding to be at Level 3 to qualify.

Coupling Innovation Theory with Process Improvement

The preceding sections have suggested that software process improvement is an important innovation, which needs some help in increasing its rate of adoption. This section proposes theory-based guidelines to enable IS managers to better understand, change and control process improvements. Key ideas and themes from the previous discussions on innovation/diffusion theory and process improvement, respectively, will be coupled.

Recalling the prior discussion of Roger's categories, the present adopters of process improvement models may be classified as early adopters, which warrants an expanded description of that category. Von Hippel (1988) introduced the category of *lead user*, which seems to encompass the innovators and part of the early adopters in Rogers' framework. This term is particularly useful in the rapidly changing world of IS. Lead users are those few who do have experience with the use of an innovation as a result of needing the innovation months or years before others in their industry. By identifying and analyzing these lead users' reactions and attitudes to the innovation under investigation, numerous insights may be achieved. Hayes and Zubrow (1995) point out that lead users of process improvement have been firms such as Department of Defense (DoD) and federal government contractors, which is consistent with the focus of the CMM. The DoD Software Technology Strategy has focused on the software development process, and declared that there must be evolutionary changes in the software process (Nordwall, 1992).

Applying Hagerstrand's seminal contention that communication and learning processes affect adoption we find another guiding principle. In the case of process improvement, the distribution of information *per se* does not constitute a barrier, but the lack of precise definitions distort relevant communication. For example, Pressman (1996) developed a Software Engineering Self-Assessment Test, which has been administered to 2,600 software developers and managers. Software developers on average answered only 49% correctly, with managers doing only marginally better (58% correct). The significant weaknesses appeared in areas of quality assurance, metrics, configuration management, and testing.

The absence of cost-benefit data also distort relevant communication. This absence can be explained, in part, by Brown's (1991) suggestion that innovations are difficult to assess properly in the early stages of the diffusion cycle. Hersch (1993) observed that process improvement efforts have encountered many of the same productivity and payoff measurement problems as IS in general. Investment payoff data are available to establish an argument for adopting the CMM as a model to guide IS investment decisions. Data, such as shown in Table 3 are emerging from early adopters. A summary of the findings by Herbsleb et al. (1994) for thirteen organizations (Bull HN, GTE Government Systems, Hewlett-Packard, Hughes Aircraft Co., Loral Federal Systems, Lockheed Sanders, Motorola, Schlumberger, Siemens Stromberg-Carlson, Texas Instruments, USAF Oklahoma Logistics Center, USN Fleet Combat Direction Systems support Activity) is shown in Table 3.

Process Improvement Category	**Range**	**Median**
Total yearly costs of SPI activities	$49,000-1,202,000	$245,000
Years engaged in SPI	1-9	3.5
Cost of SPI per software engineer	$90-2004	$1375
Productivity gain per year	9%-67%	35%
Yearly reduction in time to market	15-23%	19%
Yearly reduction in post-release defect reports	10-94%	39%
Business Value of SPI investment (dollar return for each dollar invested)	4.0-8.8	5.0
Early detection gain per year	6-25%	22%

Table 3: Summary Findings

In the case of process improvement, the final form of the model still represents an uncertainty. Innovations are difficult to assess properly in the early stages of the diffusion cycle. This is a particularly bothersome risk, as the adoption of a process improvement model that does not completely conform to the final standard may lead to an additional workload of adjusting the software development process to the new mainstream in addition to the potential cost of unlearning. Sahal (1981) suggested that innovative techniques need not supplant old techniques, and Damapour (1991) reminds us that incremental change is easier than radical change. Appropriately, the CMM is evolutionary: it is based on many small, evolutionary steps each providing the foundation for the next step. Furthermore, the CMM is flexible: it does not require any specific software technology, organizational structure, life-cycle model, or set of documents. Care has been taken to provide a complete set of valid principles that apply to a wide range of situations, not a rigid set of rules and regulations. In contrast, consider the failure of CASE tools as an example of the problems associated with attempts to make radical change.

Organizations have made huge investments in existing methodologies, models, techniques, paradigm shifts, tools, and assorted other proposed solutions to software's chronic illnesses.

The initial investment in process improvement can be low, reducing the risk for the decision makers. Nejmeh (1995) provides a useful strategic alignment matrix for comparing costs and values of process activities, adapted here, in Table 4, to include only relevant CMM activities. Note that some high-value activities have correspondingly low costs. For example, project management is acknowledged as very important, and can be implemented at very low cost. In contrast, high cost status reporting is judged as least important.

Process improvement can show substantial cost savings, even if the price/performance ratio of proprietary systems have become more favorable. Process improvement models are gaining more widespread

	Very Important	**Important**	**Less Important**
Highest Cost	Functional Specification	Requirements Definition	Prototyping
High Cost	Develop User Interface Code	Unit Test	Research
Low Cost	System Test	Project Planning	Process Quality Improvement
Lowest Cost	Project Management	Training	

Table 4: Strategic Alignment Matrix

acceptance, primarily in the form of the CMM. Diffusion of process improvement now has reached the early majority adopter stage and further rapid diffusion should be expected. (Recall Figure 1.)

Resistance to change must be overcome. Consider separately each of parameters specified by Dannemiller and Jacobs (1992). Dissatisfaction with the present situation (D) can be increased by contrasting the characteristics of mature firms with the chaos of immature ones. Vision (V) is also increased by this awareness building comparison. The first process steps toward the vision (F) are made clear by the model. Thus, by increasing each of these three parameters, their product increases substantially enough to be greater than the resistance to change.

Sampler and Galletta (1991) contend that prior to beginning a change program the IS department and organization must resolve two issues: 1) the scope of the change and 2) the depth of the change. The CMM facilitates both of these. Regarding scope—the CMM identifies which breadth items a firm may be willing to change via the CMM framework, which allows great flexibility here. Regarding depth—the industry experience with the CMM provides realistic expectations of how much of a change can be made.

Change Principle	CMM Feature
1) Develop a systematic process for change.	The CMM is a true process that defines five levels of maturity and provides step-by-step guidelines to progressively improve capability.
2) Manage equilibrium and mutual adaptation of the organization, technology and business units.	CMM subprocess categories of management, organization, and engineering.
3) Determine whether there exists enough energy for change.	CMM common feature of commitment.
4) Analyze the size of the change effort.	CMM is a common-sense application of process management and quality improvement concepts to software development; a community owned guide; a model for organizational improvement; the underlying structure for appraisal.
5) Analyze and manage stakeholder commitment.	Requirements Management Key Process area.
6) Identify and understand the champion's role.	Senior management commitment.
7) Prototype organizational response.	CMM can be introduced on a project-by-project basis.
8) Build change reviews into the management process.	CMM focus on process measurement and continuous improvement.

Table 5: Changes Principles and CMM Features

Finally and perhaps most convincingly, each of the eight change principles identified by Benjamin and Levison (1993) map directly to the CMM, as shown in Table 5.

Future Process Improvment Trends

The potential diffusion of process improvement innovations has only begun. Firms have yet to reach the higher levels, where real process improvement is expected. This is especially noteworthy due to the fact that there is a paucity of information about higher maturity levels because virtually all firms in the SEI database are Level 3 and below. Any credible discussion of the future requires firm foundations on which to base predictions. This section uses the established disciplines of economics and risk analysis to support the contention that the diffusion of process improvement innovation has only just begun.

Economics

Nejmeh (1995) discusses one future challenge of software process improvement. He states that all of the software process improvement frameworks developed are useful, but they lack a return-on-investment basis on which to evaluate and improve existing processes. A response to this challenge may be found with David's (1994) contentions that a defining characteristic of modern economic growth has been the rise of the knowledge base—reflected as investment in intangible, knowledge-intensive forms of wealth and accompanying increases in labor quality due to increased educational attainments and increases in capital quality due to the embodiment of new technologies. This is a stark contrast to innovation in the 19th century, which was based on tangible, capital-deepening growth. Technological change has tended to raise the marginal productivity of capital formed by investments in education, R&D, and information systems.

Kennedy (1989) explains that when a new production function is formulated to produce a new product it is referred to as product innovation. Alternatively, actions such as cutting production costs or boosting quality make the process more efficient and merely streamline existing production functions—this precedes innovation. In short, for a process innovation to be economically rational (e.g., beneficial) it should increase profits by decreasing input cost or increasing output value.

Risk Analysis

Even if the long term economics of process improvement are favorable, there may be an inertia barrier that blocks diffusion.

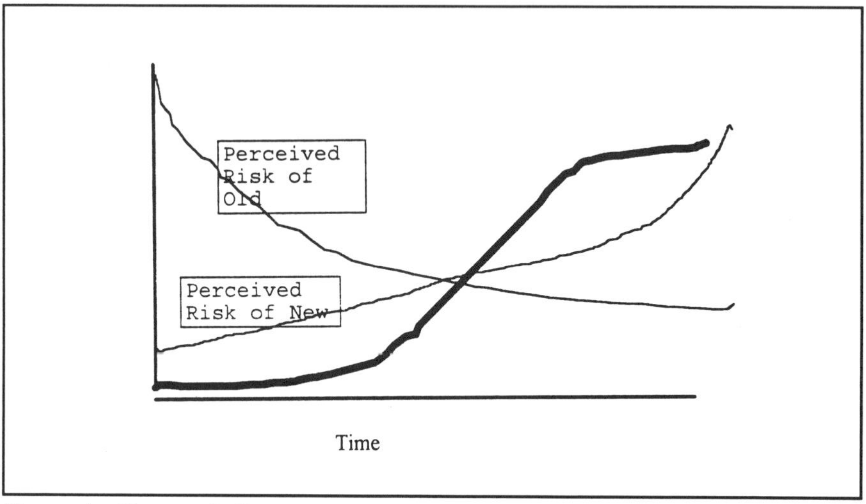

Figure 4 : Changes in Perception of Innovation Risk

Perceived risk influences the diffusion of innovations. It has been suggested that those who acquire innovations early have a particular attitude towards risk. As the innovation's life cycle evolves, the manner in which it is perceived changes. Figure 4 is an illustration of this phenomenon, with the thick, bold line representing the adoption "S" curve.

In the time period that corresponds to the early adopter phase, the perceived risk of an innovation is high, but can be expected to fall. In the this early stage, the corresponding perceived risk of the old process is low. For those who have not made a change to the process improvement innovation, a feeling of risk by keeping the old processes, which may become outdated and impede competitiveness increases. This will again stimulate demand for the new processes and further reduce the image of associated risk. Any strategic decision, such as the case of conversion to process improvement, involves an element of risk. The outcome of the decision making process will, in addition to the quasi-rational stream of reasoning, to a great extent be determined by the decision maker's attitude to risk taking.

Today, process improvement models are, at least theoretically, familiar to most IS professionals. The perceived risk aspect may therefore be reduced. If it is assumed that the level of perceived risk today as an average is at the level where the risk of keeping old technology and the risk of adopting process improvement are equal, i.e. where the two lines intersect. This would contribute to a rapid future diffusion of process improvement.

Conclusion

Solutions to the software crisis exist, but we have a gap between the state of the art and the state of the practice. Bandinelli et al. (1995) report that investment in software is increasing worldwide, and software process improvement has become one of the most important research areas for government and private industry.

This chapter has proposed the suitability of (state of the art) innovation theory techniques as a practical execution of the requirements of the higher maturity levels suggested by the (state of the practice) process improvement models. The purposes for doing so are twofold: First, to provide IS managers with yet another starting point for their process improvement efforts. The notion that process improvement is a viable method has been hampered by slow adoption and diffusion. Secondly, to strengthen the process improvement models themselves.

It appears that attempts at software process improvement (SPI) will expose IS managers to the same forces that impede innovation. For the early advocates (a.k.a. early adopters) of process improvement, innovation theory offers ways to reduce risks and increase the likelihood of successful diffusion. Returning to Gogan and Cash's IT Innovation Process Model, we can see that the CMM fits with their four-step process for continuous, incremental, IT innovation. Unlike the existing top-down models of IS planning, their approach credits ideas generated at lower levels in the organization. The CMM focus on individual projects at Level 2 implements this idea perfectly. This low-level generation of ideas is then supported by efforts to capture best practices for inclusion in organization-wide processes.

By utilizing ideas from the innovation literature, IS managers can create an environment that Pressman (1996) describes as "SPI-friendly." The principles of innovation, diffusion and change management are particularly useful for managing expectations, both your own and those of other senior managers. Process improvement, like any innovation effort loses momentum when unrealistic goals are set, resources become scarce and some people remain unenthusiastic.

References

Armenakis, A.A., Harris, S.G., and Mossholder, K.W. (1993). Creating readiness for large-scale change. *Human Relations*, 46(6):681-703.

Balzer, R. (1991). What we do and don't know about software process. In T. Katayama (Ed.), *Proceedings of the 6th International Software Process Workshop* (pp.61-65). Los Alamitos: IEEE Computer Society Press.

Bandinelli, S. C., Fuggetta, A., Lavazza, L., Loi, M., & Picco, Gian. (1995). Modeling and improving an industrial software process. *IEEE Transactions on software engineering*, 21, 440-454.

Barczak, G., Smith, C. and Wilemon. (1987). Managing large-scale organizational change. *Organizational Dynamics*, 16(2):23-35.

Basili, V. & Green, S. (1994, July). Software process evolution at the SEL. *IEEE Software*, 58-66.

Benjamin, R.I. and Levinson, E. (1993). A framework for managing IT-enabled change. *Sloan Management Review*, 34(4):23-33.

Bollinger, T.B. & McGowan, C. (1991). A Critical Look at Software Capability Evaluations. *IEEE Software* (July):25-45.

Brown, L.A. (1991). *Innovation Diffusion: A New Perspective* London: Methuen.

Constance, P. (1996 January 8). Mr. Software's Legacy Lives on at Defense. *Government Computer News* 15, 18.

Couger, J.D. (1996). *Creativity & Innovation in Information Systems Organizations.* New York:boyd & fraser publishing company.

Couger, J.D., Oppermann, E., and Amoroso, D. (1994). Changes in motivation of I.S. managers. *Information Resources Management Journal*, 7(2):5-14.

Curtis, B., & Paulk, M. (1993). Creating a software process improvement program. *Information & Software Technology*, 35, 381-386.

Damanpour, F. (1991). Organizational innovation: A meta-analysis of effects of determinants and moderators. *Academy of Management Journal.* 34(3):555-590.

Davis, A. (1995). *201 Principles of Software Development.* New York:McGraw-Hill

Dannemiller, D.K. and Jacobs, W.R. (1992) Changing the way organizations change: A revolution of communication. *Journal of Applied Behavioral Science*, 28 (December): 481-490.

Davis, D. and Wallace, D. (1992). *The Capability Maturity Model for Software*, New York: DAZIX, An Intergraph Company.

Dorling, A. (1993). SPICE: Software process improvement and capability determination. *Information & Software Technology*, 35, 404-406.

Freedman, D. (1987). Cultivating IS creativity. *Infosystems* (July):24-27.

Gogan, J. and Cash, J. (1992). IT-based innovation: Managing a disorderly process, *IEEE*, pp. 257-267.

Hagerstrand, T. (1953). *Innovation/Diffusion as a Spatial Process* Chicago: Univ. of Chicago.

Hawkins, D.I., Best, R.J. and Coney, K.A. (1989). *Consumer Behavior.* Homewood, IL:Irwin.

Hayes, W. and Zubrow, D. (1995). Moving On Up: Data and Experience Doing CMM-Based Process Improvement. *Technical Report CMU/SEI-95-TR-008.*

Heineman, G. T., Botsford, J. E., Caldiera, G., & Kaiser, G. E. (1994). Emerging technologies that support a software process life cycle. *IBM Systems Journal*, 33, 501-529.

Herbsleb, J., Carleton, A., Rozum, J., Siegel, J. and Zubrow, D. (1994). Benefits of CMM-based software process improvement: Initial results. *Technical report CMU/SEI-94-TR-13.*

Hersch, A. (1993). Where's the return on process improvement? *IEEE Software*, (July) 12.

Hippel, Eric von. (1988). *The Sources of Innovation.* New York: Oxford University Press.

Jones, C. (1994). Globalization of software supply and demand. *IEEE software*, (Nov.):17-24.

Jones, C. B. (1995). Gaps in SEI Programs. *Software Development*, (March), 41-48.

Kennedy, N. (1989). *The industrialization of intelligence: mind and ma-*

chine in the modern age. London:Unwin Hyman Ltd.

Mansfield, E. (1968). *Industrial research and technological innovation.* New York: W.W. Norton & Co.

Nabset, L. and Ray, G.F. (eds). (1974). *The Diffusion of New Industrial Processes: An International Study.* Cambridge: Cambridge University Press.

Nejmeh, B. A. (1995). Process cost and value analysis. *Communications of the ACM, 38,* 19-24.

Nordwall, B. D. (1992). Defense Department expects new strategy for improving software to save billions. *Aviation Week & Space Technology,* 136, 59-60.

Paul, D. (1994). Technological change, intangible investments and growth in the knowledge-based economy. *Proceedings of the Conference on Employment in the Knowledge-based economy,* Copenhagen, Denmark, November 7-9.

Paulk, M.C., Weber, C.V., Garcia, S.M., & Chrissis, M.B. (1993). Key Practices of the Capability Maturity Model, Version 1.1, *CMU/SEI-93-TR-25.* Software Engineering Institute, Carnegie Mellon University, Pittsburgh, PA.

Paulk, M. (1993). Capability maturity model, version 1.1. *IEEE software,* 10, 18-27.

Paulk, M. (1995) How ISO 9001 Compares With the CMM. *IEEE software,* (January) 74-83.

Pressman, R. (1996). Software process impediment. *IEEE software,* (September) 16-17.

Rogers, E.M. (1983). *Diffusion of Innovations,* 3rd ed. New York: The Free Press.

Rozum, J. (1993). Concepts on measuring the benefits of software process improvements." *Technical report CMU/SEI-93-TR-09.*

Sahal, D. (1981). *Patterns of Technological Innovation.* Reading, MA: Addison Wesley

Saiedian, H., Kuzara, R. & Hamilton, S. (1995). SEI Capability Maturity Model's Impact on Contractors. *IEEE Computer,* (January) 16-26.

Sampler, J. and Galletta, D. (1991). Individual and organizational changes necessary for the application of creativity techniques in the development of information systems. *IEEE,* pp. 404-411.

Twiss, B. (1986). *Managing Technological Innovation.* London: Pitman.

Wysocki, R.K., and Young, J. (1990). *Information Systems: Management Principles in Action.* New York: John Wiley & Sons

CHAPTER

5

The *Sim*View[1] Graphical Simulator for Participatory Decision Support: Evoking Systems Wisdom

James R. Warren
University of South Australia

Johannes N. van Dijk
Marcel J. Jobing
Delft University of Technology

Douglas Seeley
Riccardo Macri
InterDynamics Pty. Ltd.

This chapter describes the Visual Interactive Modelling (VIM) tool SimView for decision support in organizational modelling. SimView supports the Whole Systems Perspective (WSP), an approach aimed at gaining a sound understanding of the dynamic behavior of an organization in a wholistic sense in order to make effective decisions. Historically, simulation has been a technology controlled by technicians and consultants. SimView is intended to give those with direct knowledge of the organization's productive processes (e.g., managers and front-line employees) much more direct access to the simulation model and the modelling process. Application of SimView and the WSP is illustrated by a capacity study carried out for Queensland Rail (QR), a major rail carrier in Australia. The resultant SimView model augmented and clarified management's understanding of the dynamics of their sugar train system. Results obtained from modelling inspired management to cancel a planned capital investment in additional railcars. A set of policy changes was found to be sufficient to handle increased sugar produc-

tion, leading to a saving of several million dollars. It is significant that the "systems wisdom" already existed among operational managers who had a feeling for what would be the most effective policy. The simulation, especially its animation, helped management to have confidence in the solution. Modelling with SimView is now conducted by employees within QR, including decision makers, and is rapidly becoming an integral part of process improvement at QR. Our future goals for VIM technology are: (a) to use the technology to change organizational modelling methods so that pre-computing, accountancy-oriented assumptions are abandoned; and (b) to engender further diffusion of simulators into the hands of the process experts who best know the organization.

In this chapter, we consider the Visual Interactive Modelling (VIM) environment SimView. SimView is the realization of a line of research that began with one of the authors in the mid 1970's (Alemparte, Chheda, Seeley & Walker, 1975) and has been evolving as a PC-based software system (developed in C++) since 1988 (Seeley & Macri, 1991). SimView is designed to support the process of organizational improvement by using simulation modelling as the basis for improved understanding of the organization's dynamics. On the surface, a key aspect of SimView is a strict emphasis on direct manipulation to make the model visible and approachable. In fact, the VIM features of SimView exists to support an explicit philosophy of organizational modelling that we call the Whole Systems Perspective (WSP).

A number of technological trends are represented in SimView's design. These include trends that are sweeping across the entire world of computing, including the movement of processing from mainframes to microcomputers, from textual to graphical and direct manipulation interfaces, and from those who operate computers being specialist technicians to the operators being "ordinary" people with responsibilities outside computing. SimView's design also reflects trends more specific to simulation technology: the movement from general-purpose simulation software to special-purpose simulators, from programming languages to graphical specifications, and from simulation models to simulation environments that support the entire modelling process.

Also highly relevant to SimView is a trend in the business world, the pursuit of competitive advantage through innovative and sometimes radical re-design and rationalization: Business Process Re-engineering (BPR). While the SimView's developers have major philosophical differences with some of the more naive notions of BPR and "downsizing," support for critical thought about business processes is central to SimView's mission. The ultimate goal is to help managers make design decisions that are best for the business. Toward this end, a major objective of SimView is to make simulation technology much more accessible to managers and other process experts (i.e., people with substantial direct knowledge of the organization's productive processes). SimView supports pursuit of high-quality decisions by encouraging those with experience in the organization to explore alternative designs and to get rapid, comprehensible feedback on the merits of those decisions with respect to organizational performance. The result is the engendering of increased management confidence in their decisions, the emergence of innovative design proposals, and the defensibility of such proposals.

This chapter is designed to help the reader appreciate the exciting potential of today's (and tomorrow's) simulation technology to play an integral role in organizational decision making. In particular, we emphasize the ability of simulation models to act as a sort of "white board" such that people from various levels in the organization can develop a concensus on what the organizational dynamics are, and what they should be. We begin by reviewing the evolution of computer-based simulation modelling over the past 50 years that has led us to where we are now. We provide background on the developments in simulation technology that have led to SimView and describe our WSP philosophy of organizational modelling that SimView supports. The third section describes the SimView modelling environment itself, discussing the basics of model specification in SimView to show, in concrete fashion, what such an environment is like. The fourth section uses a case study to illustrate the use of simulation technology as a tool to evoke employee wisdom for organizational decision making. Therein we describe a successful application of SimView that led to the establishment of a SimView-based "modelling culture" at Queensland Rail. We then discuss future trends in visual and interactive simulation technology, including the barriers to more widespread use of simulation for organizational modelling and how they might be overcome. We conclude with a summary of the technological and conceptual innovations in management decision making that we are attempting to accomplish with SimView and related VIM technology.

Background

Innovation and Diffusion of Simulation Technology

Simulation technology matured, at least into its initial form, quickly as general purpose computers became available. In the earliest days of computing, simulation experiments were frequently referred to as Monte Carlo studies, so-called because this was the code name of random sampling studies conducted as part of the development of the atom bomb during the Second World War. By the end of the 1950's, however, simulation technology was already emerging in the form that is quite common today: programming languages for general-purpose simulation modelling. One of most popular simulation programming languages today, the General Purpose Simulation System (GPSS), was developed in the late 1950's by Geoffrey Gordon (1961), thus emerging at much the same time as the some of the classic general-purpose programming languages such as FORTRAN and COBOL. GPSS supports a discrete-event simulation model where the modeller codes transactions that move from block to block (event to event). This fairly simple formalism has proved sufficient for to provide a powerful modelling tool for a vast range of applications including manufacturing systems, health care systems, transportation systems, communication networks, defense systems, civil systems, and queuing systems in general (Schriber, 1991).

Simulation technology remained markedly stable in terms of the tools by which problems were addressed through a large portion of the era of electronic digital computing. Other simulation languages emerged to compete with GPSS, however, it is notable that three out of six simulation languages compared in a survey study conducted in 1966 (Teichroew & Lubin) were still supported in either their original or updated form in 1989 (Hoover & Perry). These three languages are GPSS, SIMSCRIPT and SLAM, with SLAM having resulted from a major revision of GASP to handle continuous and combined discrete-continuous models, as well as discrete-event models. In fact, these languages are still very popular today (1996). Another significant language innovation was the emergence of object-oriented programming for simulation, first introduced in SIMULA (Birtwistle, 1973).

Simulation technology may have been quite stable in some respects from 1966 to 1989, but it was also extremely popular. A survey of nonacademic, full-time members of The Institute of Management Sciences (TIMS) and the Operations Research Society of America (ORSA) found simulation ranking behind only economic analysis and statistical analysis in frequency of use (Watson & Baecher, 1979). Similar results have been found by others (see Watson & Blackstone, 1989, p. 14). Moreover, another survey of nonacademic, full-time members of TIMS/ORSA (Watson & Christy, 1982) found that 89% of the organizations employing members use simulation and that these members generally characterize their organizations' experiences with simulation as having yielded good results. Thus, the diffusion of simulation technology to operations research (OR) professionals, in the form of general-purpose simulation programming languages, has been quite effective.

Through the 1970's and early 1980's, simulation technology was moving into new domains and new forms following the innovation and diffusion in computer hardware. In particular, simulation languages have rapidly appeared for microcomputers. Micro-based simulators include the porting of the established general-purpose simulation languages to microcomputer platforms and also a proliferation of new simulators existing only on microcomputers. Many of these have had small impact in terms of professional use, but some have had a non-trivial history of use. For instance, SIMPAS acts as a preprocessor to Pascal extended with general-purpose simulation features, producing standard Pascal as output (Bryant, 1980).

Among the emerging languages for microcomputers, we see more *special* purpose simulation languages, aimed at use for a relatively narrow class of problems, than we see general-purpose simulation languages. These languages are formulated such that their syntax somewhat resembles the form and terminology used by knowledgeable experts in the given field. Examples include INS, designed for simulation of ambulatory care facilities (Roberts & Sadlowski, 1976), and SPEED which is used to model manufacturing facilities (Horizon Software, 1982). SPEED handles structures such as routings through job shops with relative ease. Its elements are described in terms with a manufacturing flavor, such as Station, Part and Transporter. Although these languages may be flexible enough to handle different classes of problems, obviously, domain-specific terminology may be awkward if the language is used outside of its intended area. Additionally, there is a persistent danger that the modeller may find the special-purpose constructs to be too limited to express all salient aspects of the system (Hoover & Perry, 1989). Nonetheless, with specialized constructs, time and cost for model development can be drastically reduced.

With the move of simulators onto microcomputers and into specialised domains, we see the first innovations in simulation technology that represent it "reaching out" to embrace different sorts of users beyond OR professionals. The initial movement is very tentative, embracing dedicated specialists who are knowledgeable about their application domain as well as simulation modelling, and are willing to take on the learning curve and ongoing effort of coding simulation models in a programming language (albeit a specialized one). We can see in this trend some acknowledgment that the intimate involvement of domain specialists in simulation modelling may be of value, but a tool like SPEED is still something exclusively for use by "technical types" — too difficult for use by managers and front-line workers. Further innovation is needed to bring the typical process expert into close contact with the simulation model.

Within the last ten years, we see more radical innovation in simulation modelling. This innovation, however, is not in the model itself, but in its presentation. First, many simulation languages have been extended to provide animation; that is, a visual/graphical representation of the running model. This has the obvious benefit of allowing a non-expert to at least observe the simulator in operation with some hope of understanding. Secondly, some environments (e.g.,

GPSS/PC) support graphical specification of models, beginning to free the developer from some of the burdens of programming. A further refinement is to allow interruption, and possibly modification, of the running model, supporting a more intuitive and spontaneous approach to modelling. Finally, we see the emergence of complete *simulation environments*: software products to support the entire simulation modelling process (Watson & Blackstone, 1989, p. 219). For example, The Extended Simulation Support system (TESS) by Pritsker and Associates (makers of SLAM) manages the simulation models, input data and experimental conditions. During simulation, animation is available in TESS. Simulation results are stored to a database of observations from which the system can produce statistical analyses, reports and graphics. In the 1990's, we see a proliferation of such environments, including a wide range of special purpose simulation environments by CACI (makers of SIMSCRIPT), including SIMFACTORY for manufacturing and SIMPROCESS for business process modelling. A notable feature of the special-purpose tools and they often have little or no requirement for programming, because the graphical specification tools encompass (or nearly encompass) the entire range of modelling features available. We will refer to simulation environments with animation and graphical specification as Visual Interactive Modelling (VIM) tools.

With respect to organizational modelling, we see the next innovation as an ongoing one. Simulation has followed computing hardware through its widespread proliferation, from mainframes to PC's, into graphical and direct manipulation environments, and onto the desktops of a wide variety of workers who are not aptly described as "technicians" or "engineers." Special purpose, graphical, business-process simulation environments are a technical reality, implemented for microprocessors that have more computing power than the mainframes that enabled the early successes with simulation. Now that simulation is in a position to affect how we do business process redesign, there is the responsibility to devise tools that provide the most appropriate support for this form of decision making. Simply automating pre-computing business modelling principles from accounting and related disciplines is to fail to exploit the potential of the new technology. We must have a philosophy of business modelling that explicitly accounts for the possibilities of simulation.

*Sim*View's Design Philosophy

From a mathematical perspective *Sim*View utilizes stochastic (random-number based), discrete-event simulation that is essentially the same as that which has been widely used for decades in systems such as GPSS. The emphasis of *Sim*View is on the interface design — how models are specified and viewed, how they are viewed while running, and the way their output is presented. Underlying the interface design is a philosophy of how people should analyze organizations for improvement, the Whole Systems Perspective.

The most basic aspect of the WSP is that we want to engender understanding of the "big picture," to construct a model that correctly reflects a global view of the organization. Such a view should include the organization's immediate environment (clients, suppliers, etc.) as well as the linkages between all of its internal components as they behave dynamically. Ideally, we would like a model to capture the *environmental loop* of the organization — the relationship between the quality of the organization's products and services, and the taking of market share that creates the input demand to the organization's productive process (see Figure 1). Extending consideration of the system to beyond the organizational boundaries broadens the range of possible innovative solutions. For instance, Streng and Sol (1992) discuss streamlining of the brewery-to-port delivery system for Heineken beer via modelling, and subsequent improvement, of interorganizational dynamics (including the use of Electronic Data Interchange). In general, if we are to consider improving an organization's processes, we wish to explicitly represent the chains of processes and their interdependencies, especially feedback loops as these may be key to complex and non-intuitive dynamics.

The notion of *explicit representation* suggests a role for the VIM environment as a basis for conversation among participants in organizational improvement. As has been described by Senge (1990) and earlier by Forrester (1971), decision makers often use implicit (mental) models of how their organization works for their own decision making. Because these models are not necessarily written down or explicitly articulated, they are difficult to examine. This inhibits building of consensus amongst a management team about what is the appropriate model of the organization and what assumptions can be made. The

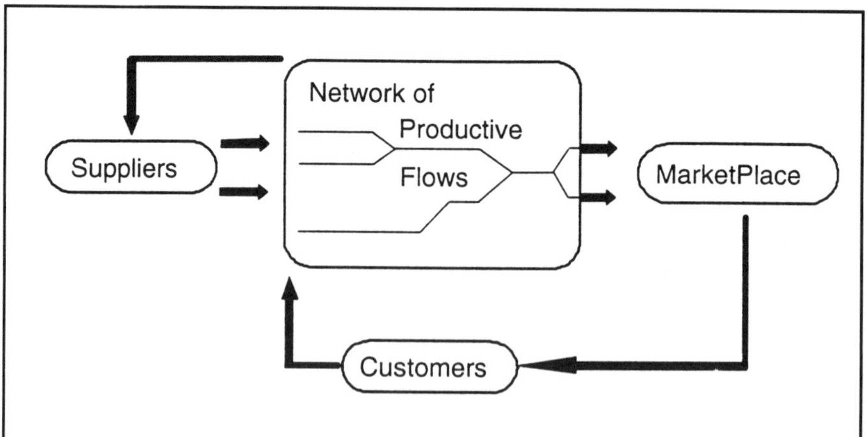

Figure 1: The "environmental loop"

simulation environment can serve to render an observable model of the organization. Irrespective of the accuracy of this model, it can serve as an objective basis for understanding and debate, where particular points can be changed to suit the perceived reality of the organization.

We believe that a key modelling objective should be to uncover *trusted linkages* among the performance measures at the various levels of the organization. That is, the performance measures used "on the shop floor" should be attainable and comprehensible at the operational level while simultaneously presenting the right targets for the achievement of organizational success at the strategic level. In the spirit of Goldratt (1992), we believe that such measures are attainable via careful consideration of organizational dynamics. Performance measures that do not fit well with the organization's productive flows may be *fragmenting*, in that the performance targets of one process can easily be unsympathetic to the performance of the next process downstream. For instance, a production target in terms of units per day gives no consideration of the smoothness of production through the course of the day. In such a situation, the downstream process may occasionally starve for lack of input and be unable to meet its target. A related problem is excessive trimming of capacity such that a system cannot meet its targets in the face of random variations in operations. See Seeley and Warren (1994) for a more detailed discussion of trusted linkages.

Finally, given the desire to model the big picture and to capture the important details of system dynamics, a hierarchical or recursive view

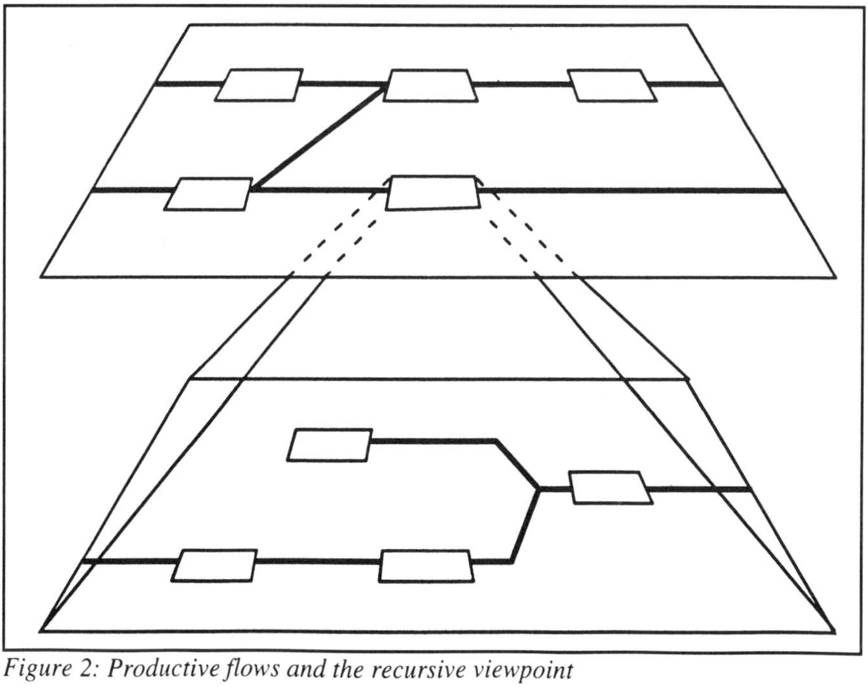

Figure 2: Productive flows and the recursive viewpoint

of the system is essential (see Figure 2). A VIM tool that readily supports views at multiple levels of detail aids the comprehension of a large model; but also, hierarchical views enable managers at different levels to participate in the modelling process effectively because they can see the VIM at the level most recognizable to them.

*Sim*View is graphical, animated and hierarchical to support the approach outlined above. *Sim*View is designed to be seen, to give live illustrations of the system model, elicit corrections from process experts and engender confidence in model validity[2]. *Sim*View is a desktop simulation package that uses a completely graphical paradigm without any underlying programming script. Graphical modelling is used for its naturalness in visualization and provision of immediate feedback of what has been described, as well as to avoid reliance on programmers for the creation of the simulation model with the consequent. Visualization encourages the participation of decision makers and others in the modelling process (Grant & Weiner, 1986), contributing to the validity of the model as the participants will normally have considerable knowledge of the modelled system (Johnson & Poorte, 1988). Participation also leads to a better understanding of the simulation model, resulting in: (a) more confidence in the model itself, (b) more confidence in the decisions based on the output of the model, and (c) better mutual understanding between staff (O'Keefe & Pitt, 1991).

It should be noted that we have not aspired to rigorously defend the derivation of our WSP approach to organizational modelling. It is as at least as much based in personal consulting experiences as research literature. We merely have presented the way of looking at business that has driven the ongoing development of *Sim*View. If one radically disagrees with our assumptions, then one would probably find a different sort of simulation tool to be more appropriate. For instance, some tools appear to be designed to support an approach where experts craft complex blocks of code that are incomprehensible and unapproachable to managers who must blindly accept the recommendations of technicians. Our objective is to support simulation modelling such that people who know the organization can build, understand and review the model; managers should come to their own decisions as supported by the model.

Model Specification in SimView

A *Sim*View simulation model consists of two elemental types of modelling entities:

- *Objects*: Entities that remain in a fixed logical position in the system and largely do not change over time.
- *Items*: Entities that move logically within or through the system and that may change during the course of the model run.

Items can represent the inputs or stimuli from the environment of the system, activity entirely within the system, or the output response to the environment of the system. Items interact with the system objects as they "flow" through the system in sequences of interactions that bring about the transformation of the items. Items are grouped in item classes each having their own graphical appearance (icon), sequence of interactions (flows), and sets of attributes. Examples of item classes are orders, invoices and goods. A special type of item is an *agent*, which interacts with some objects in a different way from ordinary items. Agents enter some objects only when they are required for an action (e.g., an operator starting or repairing a machine). Also, agents have the ability to collect and "carry" one or more other items and transport them through the system. Agents are referred to as *dynamic resources* or *item transporters* and may be used to model operating personnel, repair personnel, lift trucks, etc.

Modelling the behavior of a system is achieved by visually constructing item flows. A *path* defines a single sequence of interactions between an item of a particular item class and one or more objects in the system. A path usually consists of several steps that are movements from one object to the next. A *flow* is defined as the collection of all paths for an item class. The collection of flows in a model represents the sequences of interaction that produce the response of the system to its input.

The process of building a *Sim*View model consists of three phases: *Object Editing*, *Flow Editing* and *Interaction Editing*. These phases are normally carried out in sequence, but there is no restriction to this. In object editing, objects representing the components of the system are chosen from the object palette and are positioned on the screen (see figure 3). *Sim*View offers a large range of different objects that can be used to build the simulation model. These objects can be grouped into three categories:

1 Service Objects are objects that offer services to items. They are the resources utilized by items as they flow through the system. Examples of service objects are servers, multi-servers, assemblers, splitters and facilities.
2 Control Objects are objects that organize or control the movement and attributes of items as they flow through the system. They represent and perform the decision-making, ordering and transforming functions which occur in the system. Examples of control objects are switches, queues, entries and change-objects.
3 Modelling Objects are objects that facilitate the process of neatly organising, arranging and animating the *Sim*View screen display for the simulation model. They are not observable in the real system since they are only used to improve communication and clarity when using *Sim*View. Examples of modelling objects are portals, guides

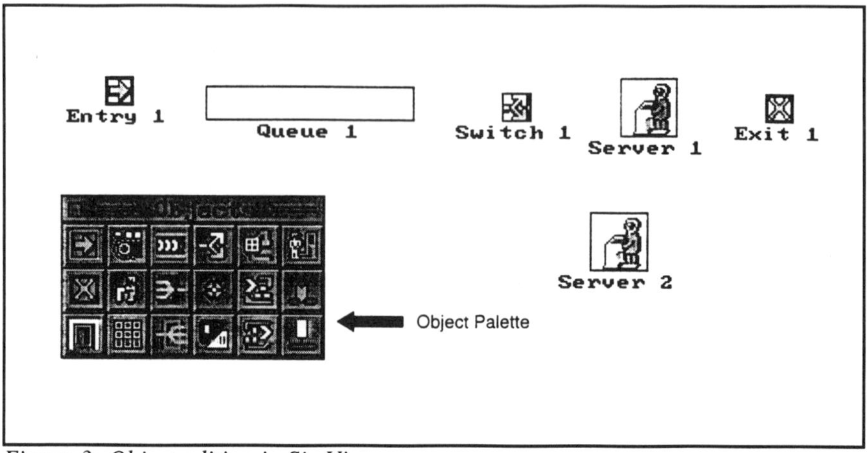

Figure 3: Object editing in SimView

and mark- and record-objects.

Objects are presented on the screen by schematic representations (bit-mapped icons). For each different state of an object (e.g. busy, unavailable or idle) different icons exist. The use of different icons for each state reflects the changing operational conditions of an object and enables operational staff to quickly understand what is happening on the computer screen. All icons may be customized for a better representation of the components in the real system under consideration, and can be kept in various user libraries of icons. Objects can be copied and pasted to create multiple occurrences of the same object instance.

In flow editing, the paths between the simulation objects are created. Before paths can be added to the model, an item class has to be created. Each item class has an icon associated with it that may also be customized. Paths are created for each item class; different item classes cannot share the same path. A distinction is made between primary paths and alternate paths. A primary path is one that proceeds from where an item enters the model to where it leaves the model. An alternate path is either a branch from a primary path, or is an additional primary path itself. The flow for a specific item class is the collection of all primary and alternate paths (see Figure 4).

In interaction editing, the interactions for each item class with objects that are included in their flow are specified (see Figure 5). These interactions occur as items move from object to object along their paths. The interaction that is specified is determined by the type of the object. In many cases, the user has to specify a distribution to quantities like process times or item inter-arrival times.

Before a model can be built, the items and objects need to be identified in the real system. Usually the entities within a system that stay in a fixed position are modelled with objects, and inputs to the

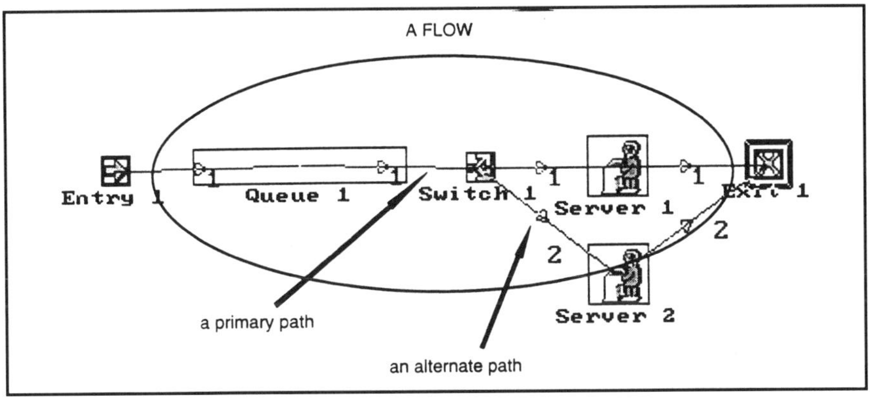

Figure 4: Flow editing in SimView

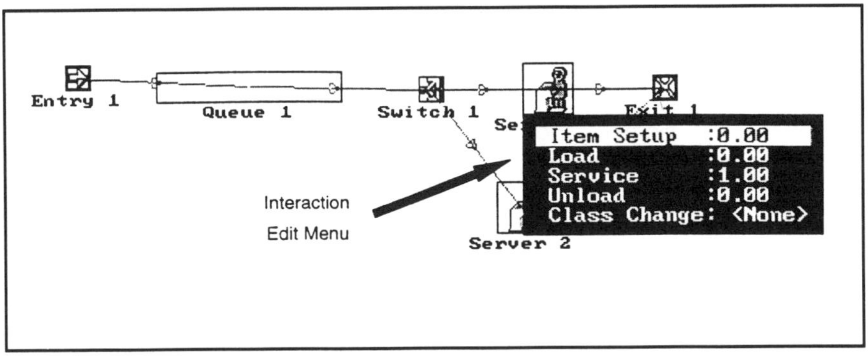

Figure 5: Interaction editing in SimView

system and outputs of the system are modelled with items. A special type of object is a *portal*. Portal objects are used to model the various levels of detail. A portal represents a component of a system that is modelled in more detail in a sub-model on a lower level. Creating sub-models forms a hierarchical model structure (as per figure 2) and enables models to be composed of a number of screens. Each portal object has a predefined entry and exit object. Items entering the portal object are 'teleported' down to the screen associated with the portal and will appear at the portal's entry in the sub-model. Items entering the portal's exit in the sub-model will be 'teleported' up and will leave the portal object on the higher-level screen.

The next section illustrates use of many specific *Sim*View objects as well hierarchical modelling with portals.

The Queensland Rail Sugar Train Case

This section illustrates a successful application of *Sim*View. First, we introduce Queensland Rail (QR) and the management decision problem that motivated them to seek the use of simulation technology.

We then describe the particulars of how *SimView* was applied to create a model of the dynamics of the QR sugar train system, demonstrating the mechanics of the modelling environment as applied to a real problem. The section concludes with the results of the simulation modelling exercise, including a detailed discussion of the decisions made. The decisions are interesting not only in that they show QR saving money, but that they reflect increased awareness of the system dynamics (or, more accurately, a coalescence and affirmation of ideas that employees already had), allowing QR to work "smarter" and in better harmony with the operational environment.

Background on Queensland Rail

Queensland Rail is the largest transport service provider in Queensland, a state in the tropical North-Eastern part of Australia which is 666,990 square miles in area (somewhat larger than Alaska). QR's activities extend from heavy haul to general freight, small freight and livestock, as well as to suburban and long distance passengers. We look at a capacity study for the rail system that transports sugar from four mills in the country to the jetty at the Townsville port. The sugar is crushed in four mills in the Burdekin region. The four mills — Giru, Pioneer, Kalamia and Inkerman — are all owned by the CSR organization. After crushing, the sugar is temporarily stored in bins at the mill. These bins have a very limited storage capacity and sugar should therefore be loaded into railcars and taken to the jetty as soon as possible. An overflow of a bin results in the loss of the spilled sugar and high costs involved in closing down and restarting mill production.

In 1992, an increase in sugar production of 50% for the next five years was expected for the Giru mill. Based on this expectation, the CSR organization expressed their concerns about the capacity of the rail system and sought QR's assurance of its ability to handle the forecasted increase in sugar tonnages. QR had anticipated the demands of greater production, and had planned a multi-million dollar investment in railcars for 1995. The CSR organization was, however, particularly concerned at the ability of the track in the Townsville area to cope with both the increase in sugar trains and the predicted increase for rail traffic in general. This concern was reinforced by the fact that the entry to the Townsville jetty consists of only a single track and that the sugar trains have to compete with other traffic to the jetty, including nickel, copper and phosphate trains.

It was decided to carry out a capacity study of the rail system for transporting sugar from the mills to the port. However, rail plays only one part of the total sugar production process. The overall process consists of the following elements:

- continuous sugar production at the mills;
- a rail system that is responsible for transporting the sugar from the

mills to the jetty;
- loading facilities and a very limited storage capacity for raw sugar at the mills;
- unloading facilities and a virtually unlimited storage capacity at the port; and
- holding roads at the mills that have a limited size.

The size of the holding roads, together with the capacity of the locomotives, determine the maximum train length that can be handled at the mill.

To meaningfully evaluate the sugar transport system, the entire sugar production chain was modelled, including the production and loading process at each of the four mills, the background rail traffic to and from the port, and the unloading operations at the port. First, a model of the old situation was made. In the old situation, there was no fixed time schedule for the trains and lengths for trains to each of the mills varied from one day to the next. After analyzing the simulation model of the old situation, a simulation model of the predicted situation for 1997 was built. In this model a proposal for change was implemented. This proposal for change included fixed train lengths for each of the mills and fixed departure times from the mills (in the old situation, trains did not leave the mill until all of their railcars were loaded). This proposal had been thought of some years previously by staff involved with the day-to-day running of the trains. At that time, however, senior management had dismissed the proposal as being infeasible.

Modelling the Sugar Train System in *Sim*View

The items identified in the Queensland Rail capacity study were in the first place the sugar trains, running from the mills to the sugar terminal at the port and back. Different item classes were created to represent full sugar trains and empty sugar trains. The full sugar trains were classified as agents so that they could 'carry' the sugar produced by the mills. The sugar itself was modelled as an item representing a railcar load of sugar. The background rail traffic was also modelled using items. *Entry* objects were used to introduce the items and agents into the simulation model; items could leave the model via *exit* objects. *Conveyer* objects and *server* objects were used to model the delays for the trains on specific parts of the track. As more than one item can enter a conveyer, conveyer objects were used to model the delays for sections that could handle more than one train at a time. Server objects (together with *queue* objects) were used to model the delays for sections that could handle only one train at a time. Server objects were also used to perform the changing of item class at the sugar terminal (from full trains to empty trains) and at the sugar mills (from empty trains to full trains). *Out tray* objects were used to model the loading of the trains at the sugar mills. The out trays represented the bins at the sugar mills. An out tray object enables a quantity of items (i.e. sugar produced by the mills) to be "picked up" by an agent (i.e., the sugar train). An *in tray* object was used to model the unloading of the sugar trains at the Port. An in tray object enables agents that are carrying items (i.e., the sugar trains carrying the sugar) to release one or more of the items.

After the identification of the items and objects in the system, the sub-systems which comprise the whole system were chosen. Portal objects were used to model these subsystems, thereby creating a hierarchical decomposition in the model. The four sugar mills were identified as sub-systems together with the unloading sub-system and the Makeup & Storage sub-system at the sugar terminal at the jetty. To model the background traffic, the unloading operations for the phosphate, nickel, copper and molasses trains were also identified as sub-systems together with the sub-systems to model the north-south traffic and the Stuart-Mt. Isa traffic. Figure 6 shows the main screen of the simulation model. It shows different portal objects representing the identified submodels, and conveyers and servers for main lines of the railroad network.

At the beginning of the *Sim*View modelling process, a conceptual model was built by schematically laying out the objects (mainly portal objects) on the main screen. The conceptual model represented the structure of the simulation model and was not yet filled with empirical data. It invited an early participation of staff directly involved in the

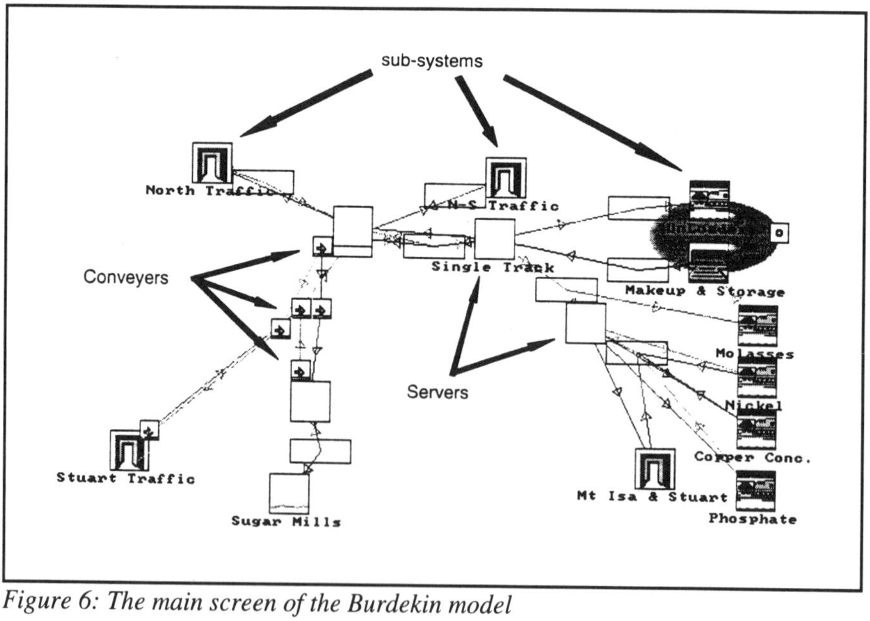

Figure 6: The main screen of the Burdekin model

day-to-day running of the system. Later, by specifying the paths for each item class and the interactions between items and objects, the model was fleshed out with more detail. The flows for the trains all consisted of a single path; none of the trains had alternate paths defined. The interaction for the entry objects for the sugar train items was specified to supply the initial sugar trains into the model. For the other trains, arrival patterns were specified for their entry objects to indicate when they entered the system. These trains exited the system after arrival at their destination. For the server and conveyer objects the associated delays were specified. For the server objects at the mills and at the Makeup & Storage, the change in item class from a full to an empty train or vice versa was specified. The servers in the Make-up & Storage sub-system are shown in Figure 7. This figure also shows the portal entry and portal exit for the sub-system. For the out tray objects at the mills, the "item capacity" and "maximum number of items to be picked up" attributes were specified. The first attribute represents the bin size at the mill and the second represents the train-length of the train servicing the mill. For the out tray at the port unloading facility, an unlimited capacity was specified representing the virtually unlimited bin size.

With the help of the animation, the underlying control logic of the simulation model was verified and modelling errors were isolated and removed. Of course, a correctly functioning animation does not necessarily imply a completely debugged model since aspects of the underlying simulation model's logic may not be evident by its animation (O'Keefe & Pitt, 1991; Verbraeck & Vreede, 1993). *Structural validation* and *replicative validation* (Dur, 1992) of the model was

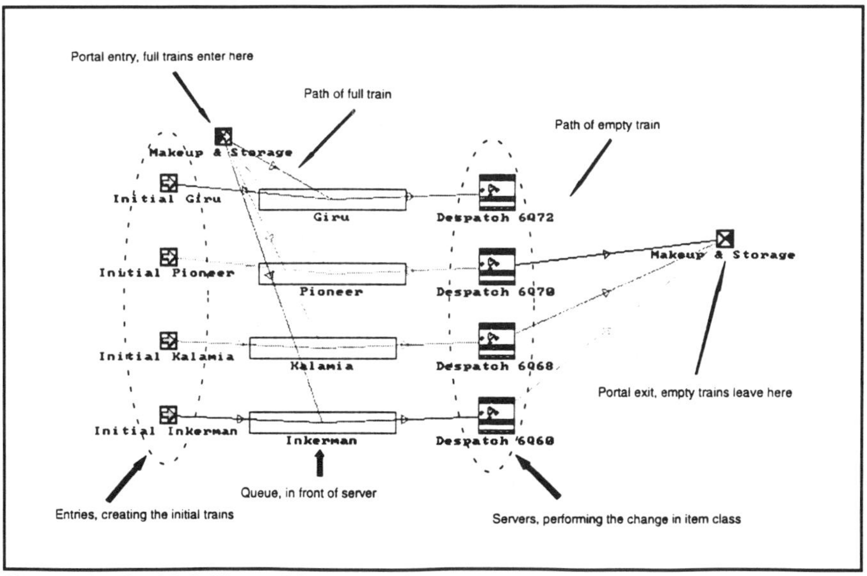

Figure 7: The MakeUp and Storage sub-system

conducted to ensure that the simulation model was a good representation of the real system. The structural validation checked that behavior of the simulation model corresponded with expectations based on experience. Together with a process expert, structural validation was carried out by running the model and observing the animation. Replicative validation checked whether the overall behavior of the simulation model corresponded with that of the real system. Data for the replicative validation was supplied by staff directly involved in the day-to-day running as well as management staff.

The *multiple runs* option of *Sim*View facilitates the performance of a series of model runs without animation. *Sim*View's facilities enable method-of-replications simulation experiments; however, it should be noted that the *Sim*View environment emphasizes the less formal modes of understanding a system via animation and graphics. Run results can be exported for modes of statistical analysis exceeding the current features of the package. The results of model runs can be examined in the Log Display System discussed in the next section.

Experiments and Results

The results of the simulation runs were analyzed using the built-in Log Display System of *Sim*View. In *Sim*View, system performance can be measured in a variety of ways. System performance statistics can be displayed dynamically on screen during a simulation run and can also be logged for viewing and analysis later. Each *Sim*View object that is part of the model (e.g., a server or a queue) can have a *statistical object* associated with it. The types of statistical measurements that are available in a statistical object depend on the object type to which it belongs. The occupancy of each object can also be set to be shown graphically during the simulation run. Different types of bar graphs, sliders and dials are available for this purpose and they can be adjusted to suit the users requirements. An example use of bar graphs is shown in Figure 8. The figure contains part of a screen from the simulation model and shows portals to three of the four mills and the rail tracks between them. The three bar graphs continually show the occupancy of the bins at the mills during the simulation run. When the *Log During Run* option of a statistical object is selected, the information produced by it is written to a log data file. The *Sim*View Log Display System is a flexible graphical interface that enables the user to study, review and report on the data that was chosen to be logged during one or more simulation runs. These data can be shown in graphical format or as textual reports. Graphs can be overlaid one on another to visually investigate any relationships or patterns that may emerge between various aspects of the system.

The simulation results of the model of the old situation clearly showed that the bins at the mills occasionally overflowed. The bins

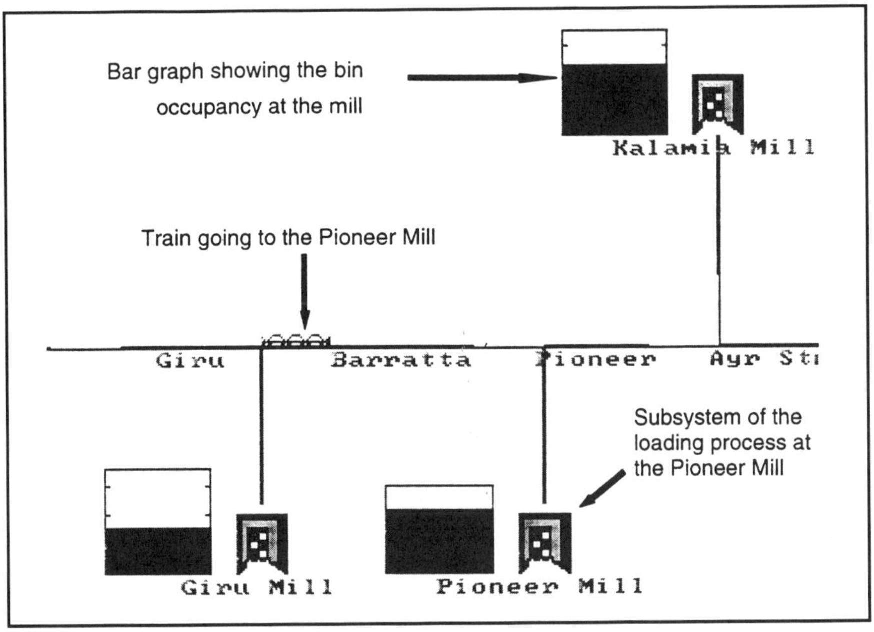

Figure 8: Example use of bar graphs

were modelled as *out trays* so the entrance of a *full*-state by one of these out trays indicated that the corresponding bin overflowed. The overflows of bins could thus easily be extracted from a status report on the out trays. From the simulation results it could also be concluded that the highly variable train lengths and the resulting shunting movements caused considerable delays in the system. The fact that trains remained at the mill until all of their railcars were filled amplified the variation in the time to fill the railcars.

In the model of the new situation, the predicted increase in sugar production at the mills and the predicted increase in rail traffic on the tracks between the mills and the jetty were modelled. In the model, two trains per day were scheduled for the Pioneer, Kalamia and Inkerman mill and three trains per day were scheduled for the Giru mill. The time between two trains departing the mills was modelled to be constant. Every step in the process was allowed a 10% variance based on advice by staff working with the system.

A major concern in evaluating the performance of the system was the level in the bins at the mills, as an overflow of one of the bins would cause a great deal of trouble. A problem, especially at the Giru mill, was the bin size. The Giru mill only had a bin size of 18 railcar loads, which was much too small to temporarily store the sugar produced during the interval between two train departures. Longer trains would not solve the problem as the trains always emptied the bins when they serviced the mill. More trains could solve the problem but QR was not capable of scheduling more trains to the Giru mill. The only alternative was to increase the bin capacity at the mill. Experiments were carried out with several bin sizes, which revealed that only with a doubled bin size could the system handle the sugar production without causing bin overflows. Figure 9 shows the occupancy of the bin at the Giru mill for one week during the crushing season. The graph can be scaled to see an overview of the entire crushing season and the data on the occupancy of the mill can also be obtained in textual format, including average occupancy and deviation. Similar experiments showed that the bin size at the Pioneer mill should be increased by seven railcar loads and that the bin sizes at the other two mills were sufficient. The number of railcars per train was at first determined by dividing the daily sugar production by the number of trains servicing the mill. However, it turned out that with these train lengths the system was not able to cope with the variation in the process. For this reason, three buffer railcars were added to the trains, providing a surge capacity for the system. A surprising result was that the total amount of railcars needed to cope with the sugar production in 1997 did not exceed the amount of railcars QR already owned! This fact alone saved several million dollars.

Another concern was the capability of the railroad network, and especially the single track entry to Townsville, to handle the increase

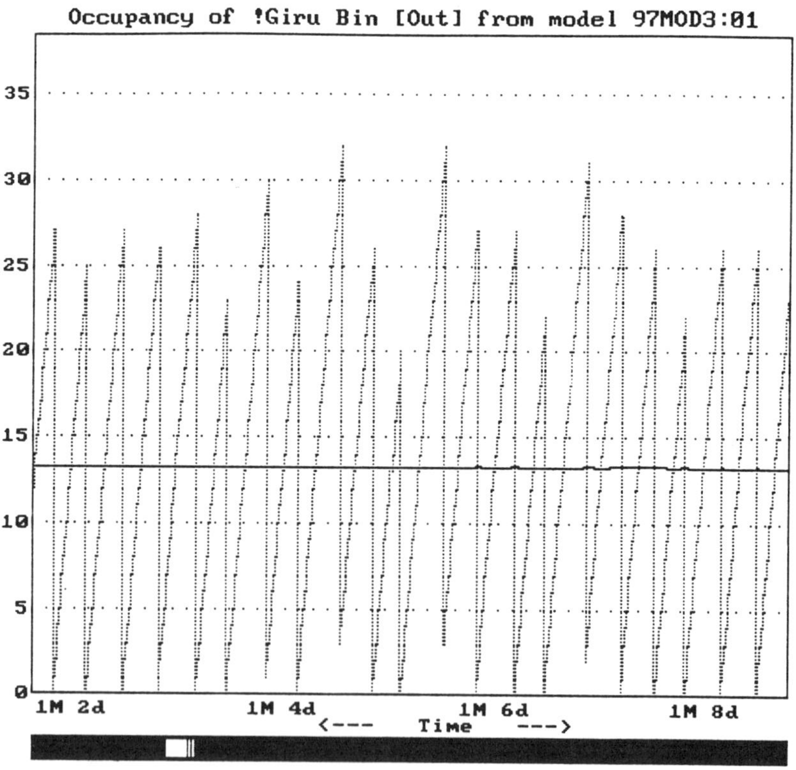

Figure 9: Occupancy of Giru sugar bin

```
==========================================================
                   Log Analysis Report
==========================================================
Model        :   97SCEN5
Object       :   Single Track [Ser]
Time         :   23d  1:06  2   to   5M 9d  0:26 49
Interval     :   4M15d 23:20

----------------------------------------------------------
    Object State Change Summary
----------------------------------------------------------
     State      Average Time       Deviation      Percentage
----------------------------------------------------------
      Idle:       0:48 39.26     0:53 10.39      ( 85.1%)
      Busy:       0:08 32.83     0:05 42.84      ( 14.9%)

----------------------------------------------------------
    Summary Of Item Activity
----------------------------------------------------------
     Class             In     Out    Avg. Occ. Time
----------------------------------------------------------
8261 Carstairs:       544     544    0:10  3.27
8271 Pioneer:         544     544    0:10 18.56
Copper Conc Full:     191     191    0:09 46.86
 8273 Giru:           816     816    0:10  7.21
8269 Kalamia:         543     543    0:09 58.66
Empty Train:          785     785    0:03 21.38
==========================================================
```

Figure 10: Occupancy of single track to Townsville

in both sugar and other rail traffic that was predicted for 1997. The occupancy of all sections of the railroad network was measured and logged using statistical objects in the *Sim*View model. These occupancies all were less than 30%. The occupancy of the single track entrance to Townsville is given in Figure 10. Though this track was the one that was expected to be a bottleneck in the system, it can be seen from the figure that the single track only had an occupancy of approximately 15%. In fact, the track that leads to the Burdekin area had the highest occupancy (29.8%). Despite initial worries, the model indicated that trains were never delayed so much that it caused congestion in the system.

Based on the experimental results, it was concluded that Queensland Rail could handle the increase in sugar tonnages, and the increase of rail traffic in general for 1997, provided that the following adjustments would be made to the system:

- Trains run on a fixed schedule. Three trains a day service the Giru mill and two trains a day service the other mills.
- The trains to the four mills have a fixed train length.
- The bin sizes at the Giru mill and Pioneer mill are increased to 36 and 29 railcar loads, respectively.
- The holding roads at the Giru and Pioneer mill are adjusted so they can handle the larger train lengths.

The fixed train lengths and fixed train schedules were implemented in the 1993 season and were found to work very well. Two significant spin-off benefits emerged from the implementation of the adjustments to the system: (a) because of its simplification, the system required less operational management and, (b) the changes to the system meant an improvement to the quality of working life for the crews who benefited from more predictable shifts. Another interesting result from the study was the identification of a performance indicator to predict when the whole system is likely to fail (i.e., when a bin will overflow), namely, the number of empty railcars returned to the port. By monitoring the number of empty railcars that are returned to the port and the average level of sugar in the bin, it is possible to predict a couple of days in advance that something is going wrong, so appropriate action can be taken.

Future Trends

Survey results from 22 *SimView* model developers (Warren, van Dijk & Jobing, in press) indicate that, in many ways, *SimView*'s use is in line with the ambitions we have for VIM environments. Ninety-five percent of respondents report participation of a process expert in the model development process, and 85% agree that participation leads to a more valid model. 95% of respondents show animation to decision makers to give them an understanding of how results are dervied, 95% rate running animation as very important in debugging a model, and 88% say people come up with corrections when watching animation. Badham, Couchman and Little (1995) discuss a human-centered approach to simulation, offering many principles in agreement with our own. For instance, that modelling should be a collaborative effort and simulations should be transparent so they can be trusted. However, they report a partial failure with *SimView* in the modelling of an assembly process. The model was successful (and potentially valuable) from a mathematical viewpoint, but in the end failed to influence management decisions. Badham et al. acknowledge that the assembly system design team was not involved in the model building, nor could they find time to view animations. Thus, *SimView* functioned as a conventional simulation modelling tool and failed to achieve its role as a communication support. The way a simulator is employed with respect to the organization's process experts and decision makers is just as vital to overall project success as are the simulator's technical capabilities.

While the developer survey indicates that *SimView*'s visual approach meets with approval in terms of ease of learning and amenability to use of animation, the results also indicate need for a number of technical enhancements, notably with respect to documentation issues. Respondents complain of the difficulty of understanding models created by others, of the problems of complex graphical structures used to circumnavigate the lack of a scripting language, and of irrelevant structures within models. Version management facilities, absent from *Sim*View, would also be desirable. Thus, even with a visual model, dealing with complex systems is still challenging. *Sim*View is constantly being extended to support the latest modelling problems indicated by users.

We see two major (and related) goals for the future development of VIM environments. First, we wish to change the nature of organizational modelling through simulation technology. *Sim*View can function as an ambassador for enlightened approaches to business process redesign, where assumptions about organizational dynamics are made explicit and put to the test in the model; and where proposals to retool, retrain, "downsize," or whatever can be evaluated in terms of their impact on the organization as a whole. To accomplish this sort of culture change, simulation technology must be brought close to decision makers and process experts. They must feel that the technology is as understandable and controllable as an account's spreadsheet. This is our second goal: the diffusion of simulation-based organizational modelling technology to (ideally) all stakeholders in business process redesign decisions. Certainly, *Sim*View has yielded some success in bringing simulation closer to key stakeholders. Managers and process experts are able to view the animations, and experienced developers can quickly alter the model in response to wisdom evoked from the experts. Moreover, *Sim*View model development is sufficiently easy that process experts can become experienced modellers with training or on their own.

Despite the success with *Sim*View to date, however, many challenges must be overcome if simulation model development is to be undertaken by a much broader population. Consider, for instance, the stochastic (random-number based) nature of simulation results. Even if a simulation model is a competely accurate representation of reality, the results of a simulation are only an estimate of the outcome that will occur in the real system. Park and Mellichamp (1990) have developed an expert system to provide support in the numerous statistical issues in simulation, such as: identifying standard distributions for input variables; determining run length and number of replications; constructing confidence intervals from estimates; and, comparing output from alternative designs. The system can help the user identify particular methods to use (or may recommend the advice of a "real" statistician). This level of help may indeed be useful to, say, an engineer who is undertaking simulation model development, but is not nearly sufficiently packaged for most process experts and managers.

Warren, Nocio, Stott and Canfield (1996) developed a simulation environment for information systems (IS) professionals to evaluate the performance of IS designs. Experiments with IS professionals as users indicated a marked and statistically significant value to simulation in terms of improvement to their ability to estimate system performance.

However, many users fell far short of the exploiting the full potential of the model by misusing the simulator. In particular, many users did not consider the inaccuracy of reported numbers in light of insufficient run length or inadequate numbers of replications. These mistakes were made in spite of context-sensitive help that provided specific recommendations to correct the problems and user training that created an awareness of the help. These results indicate that users can make substantial errors in the use of simulators and interpretation of their results even in the presence of on-line advice that is understandable to the users, known of by the users, and recommends corrective actions. More active corrective systems may be indicated — for example, a system that forces the user to conduct a sufficient number of replications — however, such methods tend to remove the user's judgement from the modelling process. Too much interference from the environment may undermine users' sense of control of the process and ultimately reduce their confidence in decisions.

Since getting the model into the hands of decision makers is so critical to having simulation results put to use, we are currently developing tools for constructing modelling environments that embed *Sim*View. These environments provide front-ends from which a high-level user can "test drive" various system configurations and observe key performance measures as "instruments." Trained users work in iterative fashion with decision makers to arrive at a solution the decision makers can back with confidence because they have "seen" it in action. It may be that, for the foreseeable future, this is the limit of diffusion of simulation technology in organizational modelling.

Conclusions

In this chapter we have explored the features and application of *Sim*View as an example of a Visual Interactive Modelling (VIM) environment to support organizational improvement. *Sim*View's design exploits a number of developments in simulation technology including the movement of simulators from mainframes to microcomputers, the use of graphics and animation, the emergence of simulation environments, and the trend toward special-purpose simulators. Moreover, *Sim*View's design takes an aggressive stance with respect to its domain; rather than merely automating existing accountancy-based models of organizational productivity, the design of *Sim*View supports a Whole System Perspective (WSP). WSP is an organizational modelling philosophy based on exploiting the power of simulation technology to overcome many of the drawbacks of pre-computing productivity measures and the often-poor decisions resulting from them. The WSP requires a hierarchical modelling tool so that one can track both the big picture and details. Equally important, however, is the visual nature of the model (*Sim*View has no scripting language) that supports substantive participation by managers and decision makers, evoking the wisdom that is innate in them from long experience with the organizational system. *SimView* models are intended to be quick and easy to modify and re-run, allowing responsiveness to challenges in model validity and investigation of ideas. *SimView* supports the insertion of various graphical and numeric counters throughout the model, visible when animations run, to encourage users in their search for performance indicators that link strategic concerns with operational realities.

SimView's application to the Queensland Rail (QR) sugar train system illustrates many of the desirable features of the tool and its appropriate use. In just a few weeks (in the hands of trained consultants) a simulation model was built that encompassed not just the sugar trains themselves, but the surrounding context of loading at the mills, unloading at the port, and the other rail traffic. The visual nature of the model helped the recommendations to be credible to management. Significantly, the model served to support an operational policy that was already suspected to be efficacious by operational staff, but which, prior to availability of evidence from the model, management felt they could not prudently support. After the initial experience with *SimView* (and resultant multi-million dollar savings), the use of the tool is moving from the consultants to the QR staff. Staff at QR have been trained in the use of *SimView* to solve problems unique to their divisions. *SimView* simulation models are now used for operational planning of coal and mineral trains. The people who build and run the simulation models are also involved in the decision making process. The use of the simulation models causes them to have more confidence in the decisions they make. Simulation modelling with *SimView* is becoming an integral part of ongoing process improvement at QR.

We believe VIM technology has now advanced to a stage where it will behoove large numbers of managers and process experts to personally participate in simulation modelling for more informed organizational redesign decision making. The ease-of-use of simulation modelling environments has not yet reached the stage where many individuals can lightly undertake the development of simulation models; but it is now practical for in-house model developers to work closely with process experts who give substantive, iterative input in terms of establishing model validity and proposing design ideas. These process experts can attain a level of familiarity with the models such that they have confidence in the simulation results. This sort of informed, meaningful involvement has great potential to transform the practice of BPR from its current stage of (almost blindly) following the recommendations of consultants to a process where stakeholders have a deep understanding of the costs and benefits of proposed changes.

Endnotes

[1]*Sim*View is a registered trademark of TechSearch, Inc., the commercial arm of the University of South Australia (UniSA). The software, which originated at the UniSA, has been developed and extended by InterDynamics Pty. Ltd.

[2] It is also notable that *Sim*View models are not compiled. The avoidance of a lengthy compilation stage allows *Sim*View to be "reactive" in a demonstration situation; it can be modified and re-run immediately in response to a correction or to follow up an idea.

References

Alemparte, M., Chheda, D., Seeley, D., & Walker, W. (1975). Interacting with discrete simulation using on-line graphic animation. *Computers and Graphics*, 1, 306-318.

Badham, R., Couchman, P., & Little, S. (1995). A human centred approach to simulation: a case study of software to support system design and development. In, *Proceeding of the 28th Hawaii International Conference on System Sciences* (vol. 4, pp. 861-870). IEEE Computer Society Press.

Birtwistle, G., Dahl, O., Myhrhaug, B., & Nygaard, K. (1973). *SIMULA Begin*. Philadelphia: Auerbach.

Bryant, R. (1980). *SIMPAS User Manual* (Computer Science Technical Report No. 391). Madison, WI: University of Wisconsin-Madison.

van Dijk, J., & Jobing, M. (1995). *Simulation for business process re-engineering*. Thesis, Department of Information Systems, Delft University of Technology, The Netherlands.

Dur, R. (1992). *Business reengineering in information intensive organisations*. Doctoral dissertation, Delft University of Technology, Delft, The Netherlands.

Forrester, J. (1971). *World Dynamics*. Cambridge: Wright-Allen Press.

Grant, J., & Weiner, S. (1986). Factors to consider in choosing a graphically animated simulation system. *Information Engineering*, August, 34-41.

Goldratt, E. (1992). *The Goal* (2nd rev. ed.). New York: North River Press.

Gordon, G. (1961). A general purpose system simulation program. In, *Proceeding of the EJCC*. New York: Macmillan Publishing, Inc.

Hoover, S., & Perry, R. (1989). *Simulation: A Problem-Solving Approach*. Reading, MA: Addison-Wesley.

Horizon Software, Inc. (1982). *Users Guide to SPEED*. Boston, MA: Author.

Johnson, M., & Poorte, J. (1988). A hierarchical approach to computer animation in simulation modelling. *Simulation*, 50(1), 30-36.

O'Keefe, R, & Pitt, I. (1991). Interaction with a visual interactive simulation, and the effect of cognitive style. *European Journal of Operational Research*, 54, 339-348.

Roberts, S., & Sadlowski, T. (1976). INS: a simulation language for analyzing operational issues in ambulatory care. In, *ORSA/TIMS Joint National Meeting*. Miami, FL.

Park, Y., & Mellichamp, J. (1990). A statistical expert system for simulation analysis. In, *Proceedings, Summer Computer Simulation Conference* (pp. 611-616). San Diego: Society for Computer Simulation.

Schriber, T. (1991). *An Introduction to Simulation Using GPSS/H.* New York: Wiley.

Seeley, D., & Macri, R. (1991). A directly-manipulable animation system for simulating productivity and quality improvements. In, *Proc. ASCILITE '91.* Launceston, Tasmania.

Seeley, D., & Warren, J. (1994). Trusted linkages and system chaos: the case for dynamic modelling of organisations. In, *Proceedings of the Fourth International Working Conference on Dynamic Modelling and Information Systems* (pp. 271-301). Noordwijkerhout, The Netherlands: Delft University of Technology.

Senge, P. (1990). *The Fifth Discipline.* New York: Doubleday/Currency.

Streng, R., & Sol, H. (1992). A dynamic modelling approach to analyze chain dynamics on the inter-organizational level. In, *Proceedings of the Third International Working Conference on Dynamic Modelling of Information Systems* (pp. 47-64). Noordwijkerhout, The Netherlands: Delft University of Technology.

Teichroew, D., & Lubin, J. (1966). Computer simulation — discussion of the technique and comparison of languages. *Communications of the ACM,* 9(10).

Verbraeck, A., & Vreede, G. (1993). Animation as a communication vehicle in simulation studies. In A. Pave (Ed.), *Modelling and Simulation ESM93* (pp. 670-674). San Diego: Society for Computer Simulation.

Warren, J., van Dijk, J., & Jobing, M. (in press). Human factors in simulation of information systems. *Simulation and Gaming* (special theme issue on Simulation of Information Systems).

Warren, J., Norcio, A., Stott, J. & Canfield, G. (1996). Experimental evaluation of a simulation environment for information system design. *Journal of Systems & Software,* 32(1), 3-20.

Watson, H., & Blackstone, J. (1989). *Computer Simulation* (2nd ed.). New York: Wiley.

Watson, H., & Christy, D. (1982). The evolving use of simulation. *Simulation and Games,* Sept., 351-363.

Acknowledgments

Special thanks are owed to the InterDynamics directors Bob Bridges and Tony Griffith, Patricia Morrow and Ross Darnell from Queensland Rail, to all the survey participants for taking time from their busy schedules to assist us in our research, and to the University of South Australia and Delft University of Technology for their commitments to international study which provided the authors with the opportunity to work together.

Section II
The Diffusion of Software Application Packages

The four chapters in this section use common organizational software application packages (e.g., Electronic Data Interchange and Local Area Networks) as the anchor to explore adoption and diffusion issues. Two distinct themes emerge out of the many issues these authors raise. First, can software application packages be clearly defined in relation to the innovation and diffusion process? Everett Rogers' work itself provides some insight into this issue. Hybrid corn and medical products were at the core of Rogers' diffusion theory with each of these defined as a distinct innovation whose salient characteristics did not change during the diffusion process. That is, the adopters could not change or make modifications to the innovation during the diffusion process. This is in direct contrast with information technology innovations. It is not easy to describe innovations of this type as distinct, separate, and unchangeable entities. The adopters of software application packages mold the innovation in accordance with their own perceptions and needs.

Second, is the diffusion of innovation theory sufficiently robust when applied to the adoption decision and the diffusion process for software application packages? The theory can only be regarded as robust when it predicts adoption/diffusion as well as non-adoption/non-diffusion better than any other approach. Traditional diffusion theory becomes stressed when it is applied to innovations (i.e. application software packages) that do not possess the key characteristics (e.g., relatively unchanging during diffusion) the original theory is based upon. The contributions included in these chapters show that the diffusion of innovation theory only partially applies to software application packages and that other approaches also require consideration.

McGowan and Madly base their analysis on Electronic Data Interchange (EDI). Their results show that the diffusion of innovation theory only partially explains the diffusion process. Apparently, relative advantage and compatibility play a minor role, while complexity was associated with usage breadth. Their results indicate that the adoption of EDI is a complex business issue. Customer influence explained the adoption decision and industry cooperation affected the acquisition of knowledge.

Niederman makes the observation that an IT innovation often is seen as primarily a technical challenge of limited complexity. He claims that because of this simplistic view, the IT adoption potential may be vastly exaggerated. Based on the observed EDI diffusion rate over the past twenty years he concludes that EDI has diffused at a steady rate, not exponentially as predicted initially. He argues that critical mass theory and perceived risk must be included in our analyses. Otherwise, the adoption process cannot be explained.

Using the adoption of local area networks as their example, Theoharakis, Wong and Powell explore the role media play in informing opinion leaders about emerging IT products. They show that media often exaggerates the IT potential. In particular, media does not convey the power of the installed IT base, that is, products in use that cover the same type of services as the IT product that catches media attention.

Palmer convincingly shows that adopters differ in their views on Quick Response IT potential. Some see it as a strategic vehicle, others look upon it as a means for increased efficiency, and others changed their perceptions during implementation. The data analysis shows that diffusion theory partially explains the diffusion process. However, Palmer also found that large IS staff size and in-house IS development play a role. He also found that among mature users the diffusion process is associated with supplier-oriented strategy and higher IT investment.

CHAPTER

6

Adoption and Implementation of Electronic Data Interchange

Matthew K. McGowan
Bradley University

Gregory R. Madey
Kent State University

This research examines the adoption and implementation of EDI (electronic data interchange) by organizations using the framework provided by diffusion of innovation theory. Recently, researchers have examined the diffusion of information technologies using the framework from diffusion of innovation theory. EDI is an information technology (IT) innovation, so it could be expected that factors which have been found to be related to other IT innovations would also explain the adoption and extent of implementation of EDI. However, EDI differs from other IT innovations because EDI is used between organizations.

The objectives of this research are: 1) to determine whether the factors which explain adoption of EDI are different from the factors that explain the extent of EDI implementation, 2) to explore the factors which cause the extent of EDI implementation to grow within an organization, and 3) to get a better understanding of the dimensions of the extent of EDI implementation. Studying the dimensions of EDI implementation can help researchers and practitioners get a better understanding of EDI implementation. Four industrial companies in the Midwest are studied to learn how the use of EDI grew in the organization.

Background

Diffusion of innovation theory is discussed in this section. We argue that diffusion of innovation theory is relevant to the study of EDI, and that EDI has unique features suggesting that EDI needs its own specific study. EDI has a technical component, similar to other IT innovations, but EDI has interorganizational elements which distinguish it from other types of innovations.

Diffusion of innovation theory has been the subject of thousands of studies (Rogers, 1983). Most of the innovation research which has been conducted to date has concerned social or scientific innovations (Nilakanta and Scamell, 1990), but there is a growing stream of research which applies this framework to information technology innovations. Zmud (1983, 1982) found that organizational size, professionalism, and task complexity help explain the level of adoption of modern software practices. Nilakanta and Scamell (1990) found that different sources of information are important at different stages of the innovation process for database design tools. Freeman (1989) found moderate support for the innovation process for fourth generation languages in organizations to be consistent with the model proposed by Rogers. Hoffer and Alexander (1992) found that management support for an innovation and existence of an advocate correlate with the adoption and extent of implementation of database machines. Rai (1990) found several factors which explain depth and breadth of penetration of CASE tools in MIS organizations.

Runge (1985) analyzed the implementation of telecommunications-based information systems and suggested that they follow a pattern similar to that explicated by diffusion of innovation theory. Grover (1990) extended Runge's work by studying the adoption and implementation of customer-based interorganizational systems. Grover (1990) recommended further study of interorganizational systems using the diffusion of innovation theory, since his research only considered one system in each organization. Bergeron and Raymond (1992) recommended the use of innovation theory as a framework for further study of EDI.

Diffusion of innovation theory describes the innovation process which consists of four components: 1) the *innovation*, which is an idea or practice which is new to the potential adopting unit; 2) *communications* through which the potential adopting unit gains knowledge about the innovation; 3) *time*, or the period over which the diffusion of the innovation occurs; and 4) the *social system*, which is a group of individuals or organizations working together to achieve a common goal (Rogers, 1983).

Much of the research based on diffusion of innovation theory has attempted to identify the characteristics of adopters which distinguish

Innovation Characteristics
- Relative Advantage
- Compatibility
- Complexity

Organizational Characteristics
- Size
- Management Support
- Champion
- Technical Expertise
- Training
- Centralization
- Functional Differentiation
- External Communications

Environmental Factors
- Customer Influence
- Supplier Influence
- Industry Cooperation
- Vendor Support

Figure 1: Classes of Factors Relevant to the Adoption and Diffusion of EDI.

them from non-adopters of an innovation. The insufficiency of a simple adopt/non-adopt variable has been criticized in the literature (Downs and Mohr, 1976; Tornatzky and Klein, 1982), and researchers have developed other measures which characterize the extent of implementation of an innovation.

This research examines three types of factors which are considered to be relevant to the adoption and diffusion of EDI. The classes of factors have been used in research on the diffusion of innovation theory and are discussed in the following sections. The three classes of factors are innovation characteristics, organizational characteristics, and environmental factors. The classification of the factors is illustrated in Figure 1.

Some of these factors may be more important at the time an organization is deciding whether to adopt EDI technology than in influencing the extent to which EDI is implemented in the organization. For example, organizations with a more centralized power structure might be less likely to adopt new ideas. Once such an organization has decided to adopt a new idea, the power structure can push for more widespread use of the technology.

Other factors may be important both in the adoption decision and in the subsequent implementation. For example, the degree to which a technology is perceived as providing relative advantages over other alternatives may influence an organization to adopt the technology. During the implementation stage, organizations which still perceive benefits from the technology (either based on prior beliefs or actual benefits) may extend the use of the technology. Many of the factors identified here are suggested by the literature which attempts to distinguish adopters from non-adopters; many of these same factors may also impact the extent to which a technology is implemented.

Innovation Characteristics

Researchers have attempted to develop classes of factors which have been associated with innovativeness (Zaltman, Duncan, and Holbek, 1973; Rogers, 1983). Tornatzky and Klein (1982) performed a meta-analysis of 75 articles on the relationship of innovation characteristics to innovation adoption and found that three characteristics were consistently significantly associated with innovation adoption: perceived compatibility, perceived relative advantage, and perceived complexity. Kwon and Zmud (1987) included these characteristics in their model of IS implementation process based on the diffusion of innovation theory. O'Callaghan, Kaufman, and Konsynski (1992) examined the adoption of EDI by agents of insurance companies and found that EDI adoption was more likely in target agencies that perceived EDI to offer efficiency or customer service advantages over alternatives to EDI. They also found that when EDI was perceived to be less compatible with the present physical system that adoption was less likely. Damanpour (1991) recommends including innovation characteristics in studies involving a single innovation. Innovation characteristics make up one class of variables included in the model developed here.

Organizational Characteristics

Researchers have frequently examined characteristics of organizations in an attempt to distinguish more innovative firms from less innovative ones. They assert that the innovation must be studied in the context of the organization, the social system which may adopt the innovation (Downs and Mohr, 1976; Bigoness and Perrault, 1981). A meta-analysis of 23 organizational level research studies found positive correlations between innovativeness and specialization, functional differentiation, managerial attitude toward change, technical knowledge resources, administrative intensity, slack resources, and external and internal communications, and a negative correlation between innovativeness and centralization (Damanpour, 1991). Organization characteristics are the second class of variables in this theoretical model.

The size of an organization has consistently been found to be correlated with innovation adoption (Rogers, 1983; Zmud, 1983; Dewar and Dutton, 1986; Rai, 1990). It has been suggested that larger organizations are more likely to have slack resources which may be used to implement the innovation. Larger organizations would tend to have a greater volume of business transactions and thus, could realize greater benefits from implementing EDI. Because of their size, larger organizations may be in a better position to require their trading partners to use EDI for conducting business with them.

Management support is another characteristic frequently found with innovation adopters (Zmud, 1983; Zmud, 1984; Ettlie, 1986; Alexander, 1989; Rai, 1990; Hoffer and Alexander, 1992). The implementation of an innovation may require significant organizational resources. The commitment of resources usually must be approved by management. Innovations which are viewed favorably by management are more likely to receive the funding required for implementation. Management support may be required to effect changes in procedures or organizational structure when implementing an innovation, particularly for a technology like EDI, which crosses organizational boundaries. Hansen and Hill (1989) cite the case of a large auto manufacturer which entirely automated the process of purchasing and disbursements through the use of EDI, a magnitude of change which could not have been possible without top management support.

Advocacy, or existence of a product champion, has frequently been found to influence the adoption of an innovation (Runge, 1985; Alexander, 1989; Grover, 1990; Rai, 1990; Hoffer and Alexander, 1992). Organizations are faced with many technologies, and a product champion can overcome barriers to adoption.

An innovation is more likely to be adopted when there is technical expertise available in the organization (Bigoness and Perrault, 1981; Dewar and Dutton, 1986). If organizations have people who are knowledgeable about an innovation, then they will be better equipped to deal with the issue of implementation. Diffusion theory suggests that firms possessing technical expertise might be better equipped to identify and evaluate information communicated to the firm regarding the appropriateness of the innovations to the firm. The finding could also be explained by an advocacy situation created by the presence of the experts, or that organizations predisposed to innovation have an organizational design conducive to innovation adoption. Since EDI has a technical component, technical expertise may also influence the extent to which the technology is implemented.

Grover (1990) found that the availability of training was positively correlated with the extent of implementation and extent of use of customer-based interorganizational systems. Rai (1990) found that availability of CASE training was a significant factor in explaining the difference between adopters and non-adopters of CASE tools, and that availability of training was also positively correlated with both depth and breadth of CASE tool usage. These results suggest that training can facilitate both the initial adoption and subsequent diffusion of IT.

In their study of hospital innovations, Kimberly and Evanisko (1981) found that centralization was negatively correlated with the adoption of technological innovations, but not related to administrative innovations. Zmud (1982) found that centralization was positively correlated with initiation of administrative innovations, but was not significant for technical innovations. Runge (1985) found that many of the organizations which bypassed or ignored normal IS planning and approval processes were more likely to implement telecommunications-based systems. In a company high in centralization, it would be more difficult to circumvent these procedures.

Functional differentiation represents the degree to which an organization is divided into subunits. Kimberly and Evanisko (1981) found that functional differentiation was positively correlated with technical innovations but there was no significant correlation with administrative innovations. Employment of many different specialists means that the organization will be exposed to a greater variety of ideas and innovations and that there will be greater opportunity for adopting innovations. Rai (1990) found that functional differentiation had a positive relation with the depth of penetration of CASE tools in IS departments.

Zmud (1983) found that the availability of communication channels was associated with greater use of modern software practices. Rai (1990) found that external communications were positively correlated with the adoption of CASE tools in IS departments. Organizations are continually identifying problems and considering alternative solutions, attempting to match solutions to problems. Organizations frequently scan the environment looking for solutions or to see how other organizations have dealt with similar problems. External communications, including books, trade journals, seminars, and conferences, provide opportunities to learn about possible solutions. Organizations which engage in more of this type of activity are more likely to find solutions to their problems.

Environmental Factors

The importance of environmental factors to the innovation diffusion process have been discussed (Kimberly and Evanisko, 1981; Kwon and Zmud, 1987) but have not been included in many empirical studies. Tushman and Anderson (1986) noted that the ultimate success of a technology is affected by market, legal, and social factors. The observation that entire industries are doing a poor job of managing technology suggests that there are environmental factors which affect the use of technology (Hage, 1986). It has been suggested that there are factors operating at the industry level which are related to the development and usage of manufacturing technologies (Leonard-Barton and Grogan, 1986) and information technologies (Dholakia,

Dholakia, and Della Bitta, 1991). Environmental factors are the third class of variables in the model developed here.

Gatignon and Robertson (1989) found that greater industry concentration ratios were positively associated with adoption of laptop computers for sales functions, and found a negative relationship between competitive price intensity and adoption. Organizations that are faced with competition may seek to adopt innovations for competitive advantage or may be forced to adopt innovations to prevent a competitor from gaining an advantage.

An organization's trading partners are part of its environment. Because EDI is interorganizational in nature, the trading partners are part of the environmental influences. The interorganizational nature of EDI means that the factors which influence the diffusion of the technology within the organization may differ from factors which have traditionally been identified in diffusion of innovation research. Holland, Lockett, and Blackman (1992) discuss several of the influences, including buyer leverage upon the organization and supplier leverage upon the organization. Some customers may require their suppliers to use EDI (Davis, 1990). Kwon and Zmud (1987) include interorganizational dependence in their model, based on studies of the innovation adoption process. Lederer and Mendelow (1990) identified customers as a source of problems for IS departments because customers may require new or modified systems.

Benjamin, de Long, and Scott-Morton (1990) suggested that industries with strong trade associations or where strong regulatory standards exist would be earlier adopters of EDI standards. Industry cooperation is necessary for the development of standards. Industry cooperation can also facilitate the spread of new ideas, such as EDI, through conferences and trade shows. The existence of industry standards makes it easier to extend the implementation of EDI because new trading partners can be linked via EDI without having to learn new standards.

Researchers have suggested that vendor support can influence the spread of an innovation. In examining the implementation of manufacturing technologies, Ettlie (1986) found that the vendor-user relationship was the most often mentioned factor identified by respondents, who were asked to identify important factors in the implementation of manufacturing technologies. Vendor involvement was suggested as an important factor in the adoption of database machines in preliminary interviews, but was not found to be significant in the survey of database machine users (Alexander, 1989). Kimberly (1986) recommends a mutually supportive relationship between the vendor and the firm for successful implementation of a technology. Lederer and Mendelow (1990) suggested vendor watching as a coping strategy for ISDs to deal with evolving technologies. Gatignon and Robertson (1989) found that vertical coordination was positively associated with adoption. That is, the stronger the links between the innovation suppliers and customers, the more likely was adoption. Value added network (VAN) service providers can provide assistance to organizations who wish to add more trading partners to their EDI activity.

The Extent of EDI Implementation

Electronic data interchange is the interorganizational exchange of business documentation (eg., invoices, purchase orders) in machine-processable form (Emmelhainz, 1990). Cash and Konsynski (1985) use the term "interorganizational system" (IOS) to refer to "an automated information system shared by two or more companies" (p. 134). EDI represents a type of IOS (Hansen and Hill, 1989).

The need to use a dependent measure other than the binary adopt/non-adopt measure has been remarked by researchers (Downs and Mohr, 1976; Tornatzky and Klein, 1982). It has been suggested that the use of technology has both a "breadth" and "depth" component (Kosnik, 1990; Rai, 1990). In considering EDI, the breadth and depth dimensions of extent of implementation seem to fit naturally with EDI usage. Breadth of implementation refers to how widely EDI is used in the organization; for example, the proportion of people using EDI. Depth of implementation refers to how extensively EDI is used in a part of the organization. For example, the billing department would be considered to have high depth of EDI implementation if it processed 90 percent of its invoices with EDI.

The sophistication of EDI implementation is the third dimension of the extent of EDI implementation. The sophistication of implementation refers to the degree to which the technical features of EDI are used in an organization. There are different ways of employing EDI technology, and some of these are more elegant than others. For example, an organization which receives orders on a personal computer, prints them out, and keys them into a mainframe order entry system is operating at a low level of sophistication compared with a company which receives orders on a mainframe and moves them electronically into its order entry system.

The information technology (IT) adoption and implementation model, developed from the theoretical framework provided in the diffusion of innovation theory (adapted from Alexander, 1989) is shown in Figure 2. The model depicts the three classes of independent variables which are suggested to influence the dependent variable, extent of EDI implementation. The model developed here was the framework used to analyze EDI activity at four industrial companies.

Methodology

A qualitative approach was used for this research in order to

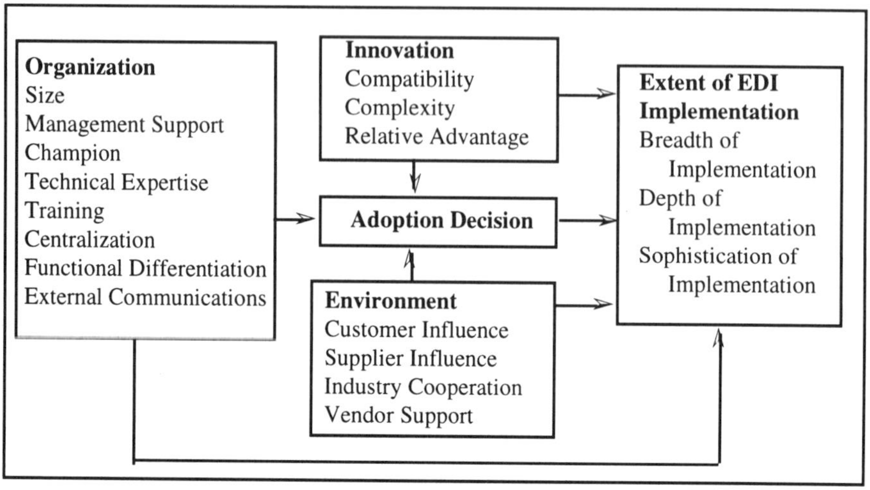

Figure 2: IT Adoption and Implementation Model. (Adapted from Alexander, 1989)

explore the factors which may cause EDI to spread within an organization. This approach also permits the study of the factors which relate to the initial adoption decision, as opposed to the subsequent implementation of EDI. A qualitative approach permits the examination of ordinary events in a natural setting (Miles and Huberman, 1994). The subject companies were selected because of their known EDI activity, not because they were considered extreme cases. AERO is a manufacturer of aircraft components with sales over $200 million. The other three companies each have sales in excess of $1 billion. AUTO is a supplier to the automotive industry, STEEL supplies the automotive and associated industries, and CHEMI is in the chemical industry. More detailed accounts of their EDI activity is provided in Appendix A.

Four industrial companies were studied. The first company, AERO, was contacted because of its known involvement with EDI and its accessibility. Numerous site visits were made during a six month period. AERO provided documents from its five years experience with EDI, including correspondence with vendors, work plans, program specifications, minutes from implementation meetings, and other internal documents. We discussed AERO's EDI implementation with the MIS manager, the supervisor responsible for EDI, the programmer who implemented the EDI software, two applications programmers who provided interfaces to the EDI software, and four users from the two functional areas involved with EDI.

The other three organizations were studied because they indicated in a preliminary survey that they were users of EDI and they were willing to subject their organizations to further study. Contact was made through the MIS manager in each organization. Preliminary contact with the interviewees was made via telephone. The purpose of the research was explained to the contacts, and they were told that they would be asked to review their company's involvement with EDI. Interviews were conducted with the person who had primary responsibility for EDI activity for the company. At AUTO there was one person whose primary responsibility was EDI activity, and that person was interviewed. At CHEMI, there was also a single individual responsible for EDI, and that person was interviewed. STEEL had a group of three people responsible for EDI. The group leader and that person's supervisor were interviewed jointly.

It was explained to the interviewees that the purpose of the session was to gain an understanding of the company's experience with EDI: how it started, how it evolved, and its current status. Several scripted questions were asked during and following the historical discourse.

The interviews began with a discussion of what the respondent considered to be EDI. Respondents generally considered the inter-company exchange of business transactions by electronic means to be EDI, whether it was done using X12 or a proprietary standard. All respondents agreed that electronic mail and facsimile transmission are not forms of EDI. One respondent specifically excluded engineering drawings from his EDI definition, noting that engineering drawings were exchanged electronically with trading partners, but that the activity was not considered to be EDI. This person also noted that intra-company exchanges of documents by electronic means are not considered EDI. Another respondent specifically excluded transfers using floppy disk or magnetic tape. The latter exclusion is noteworthy because ANSI (American National Standards Institute) specifically included floppy disk and magnetic tape media in its list of EDI communications alternatives (ANSI, 1987). The following definition of EDI was then provided: "EDI is the interorganizational exchange of business documentation in structured, machine-processable form."

The discussion regarding the definition of EDI underscores the confusion which exists regarding this new type of interorganizational communication. There is not a generally accepted nomenclature for classifying various types of data communications.

Analysis

The case studies are analyzed in this section using the model developed previously (see Figure 2.) The factors in the model are reviewed with data from the case studies to suggest whether the model is supported.

Innovation Characteristics

Each of the following sections analyzes the data from the cases with regard to one of the innovation characteristics.

Relative Advantage. All four companies believed they were getting little direct benefit from performing EDI with their customers. STEEL received some recognition from its customers for its quality service, which included an EDI component. STEEL was undertaking an effort to get its suppliers onto EDI, and believed the company would realize significantly more benefits from performing EDI with its suppliers than with its customers. AUTO believed that EDI was a competitive necessity; the organization did not benefit from EDI, but it was necessary to provide EDI support to its customers to maintain market share. AERO did not feel that EDI provided any immediate benefit, but in the long run it was a better way of conducting business, and advantages would be realized when a sufficiently large number of EDI transactions was conducted. Except for AUTO, the companies believed there were advantages to using EDI, but that their organization would not realize those advantages until some time in the future. STEEL, AERO, and CHEMI believed that increasing the percentage of transactions done via EDI would provide more benefit than simply increasing the number of different types of transactions. Relative advantage appears to be a criterion in extending the implementation of EDI, but was not a factor in the adoption decision.

Compatibility. The model suggests that companies which perceive EDI to be compatible with their data processing environment are more likely to use EDI. AERO, AUTO, and STEEL all found that EDI software was incompatible with their mainframes and started their EDI activity by using software written for personal computers. The lack of compatibility with their mainframe data processing environments did not prevent the adoption of EDI, but the implementation was the less sophisticated paper-based EDI approach. Thus incompatibility dampened the extent of EDI implementation. Each of the three companies had switched or was in the process of switching software packages to a mainframe version which was more compatible with its existing business systems.

Complexity. According to the model, companies which perceive EDI as complex will tend to have a lower extent of EDI implementation. The idea of EDI was considered a simple concept by the companies: data is merely translated from one format to another. Deciding on the details of mapping the sending firm's fields to the receiving firm's fields required detailed analysis. For example, AERO calculated a single tax amount based on the destination, but the customer wanted a breakdown into state, city, and local tax amounts. The complexity of adding new types of transactions discouraged companies from supporting additional transaction types.

All four organizations had a limited number of types of transactions being done via EDI. Companies found that it was easier to add trading partners for a given type of transaction than to add a new type of transaction. Complexity related more strongly to the breadth of EDI implementation than to depth of EDI implementation. The more complex an organization perceived EDI implementation to be, the less likely it would increase its use. Complexity would not prevent a company from adopting EDI to satisfy customers, but could keep a company from extending use of EDI to its suppliers.

Organizational Factors

Each of the following sections analyzes the data from the case studies with regard to organizational factors.

Size. Larger companies make greater use of innovations according to the model. Three of the four companies had sales in excess of $1 billion, but there did not appear to be a greater use of EDI with increase in company size. Larger trading partners, particularly customers, were frequently cited as a major reason for adopting EDI. Larger companies do seem to be expected to utilize EDI by their trading partners who might be willing to ease the requirement for smaller companies. Large companies may also possess the leverage to push their trading partners into using EDI. STEEL felt it was able to persuade its larger suppliers to adopt EDI for purchase orders. The higher volume of transactions performed at a larger organization also means greater potential for elimination of clerical labor by implementing EDI. AERO believed that increased use of EDI could lead to reductions in requirements for billing and accounts payable clerks.

STEEL and AUTO were required to use proprietary EDI standards in dealing with their larger customers. These customers supported point-to-point communications, suggesting that the customers had sophisticated EDI infrastructures. Complexity seemed to relate negatively to both breadth and level of sophistication of EDI implementation.

Management Support. We expected to observe a greater level of EDI implementation with greater levels of management support. At AERO, a directive from top management, at the urging of a large customer, caused the company to increase its EDI usage. But the concern was to satisfy the customer, not to make EDI a strategic initiative, so there was limited implementation of EDI: a PC was purchased and EDI software installed, which permitted the acceptance and printing of customer orders. The orders were then entered into the mainframe order entry system. This epitomizes the unsophisticated type of implementation which is used simply to mollify customers.

STEEL reported that the MIS function had to educate management on the merits of using EDI. Once STEEL and CHEMI obtained management approval, they undertook major EDI initiatives. Management at AUTO did not take much interest in EDI; it concomitantly had low breadth of implementation. AUTO's growth in the depth and breadth of EDI implementation has been in reaction to its customers' requests, not from any proactive approach at the organization.

Champion (advocacy). The initiative for extending the use of EDI came from within the MIS organization in the cases of AERO, STEEL, and CHEMI. Use of an innovation typically is greater in an organization which has an advocate pushing the innovation (Rai, 1990). For EDI, the role of advocate was contained in the MIS function but was not a particular individual. STEEL's MIS department convinced the purchasing group that they would benefit from getting electronic purchase orders from their customers. Support for EDI by the purchasing group grew over time. The MIS department started up a newsletter to customers, which was later taken over by the purchasing department.

AERO's MIS department worked with marketing to find out which customers had EDI capabilities, and sought to establish EDI links with these customers for customer orders and invoices, in an effort to increase the depth of EDI implementation. The MIS department also worked with its purchasing group to contact suppliers with EDI capability to begin sending purchase orders to suppliers, and increase the breadth and depth of EDI.

At CHEMI, the MIS department tried to educate several business functions on the merits of EDI, but there was limited interest. The education may have developed an awareness in the business functions for opportunities to use EDI. CHEMI's Logistics Department approached MIS to explore the feasibility of using EDI with the warehouse companies with whom CHEMI did business, but the effort was dropped when the warehouses were found to be lacking in EDI technology.

Technical expertise. Technical expertise has frequently been found to be positively associated with adoption of technical innovations. Holland, Lockett and Blackman (1992) suggest that a high level of information technology skill is required in order for a company to develop a proactive position in using EDI with its trading partners. Three companies studied here reported that problems of a business nature far exceeded technical problems during EDI implementation. Each company had at least one individual designated as an EDI expert.

Companies found it easier to establish EDI links with trading partners who were already doing EDI. These companies had already overcome the learning curve. Companies which were already doing EDI were one of the most useful sources of information about EDI for companies starting or expanding their use of EDI.

Training. The model proposes that greater levels of training will be associated with greater levels of EDI implementation. All four companies benefitted from training in their respective EDI software packages. STEEL also received education from its automotive customers on the benefits of using EDI. STEEL used that same approach for its suppliers - it provided some education on the business benefits to both trading partners from using EDI, and had a VAN service vendor provide education on VAN services. This education and training helped prospective EDI users increase their knowledge of EDI. It is an interesting approach to vendor support since the company using EDI did not have to use its own resources to train its trading partners, and the EDI vendor was provided an opportunity to win new customers. Training also helped to reduce the perceived complexity of EDI, both for technical and business issues, by providing information to better deal with those implementation issues. AERO received training in its EDI software because it hoped to do more with EDI, but it was not in a sufficiently strong position to influence its trading partners to use EDI.

Centralization. The model suggests that centralization is negatively related to the extent of EDI implementation. Respondents were asked how much influence non-management employees had on policy or decision making in the organization. Respondents were not familiar with their organizations' decision making processes, or were reluctant to share this information with the researcher. It was not possible to conjecture about this relationship from the cases.

Functional Differentiation. According to the model, greater extent of EDI implementation is expected with higher degrees of functional differentiation. Having more functional differentiation in an organization means that the specialists for a given area may be able to focus on improving their specific function. The growth in breadth of EDI implementation comes from niche applications in this case. The experiences of CHEMI offer some support for higher levels of functional differentiation being associated with greater breadth of EDI implementation. CHEMI used several different types of EDI transactions because various groups (Logistics; Export Shipping; Product Testing) were requesting these applications. CHEMI's Logistics Department requested that rail bills be received via EDI. Since railway cars had not previously been tracked at CHEMI, and since there were only eight rail carriers CHEMI used, 100 percent of the transactions were done using EDI. This represents the maximum depth of implementation score. It would suggest that having more specialized groups could also increase the depth of implementation because there is a high percentage of transactions performed in a single area.

External Communications. We expect to find a greater extent of EDI implementation with higher levels of external communications. AUTO, STEEL, and CHEMI found that industry publications were helpful in their EDI efforts. Trading partners were useful sources of EDI information for STEEL, AERO, and CHEMI. Usually trading partners have a vested interest in helping an organization increase its level of EDI knowledge. General trade journals were not considered to be as useful as journals dedicated to EDI. Industry conferences were considered important by STEEL and CHEMI; there was opportunity for asking questions of others. External communications seem to help companies improve their EDI operations, increasing the sophistication of implementation. It may also be a way for companies to learn about new types of transaction standards, permitting the organization to begin using them, and increase the breadth of EDI implementation.

Environmental Factors

Each of the following sections analyzes the data from the case studies with respect to environmental factors.

Customer Influence. The model predicts that customer influence will increase the extent of EDI implementation. Three of the companies started using EDI because of pressure from their largest customers, and all four were pressured by customers. This observation does not come as a surprise, but none of the companies took the EDI initiative. Customers have a strong influence in the EDI adoption decision, but other factors seem to influence the extent to which companies expand EDI usage. The manager at STEEL noted that "clearly, the customer is in the driver's seat when it comes to EDI." Customer influence may be indirectly related to depth of EDI implementation because a company may choose to link fewer suppliers via EDI than it had without EDI. Thus the linked suppliers would get a greater portion of the business (more transactions means greater depth of implementation). Larger companies have greater influence by virtue of their size, and they tend to use more EDI. These factors may influence the depth of EDI implementation. AERO, STEEL, and AUTO were all influenced by their largest customers to adopt EDI. Because of the large volume of transactions with these customers, it also resulted in greater depth of EDI implementation.

Supplier Influence. According to the model, supplier influence should be associated with greater extent of EDI implementation. Suppliers had very little leverage in getting organizations to use EDI. AUTO was the only company which increased its level of EDI use because of pressure by suppliers. Perhaps in a monopsony situation the supplier would have greater influence. Suppliers may be able to influence the use of EDI not through their position in the market channel, but by developing an EDI capability. STEEL began implementing EDI connections with its suppliers by first connecting to suppliers with EDI capability.

Industry Cooperation. The model suggests that greater levels of industry cooperation will be related to a greater extent of EDI implementation. Industry groups were also useful sources of EDI knowledge. AUTO and STEEL both felt they benefitted from the publications of AIAG (Automotive Industry Action Group). CHEMI participated in the industry conferences on EDI and received the CIDX (Chemical Industry Data Exchange) newsletter. Many of the standards developed for EDI resulted from intra-industry efforts. Industry groups develop standards or conventions for their type of business. Having this type of cooperation would increase the breadth of EDI implementation by making standards available for the transactions used in the industry. Having industry cooperation would also increase the depth of EDI implementation. AUTO achieved greater depth by purchasing a package which supported the AIAG standards.

Vendor Support. We expected increased vendor support to be associated with greater levels of EDI implementation. Support from the EDI software vendor and VAN service provider was of limited usefulness to the companies when they just started out using EDI. STEEL felt vendor support was useful in its efforts to increase EDI usage with its suppliers. The VAN service provider helped educate STEEL's trading partners and performed much of the initial testing with companies establishing links with STEEL. The addition of new EDI trading partners could not have moved as quickly without this support. This support relates to the depth of implementation, since more companies are being added for a given type of transaction. AERO obtained vendor support to gain a better understanding of its EDI software, and in resolving problems with adding types of transactions. Vendor support can help an organization make greater use of the features of the EDI software and assist a company in using a new type of transaction, thus increasing the levels of sophistication and breadth, respectively.

The Extent of EDI Implementation

Each organization started out doing what Emmelhainz (1990) terms "paper-based" EDI, that is, receiving transactions electronically, printing them out, and then keying them into another system. Eventually interfaces would be developed for some of the systems so that the data did not need to be entered manually. CHEMI was the only company which performed 100 percent of a transaction via EDI, and that was for a new system which only involved eight trading partners. It appears to be easier to obtain integration of EDI in new or re-engineered applications, rather than in existing applications.

Two of the companies each had their EDI software vendor get purchased by another vendor in the course of their EDI implementation because of mergers and acquisitions of one software company by another. One company felt the support from its newly created software vendor declined and ended up selecting a new vendor. Both companies considered the activity typical for software companies, and particularly for EDI software companies.

Tables 1 through 3 summarize the findings from this study. The results are discussed in the following section.

Variable	Results
Relative Advantage	Receiving little benefit now. Accommodating large customers. Expect more as suppliers added to EDI.
Compatibility	No impact on adoption. Lack of compatibility led to lower levels of implementation of EDI.
Complexity	Related only to breadth because of complexity of adding transaction types.
Extcrnal Communications	Several sources were helpful in overcoming implementation issues leading to greater breadth and depth of EDI use.
Planning	Two organizations believed this was critical to increasing breadth and depth of EDI implementation.

Table 1: Summary of Case Study Results for Innovation Characteristics.

Variable	Results
Size	All four organizations adopted EDI due to requirements of large customers. Effect on extent of implementation unknown.
Management Support	Important in adoption decision. Management support of MIS and EDI each affected extent of EDI implementation.
Champion	No champion identified, but three companies identified MIS advocacy as important factor.
Technical Expertise	Technical expertise not important; more problems due to business issues.
Training	Helped make use of sophisticated features of EDI. Led to more transactions (breadth) and more trading partners (depth) on EDI.
Centralization	No discernable affect.
Functional Differentiation	Experience of CHEMI suggests specialization may lead to greater breadth.
External Communications	Several sources were helpful in overcoming implementation issues leading to greater breadth and depth of EDI use.
Planning	Two organizations believed this was critical to increasing breadth and depth of EDI implementation.

Table 2: Summary of Case Study Results for Organizational Factors.

Variable	Results
Customer Influence	Paramount to adoption decision. Weak influence on extent of EDI implementation.
Supplier Influence	Only one company increased EDI usage due to supplier.
Industry Cooperation	Helped in adding transaction types (breadth) and adding trading partners (depth).
Vendor Support	Some association with breadth, depth, and sophistication of EDI implementation.

Table 3: Summary of Case Study Results for Environmental Factors.

Discussion

Two organizations, STEEL and CHEMI, had developed implementation plans for EDI. STEEL's plan succeeded, with numerous suppliers getting linked via EDI. CHEMI's plan was ambitious, but in the initial stage. STEEL explained that they decided "to architect an enduring plan which integrates EDI into the data processing function." The plan included new EDI software, capable of handling international standards and various options for moving EDI data. Options included printing transactions, electronic routing of text messages via electronic mail, or automatic loading of data into business systems. STEEL was implementing new business procedures to take advantage of EDI, rather than simply incorporating EDI into existing procedures. This proactive approach to EDI implementation provides a greater level of sophistication of EDI implementation. It can lead to a greater depth of implementation because it provides an infrastructure for adding EDI support for new business functions.

The development of a plan for EDI implementation appears to be associated with greater levels of EDI implementation. Improved planning for EDI implementation may improve the actual implementation process, or it may simply reflect a greater level of commitment to the technology by the organization. This factor should be added to the model developed here and addressed in subsequent research.

The factors that cause an organization to adopt EDI appear to be different from factors that explain the extent of EDI implementation for an organization. An organization may be forced to adopt EDI to prevent it from operating at a competitive disadvantage. Customer influence may make a company adopt EDI, even though EDI is perceived to be incompatible with the data processing environment. The customers' strong influence in an organization's decision to adopt EDI distinguishes this technological innovation from other innovations that have been studied. Future research should examine whether this is a normal phenomenon for interorganizational innovations in general. Organizations do not realize much benefit from adopting EDI but will extend EDI to obtain greater benefits, particularly from linking suppliers.

The growth of EDI may be multiple adoption decisions ("niche applications") or one decision to adopt and then grow it. Two of the companies were supporting multiple vendors' EDI software to accommodate different customer groups. For example, AERO used the PC-based version of the software developed by the Airline Transport Association to receive customer orders and also using EDI software on a midrange computer to invoice other customers.

Although it appears that there is not an EDI champion in the adoption of EDI, the MIS function serves that role by influencing other business areas to use the technology.

There appears to be support for the use of the theoretical framework provided by the diffusion of innovation theory to explain the extent of implementation of EDI in organizations. Further research needs to be done to test the framework on a larger sample of organizations. Future research could include several organizational users within the organization to get different perspectives and to check for intercultural reliability. The model resulting from this research is illustrated in Figure 3. The model developed here could also be applied to other types of interorganizational innovations.

It is difficult to generalize results from this qualitative study due to the small sample size. None of these organizations has made EDI a strategic initiative for their firm. All these organizations were in mature industries which may have better developed infrastructures for dealing with innovations.

Conclusions

This study examined the EDI experiences of four industrial companies from the theoretical framework provided by diffusion of innovation theory. The experiences of these organizations suggests that the adoption and subsequent diffusion of EDI behaves similarly to other technological innovations. However, the study suggests there are some distinctive differences from the traditional diffusion of innovation framework. Customer influence is paramount in the adoption decision. Industry cooperation also played a significant role in the acquisition of knowledge regarding EDI. Implications for managers are discussed in the following section.

The following recommendations for managers are based on the experiences of the industrial organizations examined in this research. Managers who want to successfully extend their level of EDI implementation should be cognizant of the developments in EDI standards. The

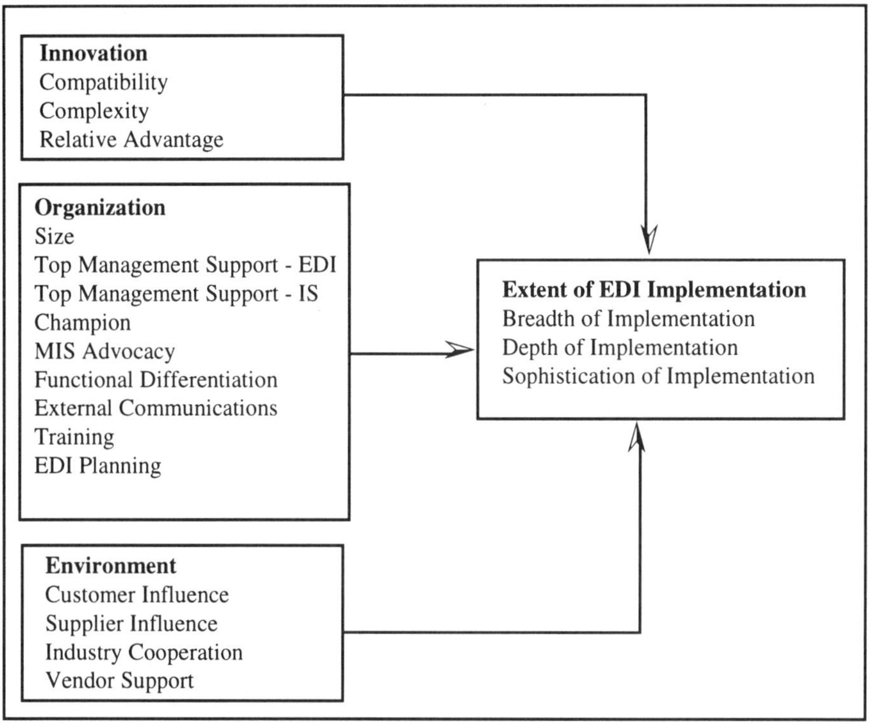

Figure 3: Model of the Extent of EDI Implementation

expansion of EDI with trading partners can be encouraged through the development of training and support, and it may be possible to enlist the aid of vendors to educate trading partners. This permits the organization to benefit from adding trading partners without having to commit its own resources to train new trading partners. Managers need to provide a leadership role for the EDI technology. Developing greater EDI capabilities may enable companies to expand their EDI connections; trading partners are more interested in adding partners who have already implemented EDI.

It may be necessary to sell top management on the benefits of EDI to gain support for it. The benefits do not come from links with customers, but from linking suppliers. The development of a plan for EDI implementation can improve the implementation process. Managers should participate in industry groups concerned with EDI and seek out experienced EDI users to gain a greater understanding of EDI. Trading partners are a good source of information.

APPENDIX A: Case Histories

AERO: Aircraft component manufacturer, $200 million revenue

AERO started its EDI activity in 1988 with the transmission of invoices to the government using a proprietary (governmental) standard, as requested by the government. Programming was done in-house to translate invoices to the specified format, and the formatted records were transmitted to a bulletin board at the government office. The benefit to AERO was that the payment cycle started with the receipt of the electronic copy of the invoice, an improvement of one to three days over the paper method. In 1989, an AERO customer requested that AERO accept customer orders via EDI. This became a top management directive to MIS, so AERO started receiving airline customer orders on a personal computer running software which worked to the Spec 2000 standard developed by the Airline Transport Association. Orders were printed from the personal computer and entered into the mainframe order entry system. There was no direct benefit to AERO, but the company felt it was necessary to comply with its customer's requests to accept orders electronically. As of 1993, there were ten customers using this system to enter orders, accounting for about ten percent of AERO's orders. In 1992, AERO purchased a software package which supports the X12 standard, and after receiving some training from the vendor, started sending invoices to two customers in 1993. The company plans to expand its EDI use with the X12 standard as much as possible, but will continue to support the Spec 2000 standard indefinitely. The government is planning to move to the X12 standard and drop its proprietary standard.

AUTO: Auto parts supplier, more than $1 billion revenue

AUTO became involved with EDI in the mid-1970's, when the big three auto manufacturers applied pressure. Initially, AUTO received material releases from the big three, then started sending advance shipment notices. The company started supporting the proprietary standards of its customers, but now its trading partners all support the X12 standard. In 1990, AUTO switched to a different EDI software vendor because the package enabled AUTO to eliminate the writing of a translation program each time a customer was added to a transaction. By 1993, the company was supporting the following EDI transactions with 12 customers and two suppliers which comprise its 14 EDI trading partners (number of trading partners in parentheses): material releases (13), advance shipment notices (10), shipping schedule (5), invoices (2), receipt advice (2), receipt advice (2), remittance advice (1), inventory advice (1), purchase order (1), text document (2), and functional acknowledgment (3). AUTO reacts to customers and suppliers requests to expand EDI activity, and has no plans to seek more EDI trading partners.

STEEL: Component manufacturer, automotive/related industries, over $1 billion revenue

STEEL started its EDI activity in 1984 at the request of the big three auto manufacturers, installing a personal computer in each of its two major product divisions to accept material releases and shipping schedules using the respective proprietary standards of the auto manufacturers. The activity involved 100 shipping locations and about 25 percent of the company's sales. In 1986 the larger division was requested to send advance shipment notices, so the company elected to develop its own software on its mainframe. In 1988, STEEL decided to form a corporate EDI group consisting of three people, and also purchased a mainframe software package which supported the AIAG standards, eliminating the PC EDI activity. The amount of activity increased to the point that about half the material release and half the advance shipment transactions representing half the sales volume was done via EDI. Although EDI activity was being done by customer requests, the company was attempting to take a proactive approach to developing its EDI capability. Some benefits began to accrue to the company as it added invoices to the group of transactions done with customers. In 1988, the company was still subscribing to four VAN (value-added network) services because it was not cost effective to pass documents between VAN providers. STEEL switched EDI software vendors in 1988 because the package it was using was no longer supported due to a buyout. In 1989, MIS was able to convince the Purchasing department to support a plan to receive supplier purchase orders (PO's) via EDI, and by the end of 1991, PO change notices, PO acknowledgments, and PO change notice acknowledgments were being done on EDI with 50 suppliers. Some customers rate STEEL on the timeliness of its EDI transmissions, so it performs some EDI transmissions in what it terms "real-time" or "event-driven" mode: the transaction is sent to a queue for immediate transmission as soon as the event occurs. As of 1993, the company figures that 80 percent of its shipment volume is being done via EDI.

CHEMI: Chemical industry supplier, over $1 billion revenue

CHEMI's MIS department decided, in 1987, that EDI technology was maturing and it would soon be necessary to support customers via EDI. An individual in the telecommunications group was designated to become the company's EDI expert, and started attending the bimonthly meetings of the Chemical Industry Data Exchange (CIDX), an industry group formed to support EDI. In 1988 the Logistics department requested that EDI links be set up with some warehouses, but investigation proved that the warehouses were not capable of supporting EDI. Also in 1988, one customer requested that CHEMI accept orders via EDI. The company purchased a software package, selected a VAN, and proceeded as far as final testing to accept the customer's orders, but conditions with the customer abruptly changed, and the application was never implemented in production. In 1989, the Logistics department requested railway bills be set up using EDI, and all eight of the company's rail carriers complied, accounting for 100 percent of this type of activity. The company attributed the success to the small number of companies involved and the fact that this was a totally new application. In 1991, the company set up the reporting of test results from independent labs, as requested by its research laboratory. That same year, CHEMI established EDI links with freight forwarders for booking information at the request of its Exporting department, and now about 25 percent of its booking transactions are done via EDI. From 1988 to 1991, the MIS group unsuccessfully attempted to garner top management support for a major EDI initiative. After receiving inquiries from many of its largest customers in 1992, CHEMI formed a project team which is pursuing plants to implement the following transactions with the company's largest customers: purchase order, PO acknowledgment, PO change notice, PO change notice acknowledgment, bill of lading, advanced ship notice, invoice, and payment/remittance advice. The transactions are to be implemented via EDI with the 25 largest customers, accounting for more than half of the business.

References

Alexander, M. B. (1989), "The Adoption and Implementation of Computer Technology in Organizations: The Example of Database Machines," Ph.D. Dissertation, Indiana University.

American National Standards Institute (ANSI) (1987), *An Introduction to Electronic Data Interchange*, American Standards Committee.

Benjamin, R. I., de Long, D. W. and Scott Morton, M. S. (1990), "Electronic Data Interchange: How Much Competitive Advantage?," *Long Range Planning*, Volume 23, Number 1, (February), pp. 29-40.

Bergeron, F., and Raymond, L. (1992), "The Advantages of Electronic Data Interchange," *Database*, Volume 23, Number 4, (Fall), pp. 19-31.

Bigoness, W. J., and Perrault, W. D., Jr. (1981), "A Conceptual Paradigm and Approach for the Study of Innovators," *Academy of Management Journal*, Volume 24, Number 1, pp. 68- 82.

Cash, J. I., Jr., and Konsynski, B. R. (1985), "IS Redraws Competitive Boundaries," *Harvard Business Review*, (March-April), pp. 134-142.

Damanpour, F. (1991), "Organizational Innovation: A Meta-analysis of Effects of Determinants and Moderators," *Academy of Management Journal*, Volume 34, Number 3, (September), pp. 555-590.

Davis, L., (1990), "Instant Orders," *Nation's Business*, April, pp. 34-36.

Dewar, R. D., and Dutton, J. E. (1986), "The Adoption of Radical and Incremental Innovations: An Empirical Analysis," *Management Science*, Volume 32, Number 11, (November), pp. 1422-1433.

Dholakia, R. R., Dholakia, N., and Della Bitta, A. J. (1991), "Acquisition of Telecommunications Products and Services: An Examination of Inter-Sector Differences," *IEEE Transactions on Engineering Management*, Volume

38, Number 4, (November), pp. 328-335.

Downs, G. W., Jr., and Mohr, L. B. (1976), "Conceptual Issues in the Study of Innovation," *Administrative Science Quarterly*, Volume 21, (December), pp. 700-714.

Emmelhainz, M. A. (1990), *Electronic Data Interchange: A Total Management Guide*, Van Nostrand Reinhold, New York.

Ettlie, J. E. (1986), "Implementing Manufacturing Technologies: Lessons from Experience," in *Managing Technological Innovation*, Donald D. Davis and Associates, Jossey-Bass Inc., San Francisco, pp. 72-104.

Freeman, J. L. (1989), "An Empirical Test of a Model of the Innovation-Decision Process in Organizations," Ph.D. Dissertation, Georgia State University.

Gatignon, H., and Robertson, T. S. (1989), "Technology Diffusion: An Empirical Test of Competitive Effects," *Journal of Marketing*, Volume 53, (January), pp. 35-49.

Grover, V. (1990), "Factors Influencing Adoption and Implementation of Customer Based Interorganizational Systems," Ph.D. Dissertation, University of Pittsburgh.

Hage, J. (1986), "Responding to Technological and Competitive Change: Organizational and Industry Factors," in *Managing Technological Innovation*, Donald D. Davis and Associates, Jossey-Bass Inc., San Francisco, pp. 44-71.

Hansen, J. V., and Hill, N. C. (1989), "Control and Audit of Electronic Data Interchange," *MIS Quarterly*, Volume 13, Number 4 (December), pp. 403-413.

Hoffer, J. A., and Alexander, M. B. (1992), "The Diffusion of Database Machines,"*Data Base*, Volume 22, Number 1 (Spring), pp. 13-19.

Holland, C., Lockett, G. and Blackman, I. (1992), "Planning for Electronic Data Interchange," *Strategic Management Journal*, Volume 13, pp. 539-550.

Kimberly, J. R. (1986), "The Organizational Context of Technological Innovation," in *Managing Technological Innovation*, Donald D. Davis and Associates, Josey-Bass Inc., San Francisco, pp. 23-43.

Kimberly, J. R., and Evanisko, M. J. (1981), "Organizational Innovation: The Influence of Individual, Organizational, and Contextual Factors on Hospital Adoption of Technological and Administrative Innovations," *Academy of Management Journal*, Volume 24, Number 4, pp. 689-713.

Kosnik, T. J. (1990), "Perennial Renaissance: The Marketing Challenge in High Tech Settings," in *Managing Complexity in High Technology Organizations*, edited by Mary Ann von Glinow and Susan Albers Mohrman, Oxford University Press, pp. 119-145.

Kwon, T. H., and Zmud, R. W. (1987), "Unifying the Fragmented Models of Information Systems Implementation," in *Critical Issues in Information Systems Research*, edited by R. J. Boland, Jr., and R. A. Hirschheim, pp. 227-251, John Wiley and Sons Ltd., New York.

Lederer, A. L., and Mendelow, A. L. (1990), "The Impact of the Environment on the Management of Information Systems," *Information Systems Research*, Volume 1, Number 2, (June), pp. 205-222.

Leonard-Barton, D., and Grogan, J. (1986), "Marketing Advanced Manufacturing Processes," in *Managing Technological Innovation*, Donald D. Davis and Associates, Jossey-Bass Inc., San Francisco, pp.187-219.

Miles, M. B., and Huberman, A. M. (1994), *An Expanded Sourcebook: Qualitative Data Analysis*, SAGE Publications, Thousand Oaks, California.

Nilakanta, S., and Scamell, R. W. (1990), "The Effect of Information

Sources and Communication Channels on the Diffusion of Innovation in a Data Base Development Environment," *Management Science,* Volume 36, Number 1, (January), pp. 24-40.

O'Callaghan, R., Kaufman, P. J., and Konsynski, B. R. (1992), "Adoption Correlates and Share Effects of Electronic Data Interchange Systems in Marketing Channels," *Journal of Marketing,* Volume 56, (April), pp. 45-56.

Rai, A. (1990), "Stimulating the Use of Computer-Aided Software Engineering in Information Systems Departments: An Empirical Test of Elements of Innovation Theory," Ph.D. Dissertation, Kent State University.

Rockart, J. F., and Short, J. E. (1989), "IT in the 1990s: Managing Organizational Interdependence," *Sloan Management Review,* (Winter), pp. 7-17.

Rogers, E. M. (1983), *Diffusion of Innovations,* Third Edition, The Free Press, New York, New York.

Runge, D. A. (1985), "Using Telecommunications for Competitive Advantage," Ph.D. Dissertation, Oxford University, England.

Tornatzky, L. G., and Klein, K. J. (1982), "Innovation Characteristics and Innovation Adoption Implementation: A Meta-Analysis of Findings," *IEEE Transactions on Engineering Management,* Volume 29, Number 1, (February), pp. 28-45.

Tushman, M. L., and Anderson, P. (1986), "Technological Discontinuities and Organizational Environments," *Administrative Science Quarterly,* Volume 31, Number 3, (September), pp. 439-465.

Zaltman, G., Duncan, R., and Holbek, J. (1973), *Innovations and Organizations,* John Wiley & Sons.

Zmud, R. W. (1984), "An Examination of 'Push-pull' Theory Applied to Process Innovation in Knowledge Work," *Management Science,* Volume 30, Number 6, (June), pp. 727-738.

Zmud, R. W. (1983), "The Effectiveness of External Information Channels in Facilitating Innovation Within Software Development Groups," *MIS Quarterly,* Volume 7, Number 2, (June), pp. 43-58.

Zmud, R. W. (1982), "Diffusion of Modern Software Practices: Influence of Centralization and Formalization," *Management Science,* Volume 28, Number 12, (December), pp. 1421-1431.

CHAPTER

7

The Diffusion of Electronic Data Interchange Technology

Fred Niederman
University of Baltimore

This chapter explores the gap between predictions of exponential growth in use of electronic data interchange (EDI) for transportation of routine business information and the actual experience of significant, steady, but not meteoric rates of adoption. Benefits of EDI focus on efficiencies gained in shifting from paper to electronic media and the potential for using more accurate and timely information for strategic decision making. Limitations of EDI include issues of standardization, of handling the varied protocols of large numbers of trading partners, of difficulties in expanding to handle new data sets, and in blending the new technology with legacy systems. Factors thought to influence the adoption of new technology based on theories of technology diffusion, in the broadest format enunciated by Tornatzky and Klein (1982), critical mass theory (Markus, 1987), and perceived risk (Teo, Tan, and Wei, 1995) are presented. The chapter concludes with practitioner and researcher issues and a discussion of how these theories may have better predicted rates of diffusion of EDI.

The Gap Between Predictions and Performance

Patterns of diffusion of information technology have varied with the particular innovation (Gurbaxani, 1990; Chin and Gopal, 1995;

Premkumar and Potter, 1995; and Ruppel and Harrington, 1995). The history of computing shows us that personal computers showed a flat, slow, steady growth among hobbyists, but accelerated in acceptance in a sharp curve once electronic spreadsheets became available. Dedicated word processing equipment grew rapidly for business applications, but diminished rapidly when word processing software became available for personal computers, thus allowing home purchasers to obtain word processing capabilities with all the other personal computer features still available. Computers hard-wired for data base applications (database machines) grew slowly and steadily in the 1970s and early 1980s but have never caught on widely. Reading the trade press, many information technologies, both hardware and software variety, are slated for rapidly accelerating diffusion, but only a subset ever achieve that potential. Other new technologies, such as personal accounting software, exemplified by Intuit's Quicken, diffuse widely before the trade press has a chance to tout it.

One technology, electronic data interchange (EDI), that was predicted in the 1980s for unprecedented growth and near universal acceptance by the mid-1990s, has shown stiff resistance to diffusion, and a steady, but hardly meteoric, growth. Sokol (1995) begins the update of her business overview of EDI technology by revising her 1989 predictions of 70% market penetration of EDI by 1993. These original projections proved to have been far off the mark. Many firms that have adopted EDI are not able to use it for the majority of business transactions. Use of EDI is largely concentrated in a few industries such as transportation and retail merchandising. On the other hand, the number of firms reporting use of EDI and amount of money spent on EDI both continue to grow at a solid rate (estimated in some cases as 10-30% annually). Sokol summarizes the reason for this significant, but slower than predicted, growth in EDI use by noting that establishing EDI is more difficult than previously recognized. Her book presents a wide variety of helpful tips to practitioners to address those difficulties and attempt to overcome them. However, she does not go very far in explaining the sizable gap between expectations and performance.

Theories of diffusion of technology have formed the backbone of numerous studies of adoption of information technologies including spreadsheets, end user computing, group support systems, imaging, budgeting software, and expert support systems (Prescott and Conger, 1995). They have also been used to study EDI technology (see O'Callaghan, Kaufmann, and Konsynski, 1992; Grover, 1993; Brouchard, 1993; Premkumar, Ramamurthy, and Nikakanta, 1994; and Iacovou, Benbasat, and Dexter, 1995).

This paper provides background on the EDI phenomenon along with discussion of the expected benefits that would lead a firm to adopt and some of the difficulties that the firm must overcome to realize those benefits. It also examines factors thought to influence the adoption of new technology based on theories of technology diffusion, in the broadest format enunciated by Tornatzky and Klein (1982), critical mass theory (Markus, 1987), and perceived risk (Teo, Tan, and Wei, 1995). In light of the actuality of EDI technology, it presents issues that continue to confront the practitioner and researcher. Finally, it concludes by exploring how such a gap developed between predicted and actual EDI diffusion and how a more realistic view of EDI diffusion might have been derived when it was first introduced.

Background

Definitions of electronic data interchange (EDI) revolve around the concept of electronic linkages between firms for the purpose of exchanging standardized business documents. At a minimum, firms substitute electronic for paper media for conveying business information such as orders, acknowledgment of orders, shipping, and payment documents. With the electronic exchange of documents, however, a wide range of secondary functionality can be actively created or passively experienced by the participating organization. Upon receipt of electronic documents from trading partners, organizations may develop computer-supported means to directly enter that information into internal data processing activities. Additionally, they can program error checking and accounting procedures, transaction monitoring by staff, routing of information to various departments (e.g. orders can directly act on inventory or production systems), and into management or decision support systems for longer term decision making.

These technically oriented possibilities, however, represent only a portion of the impact that EDI systems can have on an organization. At a minimum, these systems necessitate a standardization and discipline on the formal communications between trading partners. This sets in motion a sequence of changes in the relationship between partners. At its best, this results in more human attention on both sides to exceptional situations with the computers handling routine processing. On the other hand, it shifts the power and bargaining relationships among partners (Robinson and Stanton, 1987; Thachenkary, 1992). Internally, receipt of data through electronic channels can force the purchase or creation of new computer and telecommunications equipment, new technical staff (or training for existing staff), new procedures for traditional accounting activities such as auditing and control functions, and management possibilities to derive maximum benefit from accelerated rates of communication (Robinson and Stanton, 1987).

What motivates a firm to consider adoption of EDI?

Some firms adopt EDI for expected benefits in terms of reduced operating costs and/or strategic advantage in a particular market

(Rochester, 1989; Hill and Swenson, 1994), others respond to pressures and/or incentives from trading partners (Senn, 1992). Dearing (1990) identified three categories of benefits to firms using EDI: direct, indirect, and strategic. Scanning additional literature provides more detail regarding specific benefits expected to accrue to EDI using firms.

Direct benefits are generated from the substitution of paper with electronic transactions. Sources of benefit include: (1) cutting down on expensive paper transactions through simplification and standardization of the documentation process, speeding up information flow, and elimination of re-keying (Sadhwani and Sarhan, 1987; Canna, 1988; Janssens and Cuyvers, 1991; Sanders 1992; Scala and McGrath, 1993; Udo and Grant, 1993); (2) tracking and scheduling materials through the distribution pipeline to avoid, or quickly correct, errors and to more efficiently use physical resources (Clancy, 1990; Hall and Read, 1991); (3) providing more accuracy and discipline in transactions (Emmelhainz, 1987; Sadhwani, and Sarhan, 1987); (4) reduction in personnel (Scala and McGrath, 1993) although this is not frequently experienced according to Carter and Fredendall (1990) and more likely to be experienced as shifting personnel from activities related to documentation of transactions to more work on relationships between trading partners (Emmelhainz, 1987, Rochester, 1989; Udo and Grant, 1993); and (5) broadening the array of services to customers (Rogers, Daugherty, and Stank, 1992; Udo and Grant, 1993).

Indirect and information benefits are derived from new or improved functionality that results from having access to data electronically. Sources of benefit come from supporting: (1) just-in-time inventory and related programs; (2) shortening the ordering and receiving cycle (Emmelhainz, 1987; Sadhwani, and Sarhan, 1987; Rochester, 1989); (3) better cash-flow and related forecasting (Rochester, 1989; Boland, 1991; Hall and Read, 1991); (4) better lead times (Dearing, 1990; Rochester, 1989; Sanders, 1992; Scala and McGrath, 1993); (5) managing not just the customer's demand but also that of the customer's customer (Clancy, 1990); and (6) potentially better auditing, order control, and payment control processes (Hansen and Hill, 1989).

Strategic benefits are based on opportunities that the new technology provides to rearrange or reorganize previously arranged tasks. Benefits derive from sharing information with suppliers, (Dearing, 1990), potentially gaining market share, attracting or retaining customers/partners (Emmelhainz, 1987) and better relations with suppliers, customers, and units within the firm (Emmelhainz, 1987; Rochester, 1989; Janssens and Cuyvers, 1991; Scala and McGrath, 1993; Udo and Grant, 1993).

Asymmetrical benefits accruing to one or the other (but not both) trading partners were noted by Benjamin, de Long, and Scott Morton (1990). They found few benefits for suppliers, but customers received gains from one-time inventory reductions and better information for *inventory planning*. Customers also improved their bargaining position relative to suppliers.

What difficulties keep these benefits from being realized?

These benefits do not come without technical, managerial, and cross-organizational difficulties. Technical difficulties include the challenge of selecting cost-effective and reliable hardware and software, deciding between proprietary and standard protocols, and selecting a value-added network. Firms experiencing success in convincing many trading partners to adopt EDI can be faced with the technical problems of blending data arriving in various standards and complex reprogramming needed to generate even straightforward system enhancements (Evans-Correia, 1989; Joyce, 1989). Firms also frequently experience significant difficulties blending the system with existing transaction processing systems and controls (Evans-Correia, 1989; Cathey, 1991). This generally requires sustained high levels of top management commitment (Udo and Grant, 1993; Teo, Tan, Wei, and Woo, 1994) structured planning and effective project management (Kahn and Rosdorff, 1992) as well as the need for significant training and redefinition of roles for IS and other personnel (Carter, Clauson, Monczka, and Zenlinski, 1988; Joyce, 1989; Payne, 1989; Barber, 1992).

Firms also face issues including maintaining accounting control and the need for authorizations and audit trails (Cathey, 1991). This is related to both managerial and technical issues in maintaining security in keeping transactions private and authorized (Evans-Correia, 1989). Legal issues can also become important and tricky in resolving difficulties that arise from errors or misunderstandings (Evans-Correia, 1989; Joyce, 1989; Rochester, 1989; Cathey, 1991).

EDI participants can also anticipate some difficulties as well as benefits in changing the relationship with trading partners (Cathey, 1991; Kahn and Rosdorff, 1992). This includes issues of being "forced" to adopt EDI by trading partners (Evans-Correia, 1989) or "mission creep" by trading partners unilaterally adding new transactions or other requirements (Barber, 1992). On the other hand, it is often costly and difficulty to expand to new trading partners (Kahn and Rosdorff, 1992).

Establishing EDI is a much more complex technology than many others studied in the diffusion literature. It can be contrasted to spreadsheet software which is readily available for under $500, can be installed on a particular machine in a few minutes and immediately used. Moreover, clear direct benefits are derived from using spreadsheet software which can only be expanded as the technology penetrates the organization and spawns new applications. For EDI to work software and hardware often running above $100,000 must be purchased, additional, often substantial, programming expenses are incurred to blend the data into existing transaction processing systems, and coordination must be achieved internally between information systems, functional area personnel, and top management, and externally with trading partners and VANs. In many cases, it is not clear that direct benefits alone balance the additional costs of moving to EDI. Therefore, the firm is faced with additional capital appropriations in the hopes that indirect and/or strategic benefits will balance the costs. In some cases, it is possible for EDI to be necessary for a firm to remain in business (an arguably significant benefit), but will not in itself be cost effective.

Theoretical Influences on Technology Spread and Their Actual Manifestation with EDI Technology

The first goal of diffusion theory is to develop general understandings of individual and organizational reactions to introduction of new technologies. This general understanding includes identification of factors that can influence the processes or outcomes of the innovation project to varying degrees under varying circumstances. The second goal of diffusion theory is to provide practitioners a method of analyzing characteristics of a proposed new technology in order to (1) forecast (with precision varying with how similar the new technology is to those studied in the past) likely patterns of diffusion within and/or outside the firm and (2) anticipate problems and intervene in positive ways to overcome these.

Achieving these goals will remain probabilistic since each new technology, by virtue of its newness, will have some features that are difficult or impossible to assess in terms of their popularity with customers or users. Some factors, for example, how a new technology will interact with existing methods or the manner in which it will be introduced to potential adopters, which potentially influence diffusion are difficult to measure. Additionally, it is difficult to anticipate all the difficulties in implementing a new technology. Technologies that look straightforward have a notorious way of presenting subtle nuances that unexpectedly need much work to be aligned with existing practices. It is equally difficult to anticipate all of the innovative variations on the original concept, all of the secondary impacts on loosely related organizational systems, and all of the potential benefits which may derive from the technology.

It is important to consider not just the decision to adopt, but the degree to which the adoption is in fact implemented (Tornatzky and

Klein, 1982). Many firms that have an EDI program use it with only a subset of their trading partners. Even when there is EDI use between trading partners, it may be used for only a small percentage of transactions. Some firms will receive transactions electronically, then rekey the data into existing applications rather than add the extra component to move the data in electronic format into internal operations. Premkumar, Ramamurthy, and Nikakanta (1994) addressed this issue by measuring four aspects of EDI adoption: the decision to adopt, internal diffusion of EDI within the firm, external diffusion among trading partners, and implementation success.

Tornatzky and Klein (1982) examine ten attributes that potentially influence adoption and implementation of an innovative technology. These are: (1) compatibility; (2) relative advantage; (3) complexity; (4) cost; (5) communicability; (6) divisibility; (7) profitability; (8) social approval; (9) trialability; and (10) observability. These are listed in the order Tornatzky and Klein observed the number of times these were studied in the literature before their study. Tornatzky and Klein also noted high correlations between relative advantage and profitability and between divisibility and trialability. Because the influence of these pairs of constructs are difficult to distinguish, each pair will be treated as a single construct in the discussion which follows.

In the years since Tornatzky and Klein's study of diffusion of new technology, other studies have suggested additional potential influences on the patterns of diffusion. Particularly as regards communication where the value of the technology implies participation by multiple agents, "critical mass" theory has been used to explain some of the observed diffusion patterns. Critical mass theory (Markus, 1987) introduces the concept that the implementation decision by each firm is influenced by the penetration of that technology into the larger community. The more households with telephones in the community, the more people the new adopter can potentially call; therefore, the more value the new technology will have for them. In early stages of the new technology, the benefit of having the telephone is relatively limited, but this theory proposes that at some point, a critical mass is reached when the value of being able to access large numbers of potential message recipients outweighs the investment cost of the new technology and diffusion accelerates rapidly.

Another factor which may operate as an influence on diffusion of technology is perceived risk. Following decision theory, risk would seem to be part of the equation for determining relative advantage or cost. However, the empirical study of Teo, Tan, and Wei (1995), found it to be a distinguishable factor in the thinking of firms considering adoption of EDI technology.

For both critical mass and risk, it would be possible to hold that these are "competing" explanations for the shape of diffusion of various technologies. This would imply that these constructs and those presented by diffusion theory are somehow mutually exclusive. Another way to consider these is as part of a larger set of constructs each of which may influence the pattern of diffusion of a given technology, even if particular constructs are more likely to influence certain kinds of technologies. As a new technology has a communication or multiple participant aspect, it is more likely that critical mass may be a factor in shaping the pattern of adoption. This would not show up strongly in individually applied technologies such as spreadsheets or workstations. Similarly, risk might be a critical factor when expenditures are large or when new systems which have some possibility of not working well are replacing existing working systems.

The following discussion will focus on the eight Tornatzky and Klein factors, critical mass, and risk. Each will be discussed in general, in terms of how EDI as an innovation fares on this dimension and in terms of how firms vary regarding this dimension.

Compatibility

This variable refers to the degree to which the innovation is consistent with "the existing values, past experience, and needs of the receivers (Rogers and Shoemaker (1971)." Two aspects are normative or cognitive consistency with values and behavioral consistency with the practices of the adopters. Tornatzky and Klein (1982) reported a significant positive relationship between compatibility and adoption.

EDI potentially represents a substantial break from prior ways of operating for firms. At a minimum, EDI requires reorganizing and rethinking how a firm is connected to trading partners. It may additionally generate reorganization of purchasing and shipping functions. If extended to entry of data from external sources directly into internal transaction processing, it may begin to open new avenues of managerial decision making and evaluation of firm activities. Considered this way, firms will vary both in terms of their organizational culture being proactive in change management and technical ability to blend receipt of electronic data with legacy information systems. Premkumar, Ramamurthy, and Nikakanta (1994) differentiated technical and organizational compatibility in their study of EDI diffusion. They found technical compatibility influenced the adoption decision as well as implementation success, but not internal and external diffusion. Organizational compatibility influenced implementation success, but not adoption, internal nor external diffusion.

Eckerson (1990) has pointed out that the difficulties faced by firms integrating EDI information into legacy systems are not faced by start-up operations that can design accounting systems in coordination with EDI needs. Early EDI adopters invested heavily in proprietary networks that have constrained them in shifting to newer standardized protocols that provide for richer data sets to exchange.

Of course, these costs may be offset by benefits in utilization of EDI prior to development of standards and/or market penetration with early adoption. Holland, Lockett, and Blackman (1992) found a first mover advantage for early entrants into EDI for companies with "sufficient market share and IT experience [p. 549]" and further predicted that entering into an EDI network may be a key to firm survival in some industries as EDI becomes more universally accepted.

O'Callaghan, Kaufmann, and Konsynski (1992) provided support for system compatibility as a key factor influencing the adoption decision among those who adopted EDI technology. Those who had not adopted EDI technology, however, did not indicate this as a significant factor. Neither adopters nor non-adopters indicated organizational compatibility as a factor in adoption.

Relative Advantage/Profitability

These variables refer to the degree to which the innovation represents an improvement over prior ways of doing things. Tornatzky and Klein (1982) point out that "being better" is difficult to measure and suggest more direct inquiry into the profitability, productivity, reduction in labor costs, and other relevant potential advantages. It can be seen that expected values on this dimension might be a good predictor of intention to adopt and that initial realized values on this dimension might be a better indicator or evaluator of that decision. It should be noted that if positive return on the innovation is not realized, the firm might incur a variety of additional costs to change back to the original technology. The potential cost of changing back may or may not be factored into the original expected value. Naturally, studies of successfully adopted technologies would have little difficulty pointing out advantages gained. Perhaps in looking at patterns of diffusion, reports of results for early adopters will influence the shape of the curve for middle and late adopters. Interestingly, in spite of measurement difficulties, Tornatzky and Klein (1982) found a significant positive relationship between relative advantage and adoption.

Relative advantage can be considered in two senses: relative advantage compared to other firms and relative advantage compared to prior ways of operating. To the extent that EDI can allow a firm, the classic case being WalMart, to lower inventory costs and provide "just-in-time" stock for customers, it can potentially provide a relative advantage compared to firms in the industry not using EDI. However, it is not as clear how the trading partners of these firms benefit from EDI, other than in not losing customers who insist on EDI use. For the supplier firm, the costs of implementing EDI will be best absorbed if the totality or at least majority of its trading partners use EDI, which will diminish the advantage gained by the customer firm being the only one using EDI in its industry. Iacovou, Benbasat, and Dexter (1995) found a similar relationship of perceived benefit to adoption in five of seven small business cases (in the other two cases, adoption resulted from external pressure). Premkumar, Ramamurthy, and Nikakanta (1994) found perceived relative advantage to influence the adoption decision as well as internal and external diffusion, but did not observe a relationship to implementation success.

Firms may be differentiated in terms of relative advantage of EDI based on the information intensity of the industry (Grover, 1993; Teo, Tan, Wei, and Woo, 1994). The volume of standardized transactions that must be processed will vary between industries and between firms. Those with higher levels will have more costs for manual information processing to be deferred by EDI and a broader base to spread out EDI costs. O'Callaghan, Kaufmann, and Konsynski (1992) distinguished efficiency from service advantages and found both significantly influencing the decision to adopt EDI technology.

Complexity

This variable refers to the degree to which an innovation is perceived as "relatively difficult to understand and use (Rogers and Shoemaker, 1971, p. 154)." Complexity in organizational studies might refer to the number and difficulty of processes and activities involved in adopting and implementing the new technology. Tornatzky and Klein (1982) found a moderately statistically significant negative relationship ($p = 0.062$) between complexity and adoption or implementation.

EDI as an innovation varies in complexity based on the scope with which it is implemented. If the EDI is viewed solely as a mechanism for conveying documents, it would naturally be less complex than if it is viewed as a strategic initiative for reorganizing the interaction with trading partners. In the former case it may involve replacing mailed documents with faxed or documents downloaded from a VAN ready for keying into internal systems. Still, there is some complexity in terms of acquiring the right hardware and software for such an operation, interfacing with private or public networks, and maintaining (or creating) standardization across trading partners. As a strategic initiative, EDI is extremely complex ranging from highly technical issues at the level of telecommunication protocols, to issues of developing new mind-sets for working in new ways and rewarding different behaviors.

From the perspective of the individual firm, the level of complexity of an EDI project will also vary as a function of the firm's size and existing resources. The smaller firm in addition to handling the requirements of EDI data processing may need to hire and train new technical staff, rearrange auditing and accounting procedures, and add significant new internal data processing capabilities. The larger firm may be able to expand existing data processing and telecommunications capabilities in a relatively straightforward manner, particularly if they are not developing proprietary systems but rather using more standardized approaches.

Complexity can also be a function of the number of trading partners and the number of different protocols the firm must account for. Even assuming that at the transmission level using VANs transactions can be standardized, at the meaning level various trading partners may insist on variations in the content of the transaction messages. This can create much difficulty and added expense for maintenance and enhancement of translation software.

Cost

In Tornatzky and Klein's (1982) examination of innovation and implementation, cost is viewed as the straightforward measure of initial cost. No relationship was found between cost and innovation. In three studies cost was positively related to adoption; in two studies it was negatively related. Perhaps consideration of cost without potential return is not particularly meaningful. To the extent that innovation studies consider technologies such as new computer hardware or a particular software package, perhaps discussion of cost can be measured as the pricetag for the particular product. Initial cost may be a reasonable surrogate for total costs (including training, organizational restructuring, and cost of related equipment such as upgrading facilities) if secondary costs are proportionate.

Major EDI startup costs focus around translation and interface software where standard protocols and VANs are used. Initial costs of EDI will vary somewhat with the hardware used by the adopting firm. Costs of translation software may range from a few thousand dollars for PC based EDI to the low hundreds of thousands for mainframe based activity. Development costs for software to interface to organizational systems will also vary greatly with the scope of the implementation, how many systems will be interconnected, and the sophistication of legacy systems. In addition there are costs related to negotiating with trading partners (estimated as the largest of the costs by Clark, 1990) and training internal staff. Premkumar, Ramamurthy, and Nikakanta (1994) found cost to be correlated with the adoption decision and implementation success but not with internal and external diffusion.

Measuring costs of adopting EDI will actually be subtle and depend on where the boundary between EDI and reengineering projects are drawn. A firm that rewrites an inventory application to take full advantage of the arrival of orders in electronic format may assign all, none, or part of those costs to the EDI project. Anticipated costs will tend to be understated as firms emphasize initial cash outlay for a limited set of transactions rather than all of the long term changes cascading from the initial project and all the additional transaction sets that partners may eventually demand.

Hwang, Pegels, Rao, and Sethi (1993) present some data regarding expenditures for EDI systems. Of 280 firms reportedly surveyed, about $^1/_3$ invested between $10,000 and $50,000 and 5% more than $1 million for their system. Some 70% of firms took between 6 and 24 months to develop (10% longer than two years).

Communicability

This variable refers to the degree to which information about aspects of an innovation may be distributed to others. It is difficult to distinguish this from the complexity of an innovation (in other words does it require technical language not readily accessible to those outside a narrow specialty to convey information about the innovation). It is also difficult to differentiate this as a characteristic of the innovation rather than one of the communication media available to the potential adopter and the happenstance of how much information has actually been carried by the media and how much has actually been viewed by the potential adopter. Tornatzky and Klein (1982) found no real evidence regarding a relationship between communicability and adoption or implementation.

In considering EDI at the technology level, some specialized jargon is necessary to understand some aspects of protocols (such as use of codes internal to messages and manner of transferring information across telecommunication networks). Understanding the business implications of EDI would require use of financial, accounting, and marketing terminology. It is interesting to consider that much of the impetus for adopting EDI originates with middle management looking to streamline operations but expands to require delving into detailed technical issues and, if extended to strategic integration into firm operations, to upper management for reengineering organizational functioning. Communications regarding EDI, at least basic concepts were widely disseminated in a variety of trade publications including those in MIS, transportation, accounting, and marketing. Realistic and detailed assessments of the wide variety of issues and some of the difficult interactions among factors were probably less available in the literature. Although these more balanced looks at EDI might dissuade some firms from adopting, they would be more likely to provide the basis for firms to persist through difficulties and successfully implement (Janis and Mann, 1977). O'Callaghan, Kaufmann, and Konsynski (1992) did not find significant influence of other agents, other carriers, or industry promotion on adoption of EDI technology.

Divisibility/Trialability

These variables refer to the degree to which an innovation can be "experimented with on a limited basis" (Rogers and Shoemaker, 1971, p. 155). In other words, the degree to which an innovation can be adopted in phases, with each phase potentially leading to greater adoption. For example, new hybrid seeds can be planted on a small portion of the farm or a new teaching technique can be tried in a particular classroom or for a particular sequence of lessons. In both examples, if successful, the new product or technique can be expanded. In information technology, a divisible project may be rolled out at a particular site (e.g., a point of sale application at a particular store rather than throughout all franchises). Tornatzky and Klein (1982) found this to be positively related to adoption in three studies and negatively related in two studies.

EDI technologies are divisible in terms of the variety of documents exchanged between partners, the amount of integration within the firm, and the number of trading partners included. However, while such division might provide some smoothness to the implementation process, costs will largely be high in start-up and, therefore, more effectively spread across a variety of documents, internal processes and trading partners. In addition, some difficulties of a full-blown EDI program, such as dealing with large numbers of different protocols, will not necessarily be apparent from a trial or prototype application.

Observability

This variable refers to the degree to which the "results of an innovation are visible to others" (Rogers and Shoemaker, 1971, p. 155). It is not clear exactly how this differs from communicability, unless it differentiates between direct observation versus accepting reports from others. No significant results regarding this variable are presented by Tornatzky and Klein (1982).

One of the limitations of the diffusion model is its separate consideration of each factor. For example, firms in the Benjamin, de Long, and Scott Morton (1990) study were careful to minimize internal changes resulting from EDI to gain acceptance, but thereby lost some of the benefits anticipated from EDI use. The authors view this as a dilemma for firms considering EDI. Pilot projects tend to take a straightforward view of the technology as an external communication device and find it difficult to expand prototypes to influence production activities.

Critical Mass

"Critical mass" theory (Markus, 1987) posits that for communication technologies, some of the potential benefit to an organization considering adoption is the degree to which others have already adopted. This is easy to see in e-mail, as the first adopter has no one to communicate with. At some threshold level, according to this theory, a shift occurs such that the technology has more benefit than cost to the bulk of remaining potential adopters. An interesting secondary dynamic to this concept is the pressure asserted by those who have adopted on those who have not (Iacovou, Benbasat, and Dexter, 1995). The lack of universal acceptance of EDI among trading partners requires that firms maintain their manual system as well as using EDI for electronic transactions. This can be quite expensive and can stimulate both pressure on non-adopters and the value of subsidies to trading partners.

Bouchard (1993) showed that while general characteristics of innovation diffusion (trialability, complexity, compatibility, observability of results, and cost) were influential, critical mass factors counted more strongly in the adoption decision. Canright (1988) noted that such critical mass accumulates more quickly in those industries that use a lot of shipped materials, such as groceries, automotive, chemical, and transportation. These industries tend to have a higher volume of transactions and, therefore, more opportunity for gross cost reduction and/or spreading costs across larger bodies of transactions.

A phenomenon that may interact with critical mass theory is the added cost of adopting to improvements in a new technology if the organization became committed too soon to less effective versions. As discussed earlier, Eckerson (1990) noted difficulties faced by firms integrating EDI information into legacy systems not faced by start-up operations that can design accounting and other systems in coordination with EDI. The relationship to critical mass theory derives from firms knowing that early commitment to a non-standard solution may generate long-term costs and, thus, slow the development of a threshold level of users.

Another aspect of critical mass theory that impinges on diffusion of EDI is in terms of the existing infrastructure outside of the individual firm. Issues such as complexity, cost, and relative advantage will be a function not only of the EDI translation and application software, but also the existence of physical network, VANs, standards, and the array of mechanisms early adopters need to pioneer (generally at high cost) but upon which later adopters can piggyback. This technology appears to have required enough companies willing to absorb relatively high costs to generate the infrastructure that can bring down the price enough to create a threshold of external institutions such that firms for whom EDI benefits appear more marginal can enter.

Risk

At first consideration, risk would seem to be part of the equation comprising relative advantage and/or cost. Prior to adoption of a technology, there can be only anticipation of relative advantage. Estimation of relative advantage would have to account for probable returns minus probable costs. It would also have to account for the possibility that implementation would be operationally more difficult than expected and/or that strategically the market and opportunities may change.

However, Teo, Tan, and Wei (1995) focusing specifically on financial EDI, discovered the influence of perceived riskiness of the EDI project as a factor separate from relative advantage. They found two distinct and important kinds of risk: operational (under the control of the firm) and strategic (not subject to control by the firm). Present intention to adopt related highly to operational risk and strategic risk (as well as complexity) and to a lesser extent relative advantage and observability. Future intention to adopt was related to strategic risk (as well as complexity, and observability) and to a lesser extent trialability.

Considering the importance of perceived risk in their findings, a number of questions are raised. Does the actual level of risk vary between firms or is it constant? Are there factors which influence the degree to which members of a firm believe the new technology to be risky? Are there firms that will go ahead in spite of high risk with the inevitable result that risk will decline for other firms that learn by observing the experiences of early adopters?

Future Trends

Several practitioner questions arise from the prior discussion. *Should I adopt EDI technology (assuming the firm has not already done so)?* Factors influencing this decision, as seen from the discussion above will likely include issues such as potential relative advantage and risk, organizational and technical compatibility, and the current state of critical mass within the industry. Factors such as trialability and observability are more likely to influence whether the technology has come to the attention of the firm and the degree to which they can gather good information upon which to base a decision. More specifically, the firm will need to consider the potential for gaining customers by installing EDI (or losing them without it) and the value of that business versus the costs of installation. It must also consider the potential for the industry to evolve to one where EDI is considered essential for transacting business. Following the discussion above, this may relate to the information intensity in the particular industry. Increasingly, firms (particularly small firms linking directly to retail customers) may consider alternate technologies such as Internet.

If I adopt EDI technology, what actions will increase likelihood of generating positive returns? Firms will likely need to address a range of technical and managerial issues. Initial decisions need to focus on what manner of transporting messages are desired (VANs versus alternatives), what sort of translation and applications software are desired, and the degree to which the firm will want to use electronically received transactions as input for internal operations. Lower cost alternatives in early stages may speed adoption, but create rigidities that cost more for later extensions. High level managers, to get the most out of EDI technology, need to view it as a potentially strategic thrust that can generate changes in many aspects of how the firm does business.

If I already have EDI technology, will upgrades and extensions add increased value? To the degree that EDI becomes a sunk cost for firms, the addition of trading partners and additional transaction sets represent new projects that can change the equation of costs and benefits. Extensions offer the possibility of spreading costs for existing hardware and software; however, if hardware is near some capacity threshold, there may be new and unanticipated costs.

Conclusions

This chapter raised the issue of how such a gap developed between the predicted diffusion of EDI technology in the late 1980s and its actual diffusion through the 1990s (Sokol, 1995). This should be viewed in the context that early predictions were probably not based on diffusion of technology theory. This raises two questions: what were the early predictions based on? and where were the researchers when the predictions were being made? A cynical viewer might take the position that early optimistic projections were made by consultants and practitioners who had a stake in successful diffusion of EDI technology; a less cynical view would hold that EDI had "face validity" — it just made sense. There was a conceptual simplicity to the concept that transaction data could be moved electronically over already existing communication media. Conceptually, EDI should provide great benefits at little cost, and that the faster it diffused, the better returns at lower cost for all involved.

Information technology is well known for overly optimistic projections of the utility of new technologies (e.g. artificial intelligence replacing human decision makers). But it is also well known for unexpected technologies that grow exponentially (e.g., microcomputers, spreadsheets, connections to the Internet). Regarding the second question, perhaps it is an overly ambitious goal that diffusion of technology theory should help us to distinguish the artificial intelligences and database machines from the spreadsheets and the Internets before the curve begins to climb (or languish). In order to move toward this goal, however, a broad array of technologies, including both successes and failures, should be studied using diffusion of technology theories.

How could a more realistic view of EDI diffusion have been derived when it was first introduced? It is always easy in hindsight to see what was not viewed in the past. However, we should note a number of aspects of EDI which would militate against an exponential growth curve. EDI required multiple participants and the need to coordinate among them as well as for each to make an adoption decision. There was a natural conflict between the need for interorganizational cooperation and traditional competition. The amount of change required within the adopting firm to take advantage of EDI technology was quite high. It is generally more difficult to absorb both new technology and a new way of doing things at the same time than it is to adjust to one over the other. EDI also presents much greater challenges linking to legacy systems than was perhaps anticipated. EDI technology experiences a scaling difficulty. Adding more clients to the EDI system changes workload even when the new clients are using the same protocols (varied protocols add the need for broader capabilities in translation software) in a way that adding more personal computers or spreadsheets does not. We see this most clearly with efforts to attract trading partners to EDI. There is always a tendency to do the easy ones first which leaves it very difficult to get the last 10-20% on board. Yet that last effort is needed to completely let go of manual systems. Finally, there is difficulty with viewing EDI as a technology in the sense of a mechanism when it is more accurately pictured as a set of mechanisms and new ways of conducting business.

In a larger sense this essay raises the question — What is the contribution of diffusion theory? Diffusion theory presents the opportunity to better understand the underlying mechanisms that accelerate and decelerate technology's spread among users. Ideally, this understanding will be applied by those who wish to accelerate or inhibit diffusion early in the cycle. The goal of the research is not necessarily to accelerate adoption of all new technology but to understand underlying forces. Some technology probably ought not to be accelerated. In the case of EDI, it might be argued that the business community is better off with a smooth steady rise in use so that the amount of early protocols competing to become the standard is limited and the bulk of firms can invest in semi-mature technology and not have to discard expensive early prototypes. It is entirely possible that from an industry standpoint exponential growth may have led to intractable problems of non-robust technology. Some technologies may be of great value if adopted widely (critical mass) but each firm must make decisions before that wide adoption has come about.

In considering a wide array of technologies, it may be that the range of diffusion of technology factors all have the potential to affect rate of adoption and implementation, but that different technologies are more sensitive to some of the factors than others. Prescott and Conger (1995), for example, point out differences in diffusion of technology findings when the technologies studied were used within information technology units, within various units, within the firm, and across firms. Perhaps the level of interconnection among users of the technology will prove to be a critical differentiator of technologies. The spreadsheet user can get much value from the technology regardless of anyone else choosing to use it; however, CASE tools are most effective when part of a larger departmental coordination of analysis and design processes.

To make diffusion of technology useful as an early predictor of adoption of new technology, we need to be quite critical as we examine each element. For example, not all technologies present a positive amount of compatibility with the existing way of doing things. When they do not, we need to acknowledge much work in overcoming the inertia of current practice (and the risks that current practice is just fine as it is and that changes will bring negative unintended effects). We must account for offsetting forces. As critical mass expands and the general value of the technology increases, its opportunity to generate competitive advantage for the few who have it decreases. Finally, since there is a tendency for those firms to become proponents (for the purpose of spreading the technology and gaining from mass), we should look at early case studies very critically. Are the firms acknowledging all costs and risks? Are the firms' experiences dependent on skills or resources that are generally available? Are the firms' experiences scalable, will they experience the same returns on projects 2, 3, and 23?

Diffusion of technology theory in the broadest sense, with the addition of critical mass and risk as factors, provides a very powerful basis for differentiating the potential of technologies. It also provides a basis for proponents of particular technologies to consider tactics for stimulating its spread (e.g., designing in easier ability to test the systems and providing relief from high rates of complexity).

References

Barber, N.F. "Simultaneous Process Re-engineering," *EDI World, 2*, 8, 1992, 5-11.

Benjamin, R.I., de Long, D.W., and Scott Morton, M.S. "Electronic Data Interchange: How Much Competitive Advantage?" *Long Range Planning, 23*, 1, 1990, 29-40.

Boland, A.S. "EDI and Cash Management: Tactical Necessity or Strategic Imperative?" *EDI World, 1*, 5, May 1991, 25-27.

Brouchard, L. "Decision Criteria in the Adoption of EDI," in J. DeGross, R.P. Bostrom, and D. Robey (Eds.), *Fourteenth Annual International Conference on Information Systems*, Orlando Florida, December 1993, 365-376.

Canna, E. "Union Carbide Goes Paperless in Four Stages," *American Shipper*, August 1988, 34.

Canright, C. (1988), "Seizing the Electronic Information Advantage," *Business Marketing*, (January), 81-86.

Carter, J.R., Clauson, K.S, Monczka, R.M., and Zenlinski, T.P. "Education and Training for Successful EDI Implementation," *EDI Forum*, *1*, 1988, 89-95.

Carter, J.R. and Fredendall, L.D. "The Dollars and Sense of Electronic Data Interchange," *Production and Inventory Management Journal*, *31*, 2, 1990, 22-26.

Cathey, J.M. "Electronic Data Interchange: What a Controller Should Know, *Management Accounting*, *73*, 5, November 1991, 47-51.

Chin, W.W. and Gopal, A. "Adoption Intention in GSS: Relative Importance of Beliefs," *DATA BASE*, *26*, 2 & 3, May/August 1995, 42-64.

Clancy, D. "High Technology Prompts Efficient Inventory Flow," *Transportation and Distribution*, March 1990, 20-22.

Clark, S. "Data Interchange: How to Recruit Trading Partners," *Datamation*, *36*, 21, November 1, 1990, 75-76.

Dearing, B. "The Strategic Benefits of EDI," *The Journal of Business Strategy*, Jan-Feb 1990, 4-6.

Eckerson, W. "EDI Trailblazers Pay Price for Early Implementation," *Network World*, *7*, 4, 1990, 23-24.

Emmelhainz, M.A. "Electronic Data Interchange: Does it Change the Purchasing Process?" *Journal of Purchasing and Materials Management*, *23*, 4, Winter 1987, 2-8.

Evans-Correia, K. "EDI: The Future Frontier?" *Purchasing*, *10*, 3, 1989, 44-47.

Grover, V. "An Empirically Derived Model for the Adoption of Customer-based Interorganizational Systems," *Decision Sciences*, *24*, 3, 1993, 603-640.

Gurbaxani, V. "Diffusion in Computing Networks: The Case of BITNET," *Communications of the ACM*, *33*, 1990, 65-75.

Hall, J.D. and Read, W.F. "Focusing on the Total Order Cycle," *EDI World*, *1*, 2, February 1991, 28-30.

Holland, C., Lockett, G., and Blackman, L. "Planning for Electronic Data Interchange," *Strategic Management Journal*, 13, 1992, 539-550.

Hwang, K.T., Pegels, C.C., Rao, R., and Sethi, V. "Electronic Data Interchange Systems - State of the Art," *Journal of Systems Management*, *44*, 12), December 1993, 12-15.

Iacovou, C.L., Benbasat, I., and Dexter, A.S. "Electronic Data Interchange and Small Organizations: Adoption and Impact of Technology," *MIS Quarterly*, *19*, 4, December 1995, 465-485.

Janis, I. L. & Mann, L. *Decision Making: A Psychological Analysis of Conflict, Choice, and Commitment*. New York: The Free Press, 1977.

Janssens, G.K. and Cuyvers, L. "EDI—A Strategic Weapon in International Trade," *Long Range Planning*, 245, 2, 1991, 46-53.

Joyce, E.J. "Tales of EDI Trailblazers," *Computer Decisions*, *21*, 2, February 1989, 61-65.

Kahn, R. and Rosdorff, R. "Practical Applications of Financial EDI," *Journal of Cash Management*, *12*, 1, January 1992, 20-23.

Markus, M.L. "Toward a 'Critical Mass' Theory of Interactive Media: universal Access, Interdependence, and Diffusion," *Communication Research*, 14, 5, October 1987, 491-511.

O'Callaghan, R., Kaufmann, P.J. and Konsynski, B.R. "Adoption Correlates and Share Effects of Electronic Data Interchange Systems in Marketing Channels," *Journal of Marketing*, *56*, April, 1992, 45-56.

Payne, R.A. "EDI Implementation: A Case Study," *Journal of Systems Management, 40*, 3, March 1989, 14-20.

Premkumar, G. and Potter, M. "Adoption of Computer Aided Software Engineering (CASE) Technology: An Innovation Adoption Perspective," *DATA BASE, 26*, 2 & 3, May/August 1995, 105-123.

Premkumar, G., Ramamurthy, K., and Nikakanta, S. "Implementation of Electronic Data Interchange: An Innovation Diffusion Perspective," *Journal of Management Information Systems, 11*, 2, Fall 1994, 157-186.

Prescott, M.B. and Conger, S.A. "Information Technology Innovations: A Classification by IT Locus of Impact and Research Approach," *DATA BASE*, 26, 2 & 3, May/August, 1995, 20-41.

Robinson, D.G. and Stanton, S.A. "Exploit EDI Before EDI Exploits You," *Information Strategy: The Executive's Journal*, Spring 1987, 32-35.

Rogers, D.S., Daugherty, P.J. and Stank, T.P. "Enhancing Service Responsiveness: The Strategic Potential of EDI," *International Journal of Physical Distribution & Logistics Management, 22*, 8, 1992, 15-20.

Rogers, E.M. and Shoemaker, F.F. *Communications of Innovations: A Cross-Cultural Approach*, Free Press, New York, 1971.

Ruppel, C.P. and Harrington, S.J. "Telework: An Innovation Where Nobody is Getting on the Bandwagon?," *DATA BASE, 26*, 2 & 3, May/August 1995, 87-104.

Sadhwani, A.T. and Sarhan, M.H. "Electronic Systems Enhance JIT Operations," *Management Accounting*, December 1987, 25-30.

Sanders, N.R. "Merging EDI With JIT: The Impact on US Manufacturing," *Journal of Applied Business Research, 8*, 2, 1992, 133-137.

Senn, J.A. (1992), "Electronic Data Interchange: The Element of Implementation," *Information Systems Management*, (Winter), 45-52.

Sokol, P.K. *From EDI to Electronic Commerce: A Business Initiative*, McGraw-Hill Inc., New York, 1995.

Teo, H.H., Tan, B.C.Y., and Wei, K.K. "Innovation Diffusion Theory as a Predictor of Adoption Intention for Financial EDI," In J.I. DeGross, G. Ariav, C. Beath, R. Hoyer, and C. Kemerer (Eds.), *Proceedings of the Sixteenth Annual International Conference on Information Systems*, Amsterdam, The Netherlands, December 1995, pp. 155-165.

Teo, H-H., Tan, B.C.Y., Wei, K-K., and Woo, L-Y, "Leveraging EDIS to Improve Organizational and Interorganization Productivity: Findings from the World's Busiest Port," In J.I. DeGross, S.L. Huff, and M.C. Monro (Eds.), *Proceedings of the Fifteenth Annual International Conference on Information Systems*, Vancouver, British Columbia, December 1994, 175-185.

Thachenkary, C.S. "COINLink: The Impacts of an EDI Network on the Automobile Dealership and Lending Industries," *IEEE Transactions on Engineering Management, 39*, 4, 1992, 386-393.

Tornatzky, L.G. and Klein, K.J. "Innovation characteristics and Innovation Adoption Implementation: A Meta Analysis of Findings," *IEEE Transactions on Engineering Management, 29*, 11, 1982, 28-45.

Udo, G.J. and Grant, T. "Making EDI Pay Off: The Averitt Express Experience," *Production and Inventory Management Journal, 34* (4), 1993, 6-11.

CHAPTER

8

Strategic Implications of Hype on the Adoption of Innovative Technologies:

The Case of Local Area Networks

Vasilis Theoharakis
Veronica Wong
Philip Powell
University of Warwick

The preannouncement of a new technology can have a considerable impact on the adoption of currently available technologies. The excitement of a forthcoming technology often results in media hype which influences consumer adoption behavior and technology supplier strategy. Based on a framework for the adoption process of Local Area Network (LAN) technologies, the influence of prelaunch interest is investigated. In this paper, a methodology for quantifying media interest is applied to the LAN industry. While hype can significantly distort planning and forecasting, its measurement is an important tool for suppliers. With customers exposed to the hype generated, suppliers need to create communication programs describing future technology migration plans in order to maintain or develop a leadership image. The relative interest measurement can signal the overall threat or importance for incorporating relevant communication messages. The decision process for adopters of such technologies becomes increasingly more complex due to the abundance and evolving nature of received signals. Opinion leaders are observed to become very excited with a newly announced technology, frequently ignoring the power of the installed base.

The effects of new technology preannouncement are twofold: on consumer adoption process and supplier strategy. The impact is especially pronounced in cases where there is a long lead time from initial announcements until functional products appear in the market place. This is characteristic of high technology markets due to long development cycles. New technologies may eliminate or strengthen barriers to entry in an industry. Firms can take advantage of a new technology for entering new markets, while incumbent players are either concerned as this technology may compete with existing product lines, or view it as an opportunity to further entrench their position. During the prelaunch period, supporting suppliers seek to generate hype around the introduction of the new technology in an effort to generate awareness, accelerate adoption, and overall, minimise risk (Wind & Mahajan, 1987). One problem lies in the fact that the same marketing professionals who generate or are heavily exposed to the excitement of a new technology introduction, often perform the firm's planning and forecasting. Being immersed in the technology hype, their forecasts are subject to undesirable bias where basic marketing factors may be ignored.

The adoption process of innovative products is well researched (Abernathy & Clark 1985, Rogers, 1983). Particular focus has also been placed on buyer behaviour in industrial markets (Czepiel, 1974, Sheth 1972). With a high level of innovation and a rapid change in technology, a higher search effort is required by buyers (Weiss & Heide 1993). Given such an environment, opinion leadership is increasingly important. The competitive market environment for technology innovation has a significant influence on diffusion (Robertson & Gatignon 1986). The simultaneous technological competition and the role of externalities that lead to the dominance of a technology have also been examined (Redmond, 1991). Others have taken an economic perspective on the adoption of innovative technology (David, 1990; Farrell & Saloner, 1986, Stoneman 1995), while marketing literature has explored product diffusion modelling for new product acceptance (Mahajan, Muller & Bass, 1990). With the perspective that technologies are products that compete against each other, announcements by supporting vendors are equivalent to classical market signalling that receive reaction from the market place (Eliashberg & Robertson, 1988; Robertson, Eliashberg & Rymon, 1995). The use of technology life cycles is an important factor influencing sales predictions in industrial environments (Popper & Buskirk, 1992). Studies have also described pre-market forecasting models (Urban, Hulland & Weinberg, 1993) that suffer in environments of intense hype, which significantly lessens their predictive power.

The Local Area Network (LAN) market has experienced tremendous growth in the last decade. A LAN consists of components that form the data communications infrastructure of an organization within the geographical boundaries of a building or a campus. The advancement of semiconductor technologies has resulted in increasingly more powerful desktop PCs and improved communication devices. The adoption of these complementary technologies has created an infrastructure that enables the rapid sharing of information across the enterprise and has transformed the way business is done. These applications range from access to remote databases, to the use of electronic mail for interoffice communication to telecommuting. In general, the information systems (IS) function of an organization is responsible for the maintenance of current LANs and the planning of purchases for new equipment and technologies. New and existing applications are becoming more demanding in their requirements (bandwidth and latency) and new technologies need to be considered to meet this challenge. With networked computing becoming increasingly more pervasive through the enterprise, reliability is essential while ease of migration and the expected lifetime of the new technology are factors influencing the decision for such a transition.

Research Aims and Methodology

This study focuses on the adoption process of MIS managers, when selecting available or expected technologies. The objectives of this chapter are: i) develop an adoption framework for LAN technologies, ii) identify opinion leadership sources that influence the adoption process and, iii) measure evolution of relative interest among opinion leaders. The resulting relative interest measurement, is defined as the Interest Life Cycle (ILC). Using a chart describing the magnitude of interest for each technology, this work seeks to explore the evolution of opinion across time for a technology. More specifically, the effects of hype are discussed as it potentially influences the adoption process and requires supplier action. The research aims are achieved by initially conducting expert interviews and then by applying content analysis techniques to identified sources of opinion leadership.

A series of interviews were conducted with nine industry experts for the purpose of developing a model that describes LAN technology adoption. At the same time, the key decision influencers in the various stages of this adoption process were discussed. The selection of the experts was based on the following criteria: i) currently holding a position in a product marketing or sales organization of a leading LAN supplier, ii) have frequent customer interaction that allows observation of the adoption process for LAN technologies, and iii) have been in the industry for at least five years and have experienced the launch of several key technologies. Of the nine experts, four held product marketing positions with worldwide responsibilities. The remaining held sales positions covering most geographies (two from the United States, one from Asia, one from Europe and one from Latin America).

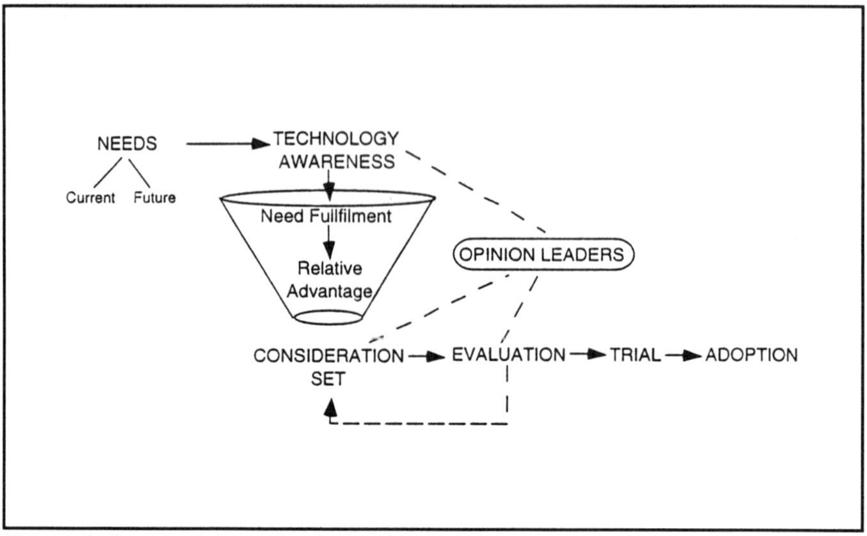

Figure 1: Stages of the technology adoption process

After identifying the key sources of opinion leadership during the interviews, computer-based content analysis is used to generate the ILC that supports the discussion of events in the industry.

LAN Technology Adoption Framework

The adoption of innovative technologies has been the subject of several studies, but the work by Rogers (1983) has been a key influence in this area. Figure 1 represents an adapted version of the innovation adoption process presented by Rogers. The adaptations to this model are guided by the series of expert interviews. This model serves as the framework for the discussion in the remainder of the paper. The following sections examine the stages of the adoption process based on discussions with selected industry experts.

The Needs That Drive Adoption

When evaluating investment in new networking technologies, current needs are considered along with future needs of the user community. Several sources drive the need for adoption of new local area networking technologies:

- *Increased use of personal computing for corporate tasks.* Information technology (IT) has penetrated the work place as the basis of the modern knowledge based business, where use of this technology can offer a competitive advantage. With the shift from manufacturing to service economies, the appropriate use of information technology translates into a key competitive advantage. Re-engineering and "right-sizing" efforts have further refuelled the use of such technologies as means to streamline business processes, resulting in an even stronger dependency on IT.

- *Client/Server computing.* The advancement of semiconductor technologies has allowed the deployment of PCs on almost every desktop. The transition from centralised processing architectures to distributed systems was realised by LAN technologies that efficiently interconnect individual computers. The combination of these technologies and advanced PC platforms has resulted in the implementation of flexible client/server systems.
- *Application requirements.* Application requirements are twofold: new applications and increased demands of existing applications. As PCs become more powerful (processing power, storage capacity) and more sophisticated (operating systems, user interfaces), new applications can be implemented. At the same time, new releases of existing applications are extended to take advantage of these new capabilities and improve their feature set. From a network perspective, the result is that more traffic needs to be transported with stricter requirements (e.g., multimedia applications generate volumes of data sensitive to network delay and jitter).
- *Transparency/Mobility.* While organizations are more dependent on information technology, particular emphasis is placed on the flexibility that IT has to offer. Geographically dispersed parts of the organization need to share information in the form of databases, or e-mail while the organization becomes more flexible with remote connectivity and telecommuting.
- *Self Interest.* Contrary to the other motivators presented where need is driven by applications and justifiable business reasons, adoption of new technology is frequently motivated because it satisfies the ego of technology-minded network managers. In this case, new technologies are sought for adoption because of the prestige or self-satisfaction that one might enjoy by being involved with leading technologies and may enhance future job opportunities.

Awareness Creation for New LAN Technologies

There are several sources that contribute to awareness creation for new technologies. *Trade journals* were consistently identified in the expert interviews as the primary source of information for the IS manager. These journals tend to report on the announcement of new products and technologies, standardization progress and contain several commentaries from leading industry figures. These media sources, seek every opportunity to discuss about new technologies that might be in their very early stages, but their opinions are perceived as being unbiased (Wind & Mahajan, 1987). The content of these journals is highly informative, as they introduce the theory of newly announced technologies, follow it through its standardization process and evaluate products that eventually implement it. It should be recognized that the authors actively seek to report on new technologies or stir controversy to attract a higher readership.

Another important source for generating awareness is *vendor presentations*. These can be either in the form of seminars or customer presentations. Only the larger customers receive the benefit of personal vendor attention when a vendor's strategy for implementing new technologies is presented. *Seminars* represent a broader vehicle and tend to be very educational. The goal of these efforts is to position sponsoring vendors as innovators and industry leaders. *Trade shows*, on the other hand, provide a more realistic view of a vendor's product line. These shows are frequently seen by users as a reality check after being bombarded with technology and product preannouncements. *Word-of-mouth* is another effective source that can come through direct personal contact or through user groups (electronic bulletin boards).

From Awareness to the Consideration Set

The first step for a technology to be considered, is need fulfilment; the technology serves the application. Depending on requirements, several technologies might be able to fulfill the application. Given several technology options, compatibility with the existing infrastructure is essential, while relative cost and expected longevity are critical decision criteria. The timing requirements for the implementation and support of an application is a determinant factor in the selection of a new technology. If it is perceived that implementation can await for the arrival of a newly announced technology, then current technology purchases will be deferred. Although the support of a critical mass of vendors is generally required for the broad diffusion of a technology, users are willing to accept a proprietary or technology at its early stages if there is a unique fit between application and technology.

For a new technology to be accepted, it must exhibit or be perceived to exhibit a relative advantage over existing solutions. These relative advantages are often reinforced by opinion leaders that tend to get absorbed in the excitement of the new technology features. In fact, a new technology may be considered even if its cost does not justify it. This subjective selection is driven by technology interest and curiosity or because familiarity with a new technology increases the value and marketability of the people involved. This implies a type of satisfying approach to acceptance; new technologies are evaluated according to a minimum set of acceptable standards, but decision makers rarely strive for optimality and investigate the full set of potential solutions.

The Evaluation Process

The evaluation process is characterised by creating a simulated environment where the application (the need) is applied on the equipment supporting the particular technology. The product evaluations that take place in trade journals greatly influence the methodology and predisposition of the evaluators. Frequently, evaluations at the early stages of the technology expose problems in its first implementations where the results often do not match expectations. This is partly due to the high level of expectation that has been set about the features and performance of this new technology as part of the prelaunch activities. If the benefits sought are not demonstrated during the evaluation, another technology from the consideration set is selected for further evaluation.

The Trial

The trial stage is the limited deployment of the selected technology. This necessitates that the evaluation has been successful in order for the network manager to introduce it into a network with real users and applications. Frequently even if an evaluation has been successful, users seek other reference sites that have implemented the technology. It is crucial for vendors to offer such sites in order to increase the comfort level of the customer by reducing the perceived risk. Although the purpose of a technology trial is to test its suitability prior to being widely adopted, it already represents a significant commitment by the network manager. Indeed, even if the initial tests are not as successful as desired, network managers have to defend their original decision. At this point, every effort is made to circumvent the problems by applying the necessary fixes.

Adoption

The main reason for the adoption of a new LAN technology, as with any other product, is need fulfillment and compatibility. Need fulfillment is always viewed in light of the cost effectiveness and expected longevity of the solution. Factors that enhance the perception for the longevity of a technology are the presence of international standards and a growing number of supporting vendors. The compatibility of the new technology with the existing infrastructure is important for a smooth migration. Besides compatibility with existing infrastructure, adoption of the new technology is facilitated if maintaining this new technology does not require extensive retraining of personnel.

Focusing on Trade Journals

In an effort to keep abreast with the latest developments in technologies and products, IS decision makers typically subscribe to several trade journals. These journals filter the massive amounts of available information and each author provides his/her opinion on the relative advantage of a new technology or product, based on their own criteria and motivations. Columnists in such journals are usually well respected consultants who use monthly columns as a further endorsement of their reputation. Their opinions are very influential since they are perceived to be unbiased and are always at the leading edge of technology. These articles provide a broad view on the state of the market and on-going product development, that IS managers seek in order to complement their own experience that remains much more focused. As vendors realize the importance of these journals, they always seek to influence the content and create relationships with the authors. Reprints of articles are frequently used to support of a vendor's technology strategy and are part of the selling process.

The rapid change of technology in this industry further intensifies the importance of opinion leaders in influencing purchase behavior. At any time the consumer is typically faced with several technologies and products offered from various vendors. Not only do IS managers find information on the broader prospects technology evolution (standards status, new technology introduction, product availability, etc.), but also on the direct comparison and testing of competing products. These product comparisons can have a significant effect on product market shares, since consumers use these test results as an important factor for product evaluation. Given the complexity and detailed knowledge required to perform these tests, it is not surprising to find consumers rely heavily on product evaluations.

The Alignment of Opinion Leadership and Trade Journals

As stated, trade journals are identified as the most influential information source. Observing the behavior of the content creators of these trade journals, they can be characterized as the opinion leaders for this market. More specifically, these figures represent the key change agents and have the characteristics of the innovative communicator (Baumgarten, 1975). The innovative communicator is described as an innovator or early adopter who is also an opinion leader. Trade journal authors are perceived as innovators as they report their experience with new technology and at the same time as peer group opinion leaders for presenting their evaluation and expert opinion. The framework defined by Rogers (1983) describing opinion leadership has been applied to demonstrate this point:

- *External Communication:* This group is always well informed and has a much greater exposure to all forms of media (trade journals, trade shows, company announcements and press releases, standards setting progress, etc.) The frequent exposure to other change agents is transferred to the reader through the contents of the published articles.

- *Accessibility:*These trade journals are made accessible by a wide worldwide distribution and by the relatively low subscription fees.
- *Socio-economic Status:* Authors are considered leading experts in the industry. They are most frequently high-profile consultants and are also perceived to be unbiased. In addition, authors include leading corporate professionals and executives responsible for large networks or overall IT strategy who present case studies of state of the art technology implementations.
- *Innovativiness:* Since this industry is characterized by constant technological innovation, change is favored. Opinion leaders are more innovative than other users, and are the first to experiment with new technology, then report on their experience.

Measuring Public Interest

After establishing the importance of trade journals, this chapter seeks to measure the relative interest for competing LAN technologies presented by this media. The measurement provides an indication of the topics discussed in the industry and is expected to be highly correlated with the topics of vendor presentations and seminars. For this purpose, computer-based content analysis is used. This methodology is selected as the paper seeks to estimate the domination and displacement of different subject matters over time and it is appropriate for measuring public attention (Krippendorff, 1985).

The first step towards the estimation of technology interest is the selection of the sampling unit used for the analysis. The unit chosen

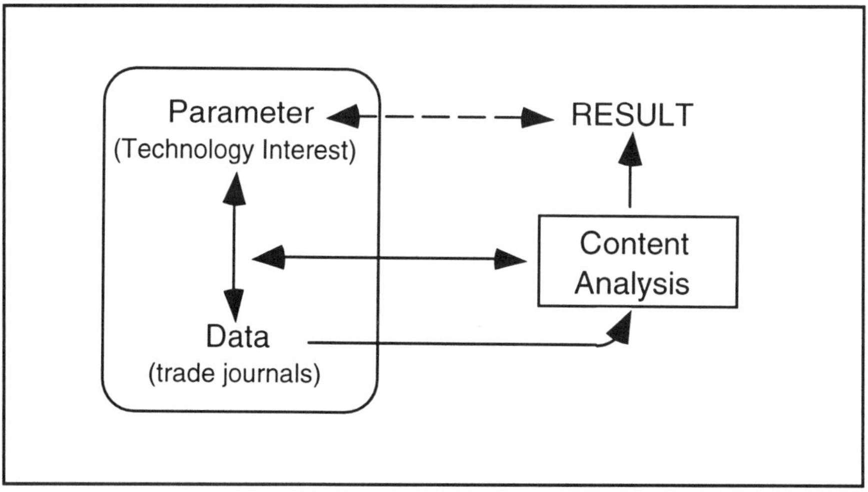

Figure 2: Content Analysis for LAN Technology Interest Estimation

for this study is simply the name (or names) that uniquely identify a particular technology. The ABI/Inform database is queried based on these keywords and the relative frequency of articles mentioning the technologies of interest tabulated on a yearly basis. The results of this tabulation create an index describing the "noise"-share of a given technology over a period of 16 years. The ABI/Inform database is deemed adequate for basing this analysis, as it includes the major trade journals related to information technologies. By using computer based content analysis, the results are reliable and any interjudge reliability issues are eliminated (Kassarjian, 1977). The database used is widely available making these results easily reproducible.

The LAN Interest Life Cycles (Figure 3) are derived based on the ABI/Inform database, display the relative media interest for a technology described by:

$$\frac{(\text{mentions for technology})_k}{\Sigma_i\ (\text{mentions for technology})_i}$$

Notice that the denominator is not equal to the total number of articles referencing all technologies of interest, as it may count an article multiple times if it mentions more than one technology. This formulation is adequate since the paper seeks a measure for the relative frequency of media interest across time. In the spirit of the concepts of product and technology life cycle (PLC/TLC), this methodology provides a new dimension: Interest Life Cycle (ILC).

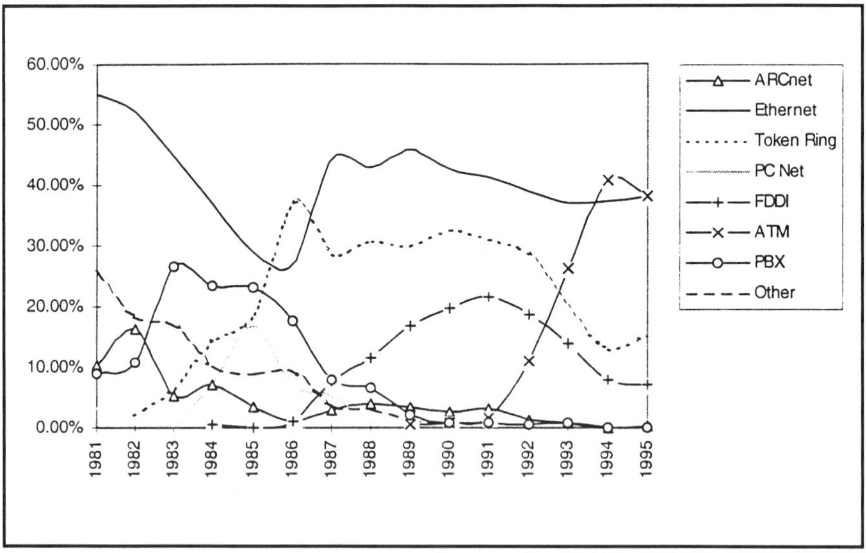

Figure 3: Relative Media Interest for LAN Technologies

The Interest Life Cycles for LAN Technologies

The content analysis-based measurement captures the evolution of opinion for printed media and provides a unique dimension in the issues described through the history of the industry. This section describes the events that shaped the Interest Life Cycle for this industry.

1981-1983: The first LAN technology introduced in the marketplace was the initially proprietary ARCnet. During this early period in the industry, although this technology had the largest installed base, attention turned to Ethernet, which was considerably more expensive, but was faster and promoted by reputable companies (Xerox, DEC, Intel). ARCnet eventually became an open architecture in 1987, and although it remained considerably cheaper than Ethernet, it could not compete in the marketplace beside its upgrade to run twice as fast as Ethernet. The anticipated growth for this market, attracts the announcement of several new technologies (marked as Other), but also presents as an opportunity for the existing communication technology within the enterprise: the PBX. PBX proponents see the role of the LAN to be simply an added function on existing or next generation PBX equipment, eliminating the need to introduce separate equipment for data and voice communications. An important event that marks this period, is the introduction of the PC that will fuel the growth of this market.

1984-1986: Although Token Ring arrived late to the market, with its major support from IBM, it surpasses all other technologies with respect to attention. While it was more expensive than Ethernet, that has a much wider vendor support, it supports Unshielded Twisted Pair (UTP) cabling. This type of cabling is very common in buildings since it is used for PBX data grade telephone systems. Token Ring technology is also viewed as being more robust, but in its first release was slower than Ethernet (4 vs. 10 Mbit/sec). At this point, with Token Ring interest at its peak, forecasts and experts predict that it would overtake the inferior Ethernet. In the meantime, due to the significant delay in the delivery of Token Ring, IBM introduced the acquired PC Network technology. Although this technology was not received well in the marketplace, it gained attention as it signified IBM's first product entry in the industry. With the introduction of Token Ring, IBM's support rapidly moved away from the PC Network, which proved to be only a fad. Ethernet's future was threatened.

1987-1992: Following the introduction of Token Ring, Ethernet vendors did not remain idle. They introduced a new standard that allowed Ethernet to communicate over UTP and guaranteed interoperability with an industry standard: 10BASE-T. The products using this technology remain much cheaper than Token Ring and have wider vendor support. Ethernet fought back and managed to recaptured attention. The introduction of 16Mbit/sec Token Ring (four times faster from the original 4Mbit/sec) maintains attention for Token Ring, but increases the entry price for this technology. The announcement of FDDI, a technology supporting much higher transmission rates (100Mbit/sec) than ever before over fiber optic cables, receives great attention. Although experts present FDDI as the technology of the future, it fails to take off at the originally anticipated rate due to its high cost and the lack of applications that would require its widespread deployment. Furthermore, in sight of the new ATM (Asynchronous Transfer Mode - not to be confused with Automatic Teller Machine) technology that competes for the same space, the proposed FDDI II enhancements to the standard were abandoned after failing to attract the necessary industry following. Early in this period, the market has decided the need of a separate network for the support of data traffic, eliminating interest for PBXs performing this function.

1993-1995: With ATM excitement at high levels, it is only the second time a new technology receives more interest than Ethernet. The market for ATM equipment remains in its very early stages while the ATM Forum (standards committee) has not yet completed the necessary elements of the standard that will enable widespread deployment. Given the maturity of Ethernet, it combats pretty well with the introduction of switching and Fast Ethernet that represent significant technology extensions. Fast Ethernet is ten times faster than plain Ethernet (100Mbit/s) and can run over the same UTP cabling, but is not alone. It competes successfully against 100VG-AnyLan proposed by Hewlett-Packard, although 100VG has some enhanced features. While Fast Ethernet maintains its Ethernet origins and enjoys wider vendor support, they both become official standards. Further, a gigabit version of Ethernet is in the works, presenting an impressive technology extension from the original 10Mbit rates. In addition to the increased line rates, the introduction of Ethernet switching revitalizes Ethernet networking. Although FDDI seems to have lost in attention, due to the continuous delays in ATM standardisation and lack of mature products, the market for FDDI products is growing at very good rate. It remains concentrated in campus backbones and still fails to reach desktops as originally proclaimed by experts.

The LAN ILC displays the great expectations experts had in the 1993-95 period for the latest and greatest technology: ATM. This technology promises support for data and real-time applications combining switching technology and high speed interface rates. With the introduction of Fast Ethernet, ethernet switching and the promise of Gigabit Ethernet, at the time of this writing (mid-1996) serious concerns are being raised about the adoption rate for this technology. Although experts still expect that one day all desktops will be using ATM (like they once did with FDDI), the adoption rate is seriously reduced by extensions of existing technologies and its application seems to be initially limited to campus and public carrier backbones. The wide industry support and vast installed base for Ethernet make this market increasingly more competitive with unmatched innovation and technology extensions. The price points achieved and continually renewed featured sets have enabled this technology to successfully compete against other technologies.

When Interest Results in Hype

The question remains: why does interest for new technologies so frequently result in hype? Hype is the expressed interest that proves to be an overreaction as the new technology fails to grow at the rates originally anticipated. With information sources creating a perceptual distortion (Sheth, 1973), it is important to understand the reasons behind the generation of this hype. While this behaviour is observed across the technology based markets, These reasons attempt to provide an explanation:

- *Opinion leaders and the media* are highly technology-oriented. Both these groups, thrive on exploring the benefits of the latest technology and with such a high technological orientation market factors are frequently ignored.
- *The standardization process and product introduction time requirements* are consistently underestimated. As technologies become more popular, their standards process becomes increasingly more political, often slowing down its progress. Suppliers at the same time tend to underestimate the time required for developing products based on a new technology, which is compounded with the pressure to preannounce. With the increased time to market, existing technologies are given the opportunity to react.
- *Incumbent technologies counterattack* as they face the threat of a new technology. Although price reductions are one way for mature technologies to compete, feature extensions prove to be even more effective.
- *The myth of rapid technology change* is observed across technology introductions. Application requirements and business needs do not change at the high rate that will be required to justify a rapid adoption rate for the new innovative technology (Schnaars, 1989)
- *Vendors supporting the new technology*, intentionally generate hype as a supportive mechanism in the prelaunch period. Such actions increase the initial rate of adoption of high risk and innovative products.

Comparing Media Interest to Actual Adoption: Can we trust the experts ?

With several LAN technologies competing to capture the same target market, there are several similarities exisy with the more popular case of competing videocassette technologies (VHS format versus Beta) (Redmond 1991). In both cases, a bifurcation point exists where one technology dominates the market. It is the combination of several distinct reasons that allow a technology to dominate: infrastructure, price, market development and technology. On one hand, ARCnet was first to market, was cheaper, but remained a proprietary technology. By the time it became an open architecture and promised performance levels well above other competing technologies, the window of opportunity had passed. Ethernet on the other hand, has enjoyed an open standard and wide vendor support resulting in considerable technology extensions. With the popularity of this technology, network managers and IS personnel have become very familiar and welcome these technology extensions that practically require no additional training. In contrast, a new technology like ATM requires considerable learning as it is significantly different from any other existing technology.

Token Ring received significant attention as it was heavily promoted by IBM and represented a more robust technology than Ethernet. In only three years of shipping, it reached 50% of the installed base of Ethernet that had been shipping for more than nine years. This is consistent with the proposition by Robertson and Gatignon (1986) that reputable supplier offerings are associated with a more rapid initial diffusion rate. As seen in Figure 3, the introduction of Token Ring was associated with a high level of interest and with forecasts showing this technology surpassing Ethernet shipments in a few years. But there were significant factors that did not allow this scenario to materialise. Although the initial growth rate was impressive, it represented purchases by loyal IBM customers. The additional technology features that Token Ring had to offer were not as important for most of the market, especially considering the associated high cost. With a much broader industry support, Ethernet improved its feature set and enjoyed considerably lower prices.

It is consistently observed that at the ILC peaks of new LAN technologies, experts predict that the specific technology will have a major impact on the market. At these instances, the following trends were expected, but never materialized:

1. PBX vendors were expected to play a key role in data networking as they "owned" the installed communications infrastructure. The requirements for data communications proved to be vastly differently than those of a telephony system.

2. PC Network was expected at its own ILC peak to capture the lower end of the LAN market, but proved to be a complete failure.
3. Token ring was consistently described as the technology to dominate the LAN environment easily bypassing Ethernet. According to the IDC Corporation, 23.7 million Ethernet network interface cards were sold in 1995, in contrast to the 3.8 million Token Ring cards sold in the same period.
4. With FDDI riding on the excitement of fiber optics, it was expected to become a pervasive technology. Although FDDI has been generally adopted, its application has been limited to backbone implementations with volumes far less those of initial forecasts.
5. ATM is the new technology that was/is expected to dominate the world. Although the verdict is still not out, the early adoption rate of this technology has been generally disappointing.

With these ongoing unmatched forecasts, should experts be trusted? As discussed earlier, experts are more technologically inclined and are easily excited with the introduction of a new technology. Caught in the early euphoria of the announcement, experts report on the technology and educate the potential adopters. In this process, they often become advocates of the new technology resulting in bias (Tyebjee 1987). Besides the bias of the experts, these are times where uncertainty dominates and a small change in market factors can significantly impact the future. At prelaunch, the new technology promises to be superior to existing technologies, but conditions frequently change as existing technologies are enhanced and therefore receive an extension in their lifetime. This can be observed in the case of ethernet that has successfully competed against newly introduced technologies. The first attack was received by Token Ring and the response included extended cabling support and an unmatched reduction in prices. The next attack is received in several fronts: 100VG-AnyLan and ATM. These attacks unite ethernet vendors that introduce Fast Ethernet standard and an array of products at a fraction of the price. In addition, Ethernet switching is introduced, relieving the bottlenecks of shared medium networks. With higher transmission rates, direct compatibility with existing infrastructures and intense price competition, ethernet has been once again successful in at least slowing the growth and need to migrate to ATM.

As described by this example, factors do change and forecasts or predictions are made with currently available data and information. The threat of a new technology introduction results in significant reaction that is frequently not considered. Given that the market is quite dynamic, it is not surprising to observe the unmet expert expectations. A common thread for the failure of these predictions appears to be in the assumptions of these early predictions: the power of the installed base is discounted as it can resist the adoption of a new technology and needs are expected to change at a much higher rate.

Discussion

The market interest measurement may at first seem simplistic as the ultimate adoption of network technologies is based on market and technology factors. However, since it adequately describes the competitive environment between technologies across time, there are several reasons that technology suppliers need to systematically monitor media interest even if it is believed to be pure hype. Since customers are exposed to the hype, they will naturally question the value of this technology. In order for a vendor to maintain or establish a leadership image in such fast-paced industries, it is necessary to communicate their strategic vision for future technology migration. This vision needs to address the incorporation or support for new technologies under development as discussed by the experts. Even if the technology being questioned does not have an immediate application for the user, association of the supplier with this technology is interpreted as technology competence and improves the competitive positioning of the supplier. With corporate network designs requiring several technologies and customers seeking to minimize the number of equipment suppliers, it is of the foremost importance to present a clear technology positioning to establish customer loyalty.

The level of the attention paid to a new technology is one indicator of planning support for such a technology. This may act as a trigger for initiating an internal investigation for new product development and exploring if the interest will translate into actual demand. The introduction of a new technology could threaten existing product line and future product plans. The strength of the hype can be interpreted as the hostility level of the signal sent for incumbent technologies. The stronger the signal, the more likely there will be a competitive and more aggressive reaction (Robertson, Eliashberg & Rymon, 1995). Monitoring the interest level can be a measure of preparing a counterattack. Executing a counterattack will depend on the market power of the supplier, the strengths and weaknesses of the competing technologies. The reaction can take several shapes: technology feature improvements, price reductions and so forth.

The danger during periods of intense interest levels for a technology is that suppliers can easily be caught in the excitement and overforecast its potential adoption. This is quite ironic given that this hype is to a great extent, created by the same suppliers in their effort to improve the chances for a new launch (Wind & Mahajan, 1987). Several sophisticated premarket forecasting models (Urban, Hulland & Weinberg, 1993) depend heavily on the prelaunch environment for making predictions for the future success or failure of a product. Models where word-of-mouth plays a key factor are rendered useless by the high level of perceptual distortion induced by the information sources and opinion leaders. This also applies to forecasters that are exposed to the interest which often shadows the true intent of the broader market for purchasing the technology.

Opinion leaders tend to become overexcited with the announcement and features of new technologies. Although this technology focus leads to the deferral of purchases for the new or expected technologies, market factors prevail in the end. Vendor support for a technology results in more feature-rich products, higher rate of incremental innovation and significantly reduced prices due to volumes and intense competition. Measuring the level of interest for technologies can result in a better understanding of the potential threat of a new technology and result in the appropriate marketing programmes strategy. Depending on vendor specific factors, there is the possibility for the following reactions: announce support for the new technology or defend and enhance existing technologies.

References

Abernathy, W. & Clark, C. (1985). Innovation: Mapping the Winds of Creative Destruction. *Research Policy,* No. 14, pp. 3-22.

Baumgarten, S. (1975). The Innovative Communicator in the Diffusion Process. *Journal of Marketing Research,* Vol. XII (February), pp. 12-18.

Czepiel, J. (1974). Word-of-Mouth Processes in the Diffusion of a Major Technological Innovation. *Journal of Marketing Research,* Vol XI (May), pp. 172-80.

David, P. and Steinmuller, E. (1990). The ISDN Bandwagon Is Coming, But Who Will Be There to Climb Aboard?: Quandaries in the Economics of Data Communication Networks. *Economics of Innovation and New Technology,* Vol.1, pp. 43-62.

Eliashberg, J. and Robertson T. (1988). New Product Preannouncing Behavior: A Market Signaling Study. *Journal of Marketing Research,* Vol. XXV (August), pp. 282-92.

Farrell, J. and Saloner, G. (1986). Installed Base and Compatibility: Innovation, Product Preannouncements, and Prediction. *The American Economic Review,* Vol. 76, No. 5, 940-55.

Kassarjian, H. (1977). Content Analysis in Consumer Research. *Journal of Consumer Research,* Vol. 4, pp. 8-18.

Krippedorff, K. (1985). *Content Analysis: An Introduction to Its Methodology,* Sage Publications, Beverly Hills.

Mahajan V., Wind Y. and Bass F. (1990). New Product Diffusion Models in Marketing: A Review and Directions for Research. Journal of Marketing, Vol. 54 (January), pp. 1-26.

Myers, James and Robertson, Thomas (1972) . Dimensions of Opinion Leadership. *Journal of Marketing Research,* Vol. IX (February), pp. 41-6.

Norton, J. and Bass F. (1992). Evolution of Technological Generations: The Law of Capture. *Sloan Management Review,* Winter, pp. 66-77.

Popper, E. and Buskirk, B. (1992). Technology Life Cycles in Industrial Markets. *Industrial Marketing Management,* Vol. 21, pp. 23-31.

Redmond, W. (1991). When Technologies Compete: The Role of Externalities in Nonlinear Market Response. *Journal of Product Innovation Management*, Vol. 8, pp. 170-183.

Rogers, E. (1983). *Diffusion of Innovations.* The Free Press, New York.

Robertson, T. and Gatignon, H. (1986). Competitive Effects on Technology Diffusion. *Journal of Marketing*, Vol. 50, pp. 1-12.

Robertson, T. S., Eliashberg, J. and Rymon, T. (1995). New Product Announcement Signals and Incumbent Reactions. *Journal of Marketing*, Vol. 59 (July), pp. 1-15.

Schnaars, S. (1989). *Megamistakes: Forecasting and the Myth of Rapid Technological Change.* The Free Press.

Sheth, Jagdish (1973). A Model of Industrial Behavior. *Journal of Marketing*, Vol. 37 (October), pp. 50-56.

Stoneman, Paul (1995). *Handbook of the Economics of Innovations and Technological Change*, Blackwell, Oxford.

Thomas, R. (1985). Problems in Demand Estimation For a New Technology. *Journal of Product Innovation Management*, Vol. 3, pp. 145-157.

Tyebjee, T. (1987). Behavioral Biases in New Product Forecasting. *International Journal of Forecasting*, Vol. 3, pp. 393-404.

Urban, G., Hulland, J. and Weinberg, B. (1993). Premarket Forecasting for New Consumer Durable Goods: Modeling Categorization, Elimination, and Consideration Phenomena. *Journal of Marketing*, Vol. 57 (April), pp. 47-63.

Weiss, A. and Heide, J. (1993). The Nature of Organizational Search in High Technology Markets. *Journal of Marketing Research*, Vol. XXX (May), pp. 220-33.

Wilton, P. and Pessemier E. (1981). Forecasting the Ultimate Acceptance of an Innovation: The Effects of Information. *Journal of Consumer Research*, Vol. 8 (September), pp.162-171.

Wind, J. and Mahajan, V. (1987). Marketing Hype: A New Perspective for New Product Research and Introduction. *Journal of Product Innovation Management*, Vol. 4, pp. 43-49.

CHAPTER

9

The Adoption and Diffusion of Quick Response Technologies in Specialty Retailing

Jonathan W. Palmer
University of Oklahoma

Quick Response (QR) has been proposed as an innovative use of technology for competitive advantage in the retail industry. QR involves a suite of technologies including electronic data interchange (EDI), point of sale (POS) systems, and several types of interorganizational systems. The paper explores the adoption and Diffusion of QR in the specialty retailing industry. Based on a study of eighty specialty retailers, QR's adoption is influenced by early adoption, firm size, public ownership, decentralized IS philosophy, and a large IS staff. Diffusion of QR is associated with early adoption, large IS staff size, and in-house IS development. At advanced stages, the diffusion is associated with a supplier-oriented strategy and higher investment in information technology (IT).

"Higher turns, fewer markdowns, better in-stock positions and lower inventory levels mean higher profitability for the retailer and its vendors," and are the result of adopting Quick Response (*Retail Information Systems News*, 1993). Quick Response is an approach in which retailers deploy a variety of interorganizational information technologies, including point-of-sale systems, bar coding, automated inventory management, electronic data interchange, and electronic invoicing. These technologies provide the capabilities of automatic replenishment and "seasonless" retailing. Consultants and industry converts point to these as innovative technologies "that will further distance today's QR innovators from their competition" (*Retail Information Systems News*, 1995).

Gains in operational efficiencies are promoted by QR advocates as providing competitive advantage to those firms implementing the technologies. What are the key influences on those firms adopting Quick Response? What are the key influences on diffusing the technologies within the organization?

The objectives of this chapter are to examine the adoption and diffusion of Quick Response technologies in specialty retailing. The chapter examines key organizational and information technology variables that may have an influence on the adoption and diffusion of these technologies. Eighty specialty retailers were analyzed regarding their use of the four levels or phases of Quick Response and the organizational and technological dimensions that influenced the adoption and diffusion of Quick Response within the firm.

The chapter is presented in five sections. The initial section develops the theoretical background including relevant aspects of the diffusion and innovation literature, Quick Response technologies, and the specialty retailing industry and its use of information technology. The second section describes the methodology and key variables. The third section presents the analysis and results. The fourth section discusses the findings and presents suggestions for future research. The final section presents conclusions and implications for theory and practice.

Theoretical Background

The innovation literature generally suggests that larger organizations are more innovative than smaller organizations (Damanpour, 1992). This relationship has been identified in groups of innovations including computer-aided software engineering (CASE) (Rai and Howard, 1994), human resource practices (Tannenbaum and Dupuree-Bruno, 1994), and telecommuting (Harrington and Ruppel, 1996).

The diffusion theory literature also identifies firm size as a key influence on diffusion of innovations and technology (Levin, Levin and Meisel, 1992). In addition, the timing of adoption has typically favored the early adopter in terms of positive organizational performance (Zmud and Apple, 1992, Levin, Levin and Meisel, 1992, Mahajan, Muller and Bass, 1990).

Several diffusion models have been used to identify the implementation of information technology. Rational choice models have typically been used to capture IT adoption (Kwon and Zmud, 1987) while political and learning models are more useful in identifying infusion of the innovation within the organization (Cooper and Zmud, 1990).

Contingency variables suggested in the innovation and diffusion literature include both organizational, technological, and strategic dimensions (Rogers, 1983, Caincara, Colombo and Mariotti, 1989). Size may affect the organizational structure, organizational resources, and environmental threats and opportunities (Levin, Levin and Meisel, 1992). A key technological contingency variable is IT sophistication (Aksoy, 1991) which might include investment in IT (Harris and Katz, 1991, Weill, 1992), level of in-house systems development, and size of IS staff. Strategic choices at the organizational and IT level can guide the adoption and use of technology (Treacy and Wiersema, 1993, 1995). In addition to Porter's (1980, 1985) generic strategies, the use of the basic value chain to identify internal, suppler, and customer oriented strategies has been proposed (Holland, Lockett and Blackman, 1992).

Earlier work in the supermarket industry discussed early adoption, diffusion and infusion of supermarket scanners (Zmud and Apple, 1990, Levin, Levin and Meisel, 1992). This chapter examines a multilevel model, Quick Response (QR), used across a wider array of specialty retail firms in terms of organizational size and product lines.

The QR model is presented in Exhibit 1. QR involves four levels or stages (*Retail Information Systems News*, 1993). The first level involves automated point-of-sale (POS), bar coding universal product codes (UPC), automatic price look-up and electronic data interchange (EDI) for order entry and inventory management. A second level involves automatic replenishment by suppliers, forecasting and electronic invoicing. EDI at this stage includes order status, invoicing and advance shipment notices. Enhancements in level three include using the information gleaned through the elements of the earlier stages for pre- and post-season planning, shipment container marking and arrangement. In the fourth level, suppliers actually take over inven-

QR Level	Technologies and applications
1	Automated point-of-sale, bar coding, universal product code, automatic price look-up, EDI for order entry and inventory management
2	Automatic inventory repleneishment, forecasting, electronic invoicing, EDI for order status, invoicing and advance shipment notices
3	Pre- and post-season planning, shipment container marking and arrangement, cross-docking
4	Suppliers take over inventory management functions, seasonless retailing, space management

adopted from: *Retail Information Systems News*, January 1993.

Exhibit 1: Levels of Quick Response

Exhibit 2: Retail Pyramid

tory management functions, this stage also includes seasonless retailing, and space management.

The retailing industry is a particularly rich environment in which to examine innovation and diffusion of information systems and technology. Specialty retailers are in a unique position in the retail pyramid (Exhibit 2), more focused, agile and flexible than the major national department stores and more able to enjoy the economies of scale unavailable to the small retail store.

Specialty retailers range in size and scope from 50 stores distributed throughout a multi-state region to nearly 2,000 stores spread across the United States and beyond. Product lines are as diverse as health foods, apparel, food, auto supplies, housewares, music and video, books, computer software and supplies, bed & bath, and specialty items.

The resulting model of diffusion of QR within specialty retailing must account for organizational, technological, and strategic influ-

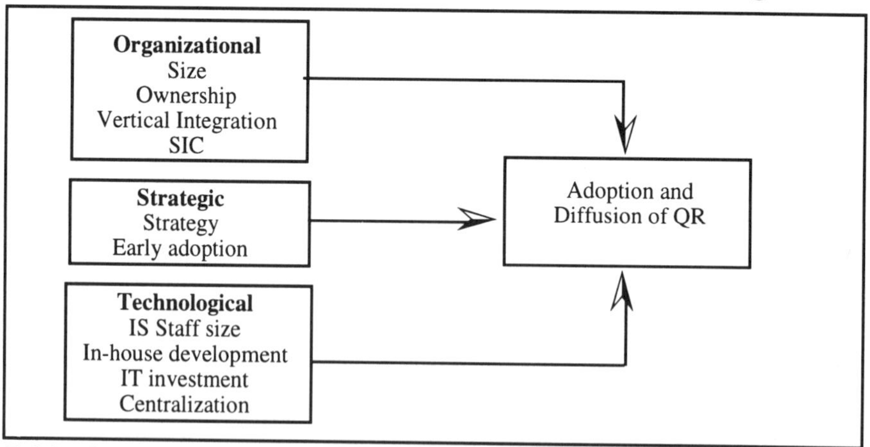

Exhibit 3: Contingency Model of QR Adoption and Diffusion

ences. These influences include organizational demographic dimensions such as size, ownership, vertical integration, and SIC; technological dimensions including size of IS staff, degree of systems developed in-house, investment in IT; and strategic dimensions including organizational and IT strategies and tendency toward early adoption. The resulting contingency model is presented in Exhibit 3. The model suggests that each of these influences can be a key to adoption and diffusion of QR.

Methodology

A survey instrument was used to gather data on the relationship between the organizational, technological and strategic variables and the adoption of Quick Response. The unit of analysis was the firm. The target audience for the questionnaire was the top IT professional in each firm. This individual was positioned to give insight into the adoption of an IT innovation. The survey was pilot tested by CIOs from three specialty retailers and reviewed by three colleagues. Changes made in a second version were again tested with CIOs from four different specialty retailers. This piloting process helped to establish content validity (Straub, 1989).

Operationalization of Variables

A review of instruments and constructs used in previous research provided some items for the survey. Other items were created by the author. The variables identified in the contingency model are described in Exhibit 4.

The dependent variables were based on respondents identification of the extent to which 11 QR technologies and 12 QR applications were used by the organization. Technologies and applications receiving a score of moderate and active were grouped as high levels of use, those receiving scores of beginning, investigating, or not using were grouped as low levels of use. Adoption of Quick Response was identified as firms with one level of QR at high levels of use. Diffusion of Quick Response was identified as firms exhibiting more than one of the Quick Response levels at high levels of use for at least two years. Descriptive statistics were used to identify levels of use and a correlation was used to explore potential relationships among the adoption of QR at various levels and the independent variables.

Selection of Respondents

RETEX, the Retail Technology Consortium, offered its membership as a sample population. The membership included firms with regional, national, and international focus; firms engaged in a variety

Variable	Measurement	Source
Organizational Dimensions		
Size	Annual Sales	Levin, Levin & Meisel, 1992, Damanpour, 1992
Degree of vertical integration	High, Medium, Low	
Ownership	Public/private	
Product Line	SIC	
Strategic Dimensions		
Organizational & IT Stratcgy	% of effort on each of internal, supplier, customer	Holland, Lockett & Blackman, 1992
Early Adoption	Year in which technology initially implemented	Zmud & Apple, 1990
Technological Dimensions		
Size of IS Staff	Numer of IS personnel	Aksoy, 1991
Degree of centralization	High, Medium, Low	
Level of IT investment	% of annual sales	Harris & Katz, 1991 Weill, 1992
In-house IS development	% of IS developed	

Exhibit 4: Independent Variables/Potential Influences on Adoption of QR

of different product lines (fashion, health, food, specialty items, entertainment products); and firms with differing approaches to IT. Corroborating organizational and IT investment information, with industry sources provided an additional level of information and helped to avoid respondent bias.

Analysis

The questionnaire was mailed to 175 RETEX members. Eighty usable responses were received, for a usable response rate of 45.7%. The general demographic information on the respondents compared to the specialty retailing industry is presented in Exhibit 5. The 80 respondents are representative of the specialty retailing industry, with a mix of products, and public/private ownership. The respondents are performing better than the industry averages in terms of sales growth and comparable store sales.

Several key dimensions in the respondent profile are highlighted in Exhibit 6, including size, level of IT investment, and initial implementation of information technology. Most respondents began their initial IT investment in the mid 1980s and currently spend the equivalent of 1.82% of their annual sales on IT.

The independent variables included organizational, technological and strategic variables. To analyze the relationships between the

Variable	Sample	Industry
Ownership	33% public	25% public
Size	57% under $100 million	65% under $100 million
SIC/Product		
Hardware	15%	10%
Specialty	21%	25%
Apparel	27%	25%
Furniture/Appliance	19%	15%
Book/Sport/Drug	19%	25%
Sales Gowth	just over 10%	8%
Comparable Store Sales	just under 10%	7%
Sales/s.f.	$200-250	$150-175
Stock Turn	4-7	5-6

Exhibit 5: Comparison of Sample and Industry Means

independent variables, correlations were run (Exhibit 7). The organizational variables showed significant correlations in size, ownership, and early adoption. Both SIC and vertical integration exhibited no significant correlations, and suggest different constructs. The technological variables showed significant correlations in IS staff size, in-house development and centralization. IT investment exhibited no significant correlation. The supplier oriented strategy was significantly correlated with in-house development. The correlations reflect earlier findings on size and early adoption (Zmud and Apple, 1990,

Firm Size	N=	Avg Annual Sales ($ mil 1993)	Avg Employees 1993	Avg IT Invest (% of Sales)	Avg Year IT Began
Small (under $25 million)	42	$20	250	1.92	1986
Medium ($25 -250 million)	19	$182	2064	1.73	1984
Large (over $250 million)	19	$1329	10701	1.63	1983
Total	80	$345	2969	1.82	1985

Exhibit 6: Profile of Respondent Firms (Annual Sales, Employees, IT Investment, IT Adoption)

	Early Adoption Invest	Size	Public Ownership	Vertical Integration	SIC	In-house	IS Staff Develop	IT Size
Size	.3922**							
Public Ownership	.2212*	.6460**						
Virtual Integration	.1366	.1291	-.0643					
SIC	.1155	-.0372	.2204*	.0292				
In-house development	-.0623	.1208	-.1608	.1292	.1001			
IS Staff size	.3490**	.6979**	.6952**	.1805	.1325	.2161*		
IT Investment	-.0096	.0368	-.0106	.1049	.0112	-.0280	.0560	
Centralization	-.0373	.2138*	-.1909	.1440	.0744	.0230	.2387*	-.1192

Kendall Tau-b, two-tailed, * <.05 ** < .01

Exhibit 7: Independent Variables Correlations N = 80

Levin, Levin and Meisel, 1992). The correlations of the IS size, development and philosophy of decentralization capture the importance of size on the technological dimensions. The SIC, vertical integration and IT investment variables suggest different influences. It is important to remember in interpreting the correlations that two variables may be correlated with each other, only because they share correlation with a common other variable.

Adoption of QR was mixed. Half of the respondents had adopted at least one level of QR (Exhibit 8). This seems low, given the strong support for the QR concept in the practitioner literature (*Retail Information Systems News*, 1993, 1995). Surprisingly, adoption occurred at levels other than level 1 for several firms. The finding does call into question the QR Levels model. It appears that while many of the firms adopted at multiple levels, it is possible to adopt at "higher" levels without the preceding levels being adopted.

QR Level	Technologies and Applications	# (and %) of respondents using QR technologies	% of technology users also using previous level
1	POS, bar coding, UPC labeling, EDI	40 (50)	---
2	Automated inventory replenishment, inventory and sales forceasting, invoiceless playments, EDI	36 (45)	63.9
3	Cross-docking, pre and post-season	17 (21.3)	47.1
4	Supplier space management, seasonless retailing	9 (11.3)	44.4

Exhibit 8: The QR Program Levels—Number (and Percentage) of Respondents Using QR

	QR Adoption
Early Adoption	.2160*
Firm Size	.2805*
Public	.3314**
Vertical Integration	.0864
SIC	.0974
In-house IS Development	.0742
Size of IS staff	.332**
IT Investment	-.0896
Centralization of IS	.3548**
Supplier Focused Strategy	.1095

Kendall Tau-b, two-tailed, * <.05 ** < .01

Exhibit 9: Adoption of Quick Response and Intervening Variables Correlations—N =80 (non QR and QR adopters)

To analyze the influences on adoption, correlations were run for the key variables presumed to influence adoption (Exhibit 9). The organizational variables of early adoption, size, ownership and centralization were significantly correlated with the adoption of QR. These results support earlier work (Zmud and Apple, 1990; Levin, Levin, 1992; and Damanpour, 1992). In addition, firms adopting QR had significantly larger staff sizes than those that did not, but did not perform significantly more of the IS development in-house. Surprisingly, those adopting QR did not invest significantly more on IT. Nor were those adopting QR more likely to be following a supplier-oriented strategy.

To examine the level of diffusion of QR, the forty firms adopting QR were analyzed separately. To identify diffusion, only those firms with multiple levels of QR for at least two years were considered to have diffused QR. As mentioned earlier, there were a number of firms who adopted levels of QR that did not follow the suggested hierarchy of adoption. Of the forty firms adopting QR, 29 were considered to have diffused QR (Exhibit 10). These 29 firms had two or more levels of QR represented and had at least two years since initial adoption.

Correlations for the 29 firms diffusing QR are presented in Exhibit 11. Firms with two levels were significantly correlated with early adoption, firm size and in-house IS development. Firms with three

	One QR Level	Two QR Levels	Three QR Levels	Four QR Levels
Number of Firms	9	19	8	4
Number of Firms with Adoption> 2 years	7	17	8	4

Exhibit 10: Diffusion of QR in Firms with QR Adoption—N = 40

	2 QR Levels n=17	3 QR Levels n=8	4 QR Levels n=4
Early Adoption	.2891	.3894**	.0963
Firm Size	.2964*	.2412	.2824*
Public	.1095	.1080	.3958**
Vertical Integration	.0599	.1137	.0158
SIC	.0715	-.0614	.0915
Centralization	.0108	.1519	.1106
In-house Development	.4455**	.3711*	.0023
IS Staff	.1780	.3294*	.2848*
IT Investment	-.1625	.1017	.3231*
Supplier Focused Strategy	.0504	.1353	.3026*

Kendall Tau-b, two-tailed, * <.05 ** < .01

Exhibit 11: Diffusion of Quick Response and Intervening Variables Correlations —N = 29

levels were correlated with early adoption, in-house IS development, and size of IS staff. Firms with four levels of QR were different, correlating with public ownership, IS staff size, higher investment in IT, and supplier-oriented IT strategies. Vertical integration, SIC, and decentralization of IS were not associated with the diffusion of QR.

These findings on diffusion suggest several possibilities. One is that the QR model is somewhat flawed. The other is that some of the technologies at the higher levels do not require the support of "lower level" technologies. In only four cases did firms only adopt Level 1 and in only four other cases were firms only adopting a single level of QR. The 29 firms that adopted more than one level suggest that these firms had at least begun to diffuse the innovation throughout the organization. The four firms that had implemented all four levels of QR merit some additional analysis in terms of their profile. The significant differences were that these firms, in addition to being public, large, and having large IS staffs, they had supplier-oriented strategy and higher IT investment.

The findings suggest several significant influences on the adoption and diffusion of QR in specialty retailing. Early adoption and firm size were significant in both cases. Adoption was also influenced by public ownership and two IT related variables, decentralized IS and large IS staff. Diffusion was also influenced by in-house IS development, a large IS staff, and at level 4 by greater IT investment and a supplier-oriented strategy. A summary of the findings are presented in Exhibit 12.

Discussion

The adoption and diffusion of QR within specialty retailing appears to be influenced by several factors. These factors include both technological and organizational ones, reflecting both the IT approach

	Adoption of QR	Diffusion of QR
Early adoption	Yes	Yes
Firm Size	Yes	Yes
Public	Yes	Mixed (Yes on 4 Levels)
Vertical integration	No	No
SIC	No	No
In-house IS development	No	Yes
IS staff size	Yes	Yes
IT investment	No	Mixed (Yes on 4 Levels)
Centralization	Yes	No
IT strategy	No	Mixed (Yes on 4 Levels)

Exhibit 12:Summary of Results

and expertise within the firm and the organizational strategies being implemented. The factors also appear to change as firms become more sophisticated in their use of QR, adopting more advanced levels. The following discussion examines the adoption and diffusion findings.

Adoption

The findings on adoption supported much of the existing theory regarding early adoption (Zmud and Apple, 1992, Mahajan, Muller and Bass, 1990) and firm size (Levin, Levin, and Meisel, 1992, Damanpour, 1992, Harrington and Ruppel, 1996).

This finding differs somewhat from Levin, Levin and Meisel (1992), who found that larger market share firms adopted innovations more quickly initially, but diffused the innovation more slowly; whereas those firms lagging in initial adoption diffused the innovation more quickly.

The findings suggest the likely candidate firms for the adoption of QR as an innovation are large, public firms which are early adopters of technology. In addition, the likely candidate firms have large IS staffs operating in decentralized environments.

The limited adoption of QR technologies and applications may also provide some insights. Few companies use the entire range of QR technologies and applications. Firms' concentration on different elements of the retail pipeline may drive choices of what parts of the QR program to adopt or not adopt. It may also be that the larger the number of suppliers, the more difficult it becomes for a firm to develop an economic justification for QR (Bouchard, 1992). Finally, many retailing firms appear to wait until technologies are ubiquitous and commoditized before they purchase and adopt them.

Diffusion

The model developed in the initial theory section appears to have captured some important new elements associated with QR diffusion. In addition to firm size and early adoption, the model identifies three types of influences: organizational, technological and strategic, all of which were shown empirically to have an influence on adoption and diffusion.

The technological issues become more important in moving from adoption to diffusion of QR. In addition to larger IS staffs, the firms diffusing QR had typically more emphasis on in-house systems development and, at the higher levels of QR, had made larger investments in IT. Existing models have not typically captured the implementation issues of technological innovation. The findings on organizational and IT strategy suggest that advanced QR is strongly correlated with supplier-oriented strategies. This is in keeping with the transaction oriented nature of the QR implementation. Surprisingly, the correlation is only significant at higher levels of QR which require greater connectivity with suppliers. The identification of appropriate strategies linked to the innovation and the technological implementation issues are potentially significant additions to the discussion of the diffusion of IS/IT innovations in organizations.

The emerging trend appears to be greater levels of influence from internal sources of IS expertise (levels of in-house development, size of IS staff and investment in IT) necessary to fully diffuse the innovation, particularly a the most sophisticated levels. While general strategies of being early adopters of innovation play a role in both the adoption and the diffusion process, more specific strategies aimed at specific uses of the innovation (in this case supplier-oriented strategies) become more important in the diffusion of the innovation.

Limitations

One possible explanation for the results may lie in the limitations of this particular study. The study involves the specialty retailing industry; this segment may differ from retailing in general and other industry in its adoption and diffusion on innovation in general and information technology in particular. Since the respondents were members of a technology-oriented consortium, they may differ in their greater use of IT. Thus some questions about the findings will only be settled when the study has been replicated in other industry environments.

The implications of the study for practice are more challenging. The study suggests that larger firms will be early adopters of innovations. This firm size is also reflected in the apparent need for larger IS staff to undertake QR. At the more advanced stages, it appears that both increased IT investment and a willingness to support in-house systems development. Finally, at the advanced stages, the diffusion of the innovation is associated with specific firm strategies.

Conclusion

It appears that the existing diffusion models do not fully account for the element of implementation and execution in the diffusion of the innovation. The results suggest that the same equipment and technology can be used in different ways and that in the right hands, these systems can become strategic. In addition to the moderating variables of early adoption and firm size, in the adoption and diffusion of information technology innovations, the degree of in-house IS development, the size of the IS staff, the level of IT investment, and the supplier-oriented strategy were key to diffusing the innovation in specialty retailers.

The move to implementation and technological issues as adoption moves to diffusion suggest that firms interested in effectively diffusing information technology innovations need to support these innovations with increased staff, in-house development capabilities and IT investment. In addition, linking the innovations with specific strategic emphasis within the organization can be important to successful diffusion.

References

Aksoy, Asu, "Computers Are Not Dynamos: Frontiers in the Diffusion of Information Technology," *Futures*, Volume 23, no. 4, May 1991, p. 402-414.

Antonelli, Cristiano, "Information Technology and the Derived Demand for Telecommunication Services in the Manufacturing Industry," *Information Economics & Policy* (Netherlands), Volume 4, no. 1, 1989/1990, p. 45-55.

Bouchard, Lyn, "EDI: Status Report," Anderson Graduate School of Management, UCLA, September 1992.

Caincara, Gian Carlo; Colombo, Massimo G.; Mariotti, Sergio, "An Evolutionary Pattern of Innovation Diffusion. The Case of Flexible Automation," *Research Policy* (Netherlands) Volume 18, no. 2, April 1989, p. 59-86.

Cooper, Randolph B.; Zmud, Robert W., "Information Technology Implementation Research: A Technological Diffusion Approach," *Management Science*, Volume 36, no. 2, February 1990, p. 123-139.

Damanpour, F., "Organization size and innovation," *Organization Studies*, Volume 13, no. 2, 1992, p. 375-402.

Harrington, Susan J. and Ruppel, Cynthia P., "The Contingency Effect of Organization Size on Programmer/Analyst Telecommuting," *Proceedings of the 1996 ACM SIGCPR/SIGMIS Conference*, April 11-13, 1996, Denver, CO., p. 234-243.

Harris, Sidney E. and Joseph L. Katz, "Organizational Performance and Information Technology Investment Intensity in the Insurance Industry," *Organization Science*, Volume 2, no 3, August 1991; 263-295.

Holland, C., G. Lockett and I. Blackman, "Planning for Electronic Data Interchange," *Strategic Management Journal*, 1992, 13, 539-550.

Levin, Sharon G.; Levin, Stanford L.; Meisel, John B., "Market Structure, Uncertainty, and Intrafirm Diffusion: The Case of Optical Scanners in Grocery Stores," *Review of Economics & Statistics*, Volume 74, no. 2, May 1992, p. 345-350.

Mahajan, Vijay; Muller, Eitan; Bass, Frank M., "New Product Diffusion Models in Marketing: A Review and Directions for Research," *Journal of Marketing*, Volume 54, no. 1, January 1990, p. 1-26.

Porter, Michael E. *Competitive Advantage: Creating and Sustaining Superior Performance*. New York: Free Press; London: Collier Macmillan, 1985.

Porter, Michael E., *Competitive Strategy: Techniques for Analyzing Industries and Competitors.* New York: The Free Press, 1980.

Rai, A. and Howard, G.S., "Propagating CASE usage for Software Development: an Empirical Investigation of Key Organizational Correlates," *OMEGA International Journal of Management Science*, Volume 22, no. 2, 1994, p. 133-147.

Retail Information Systems News, "Customer Driven Retailing," Special Supplement, January 1995.

Retail Information Systems News, "Uncovering the Keys to Quick Response," Special Supplement, January 1993.

Rogers, Edward M., *Diffusion of Innovations*, Free Press, 1983.

Straub, Detmar, "Validating Instruments in MIS Research," *MIS Quarterly*, Volume 13, no. 2, 1989, p. 147-170.

Tannenbaum, S.I. and Dupuree-Bruno, L.M., "The Relationship Between Organizational and Environmental Factors and the Use of Innovative Human Resource Practices," *Group and Organization Management*, Volume 19, no. 2, 1994, p. 171-202.

Treacy, Michael and Fred Wiersema, "Customer Intimacy and Other Value Disciplines," *Harvard Business Review*, January/February 1993, p. 84-93.

Treacy, Michael and Fred Wiersema, *The Discipline of Market Leaders*, HarperCollins, 1995.

Weill, Peter, "The Relationship Between Investment in Information Technology and Firm Performance: A Study of the Valve Manufacturing Sector," *Information Systems Research*, Volume 3, no. 4, December 1992, p. 307-333.

Zmud, Robert W.; Apple, L. Eugene, "Measuring Technology Incorporation/Infusion," *Journal of Product Innovation Management*, Volume 9, no. 2, June 1992, p. 148-155.

Section III: Facilitation of Technology Diffusion

This section contains five chapters that describe and explore different facilitation mechanisms that can assist organizations in the information technology diffusion and adoption process. The facilitation mechanisms discussed by the authors in this section range from *de facto* industry standards to interorganizational networks to a community center designed to provide information and skills training. Other chapters discuss the role of an organizational learning approach to facilitate change and contingency factors affecting the diffusion process in rural enterprises. The chapters in this section all address drivers, levers, enablers, and facilitators of the diffusion process. These mechanisms all serve to focus the organization's attention on the information technology or process that needs to be adopted or changed. This focus serves to marshal resources and sponsorship and thus facilitate the diffusion effort. All of these chapters offer practical recommendations and advice for information technology diffusion.

Paulk explains the rationale behind and the structure of the Software Engineering Institute's Software Capability Maturity Model (SW-CMM), one of the *de facto* standards for improving quality in the software industry. Paulk, who is one of the principal contributors to the SW-CMM, provides recommendations on how to implement some of the practices in this well-known model. He notes that the SW-CMM is frequently viewed as a driver or lever within an organization to focus management attention and sponsorship on adopting effective process, methods, and tools. The SW-CMM, because of its visibility and stature within the software community, can often serve as the impetus for innovation in the form of new or revised software processes and practices. These innovations may be rapidly diffused and adopted throughout an organization given appropriate top management support, an essential component of the model itself.

Newell, Swan, and Robertson present another example of the effect of a facilitation mechanism on the innovation and diffusion process within an organization. In this chapter, they discuss two case studies of organizations which were using interorganizational networks to facilitate the diffusion of new ideas into the firms. The firms discussed in the case studies are both manufacturing firms in the UK which were attempting to introduce new information technology to support their operations management. Both firms were networked to professional associations but made different use of the networks and experienced different outcomes. The two case studies discussed in this chapter demonstrate the important role that professional associations can play in diffusing knowledge about a particular technology throughout an organization. These case studies also illustrate the limitations of this role and notes other activities which firms should undertake to further facilitate the diffusion process for introducing new information technology.

Another example of an intervening variable being used to facilitate the diffusion process is presented in the chapter by Charlton, Gittings, Leng, Little, and Neilson. These authors first employ mathematical analysis to describe the diffusion of the Internet for purposes of electronic commerce. They find this model lacking and then undertake a descriptive micro-analysis of the key factors that have influenced the adoption of Internet related technology at the local community level (the Merseyside region in the UK). Within this region, the Internet Centre for Merseyside Business (Connect) has been created to facilitate the diffusion of Internet awareness and skills throughout the local community. This chapter reviews the impact of Connect's activities on the process of diffusion and examines the many interacting factors which influence the diffusion process and the synergetic nature of these interactions. They conclude that classical models of innovation and diffusion are overly static when applied to an ongoing and evolving process of diffusion as is typified by Internet applications such as electronic commerce. They call for a new model of diffusion that can readily handle synergy and interaction between many contributing factors to handle this type of innovation.

Selzer, Mellis and Herzwurm offer their view of key factors that facilitate or impede the success of software process improvement initiatives. These factors have been identified through two empirical studies into success factors and are explored in this chapter from a technology diffusion perspective. The role of organizational learning as a facilitator to the diffusion process is emphasized. A total of 12 success factors were derived from two empirical studies of software process improvement initiatives (one study focused on an effort based on ISO 9000 and the other study focused on an effort based on the Software Engineering Institute's Capability Maturity Model). Each of these 12 factors is discussed using Rogers' diffusion theory which the authors conclude provides an adequate model for explaining the success and failure factors. The authors use these success factors to derive requirements for implementing these factors, then examine whether an organizational learning approach will facilitate this implementation. Their conclusions provide a model whereby organizations can better diffuse software process improvement innovations throughout their software development environments.

Jansen explores the diffusion of information technology in rural areas in his chapter and finds that small enterprises located in rural economies face different challenges in the information technology diffusion process than do larger, centrally located organizations, when trying to implement new information technology solutions. He proposes a new framework for analyzing success and failure efforts in rural settings that integrates three contingency factors: technology, organization, and local infrastructure. This contingency model assumes that the outcome of successful information technology implementations in small, rural organizations is dependent on internal and external factors. The author's research reveals that organizational factors play a larger role in successful technology diffusion and adoption than do technical factors for rural firms. He concludes that rural enterprises have different patterns for technology diffusion than do larger, more centrally located firms and thus have correspondingly different needs for assistance and support. The contingency model developed by the author proves useful in examining these differences.

CHAPTER

10

Using the Capability Maturity Model for Software to Drive Change

Mark C. Paulk
Carnegie Mellon University

This chapter provides an overview of the Capability Maturity Model[SM] for Software, SW-CMM[SM] version 1.1 and discusses its use for software process improvement. The SW-CMM describes the software engineering and management practices that characterize organizations as they mature their processes for developing and maintaining software. The model does not prescribe the "best" software engineering or management methods or tools in an absolute sense, but it can be used as a driver in introducing effective methods and tools. The SW-CMM tells software organizations "what" to do to improve, and I have added recommendations on "how" to effectively implement some of the practices in the model.

In many organizations, software projects are often late and over budget. This state of affairs is sometimes referred to as the "software crisis." In November 1986, the Software Engineering Institute (SEI), with assistance from the Mitre Corporation, began developing a process maturity framework that would help organizations improve their software process. In September 1987, the SEI released a brief description of the process maturity framework, which was later expanded in Watts Humphrey's book, *Managing the Software Process* (Humphrey 1989). Two methods, software process assessment[1] and software capability evaluation[2] were developed to appraise software process maturity.

The SEI evolved the maturity framework into the Capability

Maturity Model for Software (CMM or SW-CMM[3]) (Paulk 1995). The SW-CMM presents sets of recommended practices in a number of key process areas that have been shown to enhance software process capability. It provides software organizations with guidance on how to gain control of their processes for developing and maintaining software and how to evolve toward a culture of software engineering and management excellence. By focusing on a limited set of activities and working aggressively to achieve them, an organization can steadily improve its organization-wide software process to enable continual and lasting gains in software process capability.

An observation from some workers involved in software process improvement has been that they could focus management attention and sponsorship on making needed changes because of the visibility that the SW-CMM provided. The problems facing an organization are usually known. The challenge is diverting resources from "fighting fires" to "preventing fires." The visibility of CMM-based process improvement can provide a lever for managers and technical staff who wish to adopt effective processes, methods, and tools. The purpose of this chapter is to provide an overview of what the SW-CMM recommends and to provide some suggestions on how to implement those recommendations. The intended result is improved organizational capability in building software.

Background

Setting sensible goals for process improvement requires an understanding of the difference between immature and mature software organizations. In an immature software organization, software processes are generally improvised by practitioners and their management during the course of the project. Even if a software process has been specified, it is not rigorously followed or enforced. The immature software organization is reactionary, and managers are usually focused on solving immediate crises (better known as firefighting). Schedules and budgets are routinely exceeded because they are not based on realistic estimates. When hard deadlines are imposed, product functionality and quality are often compromised to meet the schedule.

In an immature organization, there is no objective basis for judging product quality or for solving product or process problems. Therefore, product quality is difficult to predict. Activities intended to enhance quality, such as reviews and testing, are often curtailed or eliminated when projects fall behind schedule.

On the other hand, a mature software organization possesses an organization-wide ability for managing software development and maintenance processes. The software process is accurately communicated to both existing staff and new employees, and work activities are carried out according to the planned process. The mandated processes are usable and consistent with the way the work actually gets done. These defined processes are updated when necessary, and improvements are developed through controlled pilot tests and/or cost benefit analyses. Roles and responsibilities within the defined process are clear throughout the project and across the organization.

In a mature organization, managers monitor the quality of the software products and the process that produced them. There is an objective, quantitative basis for judging product quality and analyzing problems with the product and process. Schedules and budgets are based on historical performance and are realistic; the expected results for cost, schedule, functionality, and quality of the product are usually achieved. In general, a disciplined process is consistently followed because all of the participants understand the value of doing so, and the necessary infrastructure exists to support the process.

Fundamental Concepts Underlying Process Maturity

A *software process* can be defined as a set of activities, methods, practices, and transformations that people use to develop and maintain software and the associated products (for instance, project plans, design documents, code, test cases, and user manuals). As an organization matures, the software process becomes better defined and more consistently implemented throughout the organization.

Software process capability describes the range of expected results that can be achieved by following a software process. An organization's software process capability is one way of predicting the most likely outcome to expect from the next software project the organization undertakes.

Software process performance represents the actual results achieved by following a software process. Thus, software process performance focuses on the results achieved, while software process capability focuses on results expected.

Software process maturity is the extent to which a specific process is explicitly defined, managed, measured, controlled, and effective. Maturity implies a potential for growth in capability and indicates both the richness of an organization's software process and the consistency with which it is applied in projects throughout the organization.

As a software organization gains in software process maturity, it institutionalizes its software process via policies, standards, and organizational structures. Institutionalization entails building an infrastructure and a corporate culture that supports the methods, practices, and procedures of the business so that they endure after those who originally defined them have gone.

The Five Levels of Software Process Maturity

Continual process improvement is based on many small, evolutionary steps rather than revolutionary innovations, although revolutionary innovations may be part of a process improvement program. The staged structure of the SW-CMM is based on principles of product quality espoused by Walter Shewart, W. Edwards Deming, Joseph Juran, and Philip Crosby. The SW-CMM provides a framework for organizing these evolutionary steps into five maturity levels that lay successive foundations for continual process improvement. These five maturity levels define an ordinal scale for measuring the maturity of an organization's software process and for evaluating its software process capability. The levels also help an organization prioritize its improvement efforts.

A *maturity level* is a well-defined evolutionary plateau toward achieving a mature software process. Each maturity level comprises a set of process goals that, when satisfied, stabilize an important component of the software process. Achieving each level of the maturity framework establishes a higher level of process capability for the organization.

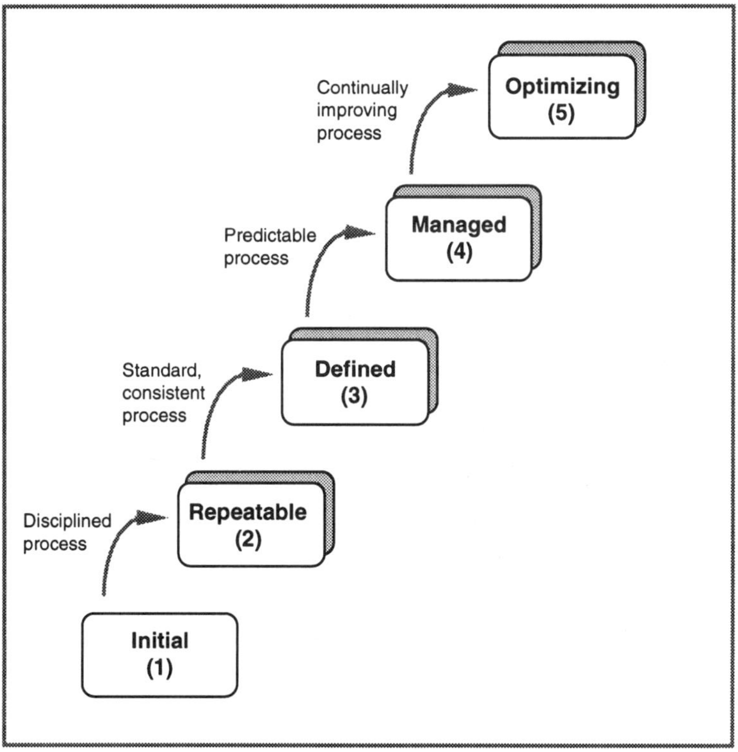

Figure 1: The Five Levels of Software Process Maturity

Organizing the SW-CMM into the five levels shown in Figure 1 prioritizes improvement actions for increasing software process maturity. The labeled arrows in Figure 1 indicate the type of process capability being institutionalized by the organization at each step of the maturity framework.

The five levels can be briefly described as:

1) *Initial.* The software process is characterized as ad hoc, and occasionally even chaotic. Few processes are defined, and success depends on individual effort and heroics.
2) *Repeatable.* Basic project management processes are established to track cost, schedule and functionality. The necessary process discipline is in place to repeat earlier successes on projects with similar applications.
3) *Defined.* The software process for both management and engineering activities is documented, standardized and integrated into a standard software process for the organization. All projects use an approved, tailored version of the organization's standard software process for developing and maintaining software.
4) *Managed.* Detailed measures of the software process and product quality are collected. Both the software process and products are quantitatively understood and controlled.
5) *Optimizing.* Continual process improvement is enabled by quantitative feedback from the process and from piloting innovative ideas and technologies.

These five levels reflect the fact that the SW-CMM is a model for improving the capability of software organizations. The priorities in the SW-CMM, as expressed by these levels, are not directed at individual projects. A project that is in trouble might justifiably set its priorities for corrective action differently from those in the SW-CMM. Its solutions, however, might be of limited value to the rest of the organization, because other projects might have different problems or because other projects could not take advantage of its solutions if they lack the necessary foundation to implement the solutions. The SW-CMM focuses on processes that build organizational capability.

Behavioral Characterization of the Maturity Levels

Maturity Levels 2 through 5 can be characterized through the activities performed by the organization to establish or improve the software process, by activities performed on each project, and by the resulting process capability across projects. A behavioral characterization of Level 1 is included to establish a base of comparison for process improvements at higher maturity levels.

Level 1 - The Initial Level

At the Initial Level, the organization typically does not provide a stable environment for developing and maintaining software. Over-commitment is a characteristic of Level 1 organizations, and such organizations frequently have difficulty making commitments that the staff can meet with an orderly engineering process, resulting in a series of crises. During a crisis, projects typically abandon planned procedures and revert to coding and testing. Success depends on having an exceptional manager and a seasoned and effective software team. Occasionally, capable and forceful software managers can withstand the pressures to take shortcuts in the software process; but when they leave the project, their stabilizing influence leaves with them. Even a strong engineering process cannot overcome the instability created by the absence of sound management practices.

In spite of this ad hoc, even chaotic, process, Level 1 organizations frequently develop products that work, even though they may exceed the budget and schedule. Success in Level 1 organizations depends on the competence and heroics of the people in the organization and cannot be repeated unless the same competent individuals are assigned to the next project. Thus, at Level 1, capability is a characteristic of the individuals, not of the organization.

Level 2 - The Repeatable Level

At the Repeatable Level, policies for managing a software project and procedures to implement those policies are established. Planning and managing new projects are based on experience with similar projects. Process capability is enhanced by establishing basic process management discipline on a project-by-project basis. Projects implement effective processes that are defined, documented, practiced, trained, measured, enforced, and able to improve.

Projects in Level 2 organizations have installed basic software management controls. Realistic project commitments are made, based on the results observed on previous projects and on the requirements of the current project. The software managers for a project track software costs, schedules, and functionality; problems in meeting commitments are identified when they arise. Software requirements and the work products developed to satisfy them are baselined, and their integrity is controlled. Software project standards are defined, and the organization ensures they are faithfully followed. The software project works with its subcontractors, if any, to establish an effective customer-supplier relationship.

Processes may differ among projects in a Level 2 organization. The organizational requirement for achieving Level 2 is that there are policies that guide the projects in establishing the appropriate management processes.

The software process capability of Level 2 organizations can be summarized as disciplined because software project planning and tracking are stable and earlier successes can be repeated. The project's process is under the effective control of a project management system, following realistic plans based on the performance of previous projects.

Level 3 - The Defined Level

At the Defined Level, a standard process (or processes) for developing and maintaining software is documented and used across the organization. This standard process includes both software engineering and management processes, which are integrated into a coherent whole. This standard process is referred to throughout the SW-CMM as the *organization's standard software process*. Processes established at Level 3 are used (and changed, as appropriate) to help the software managers and technical staff perform more effectively. The organization exploits effective software engineering practices when standardizing its software processes. A group such as a software engineering process group or SEPG is responsible for the organization's software process activities. An organization-wide training program is implemented to ensure that the staff and managers have the knowledge and skills required to fulfill their assigned roles.

Projects tailor the organization's standard software process to develop their own defined software process, which accounts for the unique characteristics of the project. This tailored process is referred to in the SW-CMM as the *project's defined software process*. It is the process used in performing the project's activities. A defined software process contains a coherent, integrated set of well-defined software engineering and management processes. A well-defined process includes entry criteria, inputs, standards and procedures for performing the work, verification mechanisms (such as peer reviews), outputs, and exit criteria. Because the software process is well-defined, management has good insight into technical progress on the project.

The software process capability of Level 3 organizations can be summarized as standard and consistent because both software engineering and management activities are stable and repeatable. Within established product lines, cost, schedule, and functionality are under control, and software quality is tracked. This process capability is based on a common, organization-wide understanding of the activities, roles, and responsibilities in a defined software process.

Level 4 - The Managed Level

At the Managed Level, the organization sets quantitative quality goals for both software products and processes. Productivity and quality are measured for important software process activities across all projects as part of an organizational measurement program. An organization-wide software process database is used to collect and analyze the data available from the projects' defined software processes. Software processes are instrumented with well-defined and consistent measurements. These measurements establish the quantitative foundation for evaluating the projects' software processes and products.

Projects achieve control over their products and processes by narrowing the variation in their process performance to fall within acceptable quantitative boundaries. Meaningful variations in process performance can be distinguished from random variation (noise), particularly within established product lines. The risks involved in moving up the learning curve of a new application domain are known and carefully managed.

The software process capability of Level 4 organizations can be summarized as being quantified and predictable because the process is measured and operates within quantitative limits. This level of process capability allows an organization to predict trends in process and product quality within the quantitative bounds of these limits. Because the process is both stable and measured, when some exceptional circumstance occurs, the "special cause" of the variation can be identified and addressed. When the pre-defined limits are exceeded, actions are taken to understand and correct the situation. Software products are of predictably high quality.

Level 5 - The Optimizing Level

At the Optimizing Level, the entire organization is focused on continual process improvement. The organization has the means to identify weaknesses and strengthen the process proactively, with the goals of preventing defects and improving efficiency. Data on process effectiveness are used to perform cost/benefit analyses of new technologies and proposed changes to the organization's software process. Innovations that exploit the best software engineering practices are identified and transferred throughout the organization.

Software teams in Level 5 organizations analyze defects to determine their causes, evaluate software processes to prevent known types of defects from recurring, and disseminate lessons learned throughout the organization.

There is chronic waste, in the form of rework, in any system simply due to random variation. Organized efforts to remove waste result in changing the system by addressing "common causes" of inefficiency. While efforts to reduce waste occur at all maturity levels, it is the focus of Level 5.

The software process capability of Level 5 organizations can be characterized as continually improving because Level 5 organizations are continually striving to improve the range of their process capability, thereby improving the process performance of their projects. Improvements occur both by incremental advancements in the existing process and by innovations using new technologies and methods. Technology and process improvements are planned and managed as ordinary business activities.

Process Capability and the Prediction of Performance

An organization's software process maturity helps predict a project's ability to meet its objectives. Projects in Level 1 organizations experience wide variations in achieving cost, schedule, functionality, and quality targets. Figure 2 illustrates the kinds of improvements expected in predictability, control, and effectiveness in the form of a probability density for the likely performance of a particular project with respect to targets, such as cycle time, cost, and quality.

The first improvement expected as an organization matures is in predictability. As maturity increases, the difference between targeted results and actual results decreases across projects. For instance, Level 1 organizations often miss their originally scheduled delivery dates by a wide margin, whereas higher maturity level organizations should be able to meet targeted dates with increased accuracy.

The second improvement is in control. As maturity increases, the variability of actual results around targeted results decreases. For instance, in Level 1 organizations delivery dates for projects of similar size are unpredictable and vary widely. Similar projects in a higher maturity level organization, however, will be delivered within a smaller range.

The third improvement is in effectiveness. Targeted results improve as the maturity of the organization increases. That is, as a software organization matures, costs decrease, development time becomes shorter, and productivity and quality increase. In a Level 1 organization, development time can be quite long because of the amount of rework that must be performed to correct mistakes. In contrast, higher maturity level organizations have increased process effectiveness and reduced costly rework, allowing development time to be shortened.

The improvements in predicting a project's results represented in Figure 2 assume that the software project's outcomes become more predictable as noise, often in the form of rework, is removed from the software process. Unprecedented systems complicate the picture since new technologies and applications lower the process capability by increasing variability. Even in the case of unprecedented systems, the management and engineering practices characteristic of more mature organizations help identify and address problems earlier than for less mature organizations. In some cases a mature process means

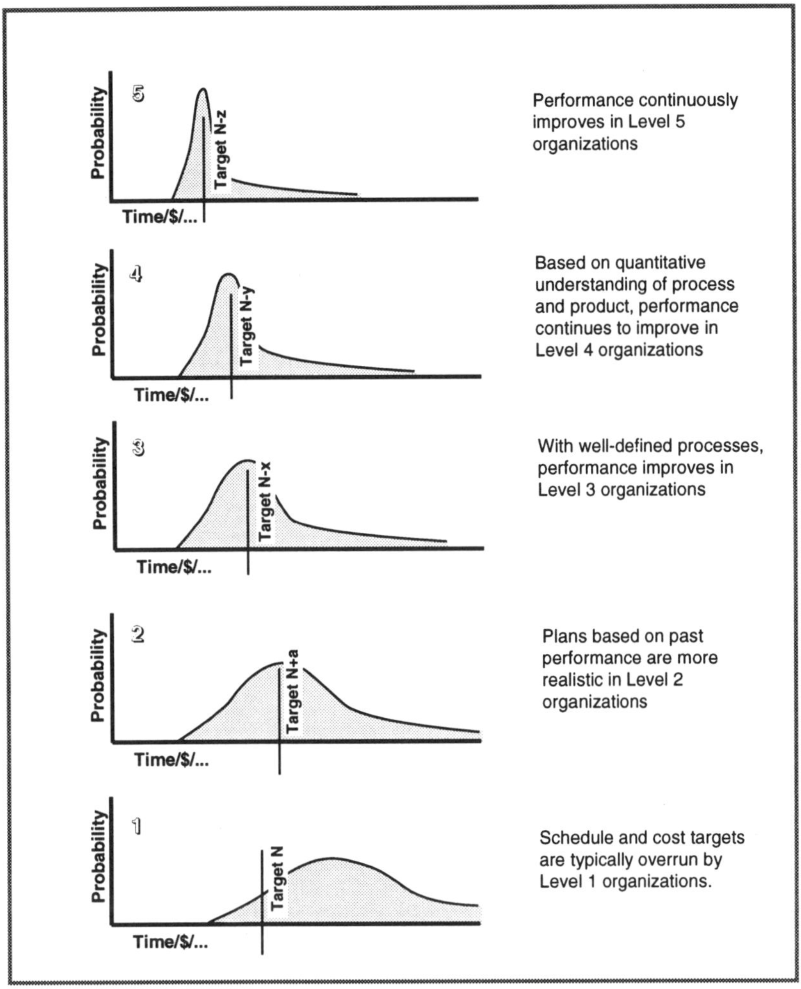

Figure 2. Process Capability as Indicated by Maturity Level.

that "failed" projects are identified early in the software life cycle and investment in a lost cause is minimized.

The documented case studies of software process improvement indicate that there are significant improvements in both quality and productivity as a result of the improvement effort (Herbsleb 1994; Lawlis 1995). The return on investment seems to typically be in the 4:1 to 8:1 range for successful process improvement efforts, with increases in productivity ranging from 9-67 percent and decreases in cycle time ranging from 15-23 percent reported (Herbsleb 1994).

An Overview of the Key Process Areas

The SW-CMM is a framework representing a path of improvements recommended for software organizations that want to increase their software process capability. The intent is that the SW-CMM is at a sufficient level of abstraction that it does not unduly constrain how the software process is implemented by an organization. The SW-CMM describes what we would normally expect in a software process, regardless of how the process is implemented.

Each maturity level, with the exception of Level 1, has been decomposed into constituent parts. The decomposition of each maturity level ranges from abstract summaries of each level down to their operational definition in the key practices. Each maturity level is composed of several key process areas, which indicate where an organization should focus to improve its software process. Each key process area is organized into five sections called common features. The common features specify the key practices that, when collectively addressed, accomplish the goals of the key process area. The common features are Commitment to Perform, Ability to Perform, Activities Performed, Measurement and Analysis, and Verifying Implementation.

Each *key process area* identifies a cluster of related activities that,

Level	Focus	Key Process Areas
5 Optimizing	*Continuous process improvement*	Defect Prevention Technology Change Management Process Change Management
4 Managed	*Product and process quality*	Quantitative Process Management Software Quality Management
3 Defined	*Engineering processes and organizational support*	Organization Process Focus Organization Process Definition Training Program Integrated Software Management Software Product Engineering Intergroup Coordination Peer Reviews
2 Repeatable	*Project management processes*	Requirements Management Software Project Planning Software Project Tracking & Oversight Software Subcontract Management Software Quality Assurance Software Configuration Management
1 Initial	*Competent people and heroics*	

Figure 3. The Key Process Areas by Maturity Level.

when performed collectively, achieve a set of goals considered important for enhancing process capability. The key process areas have been defined to reside at a single maturity level as shown in Figure 3. The path to achieving the goals of a key process area may differ across projects based on differences in application domains or environments. Nevertheless, all the goals of a key process area must be achieved for the organization to satisfy that key process area.

The adjective "key" implies there are process areas (and processes) that are not key to achieving a maturity level. The SW-CMM does not describe in detail all the process areas that are involved with developing and maintaining software. Certain process areas have been identified as key determiners of process capability, and these are the ones described in the SW-CMM.

The key process areas are the "requirements" for achieving a maturity level. To achieve a maturity level, the key process areas for that level and the lower levels must be satisfied (or not applicable, such as Software Subcontract Management when there are no subcontractors).

The specific practices to be executed in each key process area will evolve as the organization achieves higher levels of process maturity. For instance, many of the project estimating capabilities described in the Software Project Planning key process area at Level 2 must evolve to handle the additional project data available at Level 3, as described in Integrated Software Management.

The key process areas at Level 2 focus on the software project's concerns related to establishing basic project management controls.

- Requirements Management: establish a common understanding between the customer and the software project of the customer's requirements that will be addressed by the software project. This agreement with the customer is the basis for planning and managing the software project.

- Software Project Planning: establish reasonable plans for performing the software engineering and for managing the software project. These plans are the necessary foundation for managing the software project.

- Software Project Tracking and Oversight: establish adequate visibility into actual progress so that management can take effective actions when the software project's performance deviates significantly from the software plans.

- Software Subcontract Management: select qualified software subcontractors and manage them effectively.

- Software Quality Assurance: provide management with appropriate visibility into the process being used by the software project and of the products being built.

- Software Configuration Management: establish and maintain the integrity of the products of the software project throughout the project's software life cycle.

The key process areas at Level 3 address both project and organizational issues, as the organization establishes an infrastructure that institutionalizes effective software engineering and management processes across all projects.

- Organization Process Focus: establish the organizational responsibility for software process activities that improve the organization's overall software process capability.

- Organization Process Definition: develop and maintain a usable set of software process assets that improve process performance across the projects and provides a basis for defining meaningful data for quantitative process management. These assets provide a stable foundation that can be institutionalized via mechanisms such as training.

- Training Program: develop the skills and knowledge of individuals so they can perform their roles effectively and efficiently. Training is an organizational responsibility, but the software projects should identify their needed skills and provide the necessary training when the project's needs are unique.

- Integrated Software Management: integrate the software engineering and management activities into a coherent, defined software process that is tailored from the organization's standard software process and related process assets. This tailoring is based on the business environment and technical needs of the project.

- Software Product Engineering: consistently perform a well-defined engineering process that integrates all the software engineering activities to produce correct, consistent software products effectively and efficiently. Software Product Engineering describes the technical activities of the project, for instance requirements analysis, design, code, and test.

- Intergroup Coordination: establish a means for the software engineering group to participate actively with the other engineering groups so the project is better able to satisfy the customer's needs effectively and efficiently.

- Peer Reviews: remove defects from the software work products early and efficiently. An important corollary effect is to develop a better understanding of the software work products and of the defects that can be prevented. The peer review is an important and effective engineering method that can be implemented via inspections, structured walkthroughs, or a number of other collegial review methods.

The key process areas at Level 4 focus on establishing a quantitative understanding of both the software process and the software work products being built.

- Quantitative Process Management: control process performance of the software project quantitatively. Software process performance represents the actual results achieved from following a software process. The focus is on identifying special causes of variation within a measurably stable process and correcting, as appropriate, the circumstances that drove the transient variation to occur.

- Software Quality Management: develop a quantitative understanding of the quality of the project's software products and achieve specific quality goals.

The key process areas at Level 5 cover the issues that both the organization and the projects must address to implement continual and measurable software process improvement.

- Defect Prevention: identify the causes of defects and prevent them from recurring. The software project analyzes defects, identifies their causes, and changes its defined software process.

- Technology Change Management: identify beneficial new technologies (such as tools, methods, and processes) and transfer them into the organization in an orderly manner. The focus of Technology Change Management is on performing innovation efficiently in an ever-changing world.

- Process Change Management: continually improve the software processes used in the organization with the intent of improving software quality, increasing productivity, and decreasing the cycle time for product development.

Goals summarize the key practices of a key process area and can be used to determine whether an organization or project has effectively implemented the key process area. The goals signify the scope, boundaries, and intent of each key process area. Satisfaction of a key process area is determined by achievement of the goals.

Key practices describe the activities and infrastructure that contribute most to the effective implementation and institutionalization of the key process area. Each key practice consists of a single sentence, often followed by a more detailed description, which may include examples and elaboration. These key practices, also referred to as the top-level key practices, state the fundamental policies, procedures, and activities for the key process area. The components of the detailed description are frequently referred to as subpractices. The key practices describe "what" is to be done, but they should not be interpreted as mandating "how" the goals should be achieved. Alternative practices may accomplish the goals of the key process area. The key practices should be interpreted rationally to judge whether the goals of the key process area are effectively, although perhaps differently, achieved.

The IDEAL Approach

Effective software process improvement, whether based on the SW-CMM or some other model, occurs in a systematic fashion. The SEI has developed the IDEAL[SM] model, shown in Figure 4, to depict the

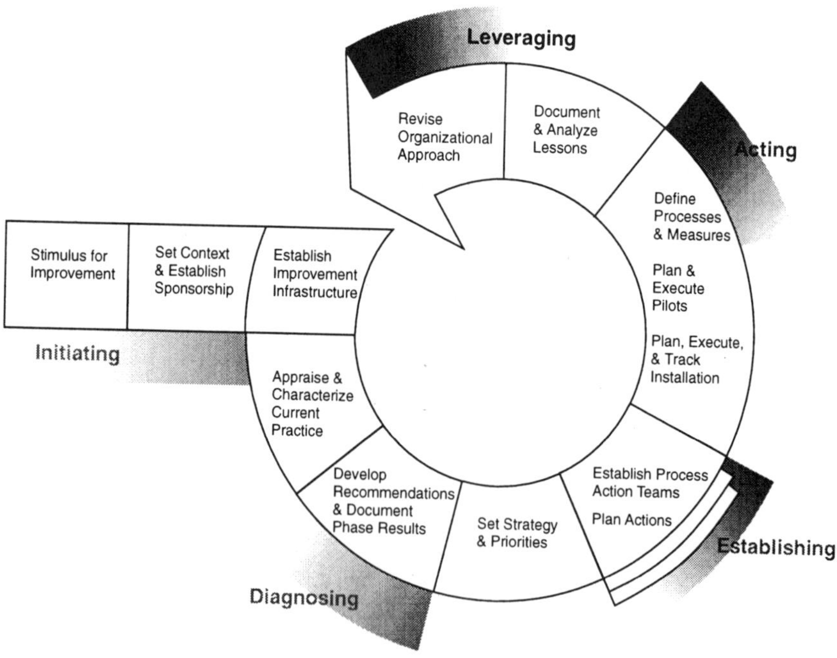

Figure 4. The SEI's IDEAL Approach to Software Process Improvement.

activities of an improvement program based on the SW-CMM. It consists of five phases:

I Initiating (the improvement program)
D Diagnosing (the current state of practice)
E Establishing (the plans for the improvement program)
A Acting (on the plans and recommended improvements)
L Leveraging (the lessons learned and the business results of the improvement effort)

The *Initiating* phase establishes the business reasons for undertaking a software process improvement effort. It identifies high-level concerns in the organization that can be the stimulus for addressing various aspects of quality improvement. Communication of these concerns and business perspectives is needed during the Initiating phase in order to gain visible executive buy-in and sponsorship at this very early part of the improvement effort.

The *Diagnosing* phase is used to baseline, through an assessment, the organization's software process maturity. This baseline will build a common understanding of the current processes of the organization, especially the strengths and weaknesses of those current processes. It will also help identify priorities for improving your software processes. This diagnosis is based on the SW-CMM (or one of the other CMMs).

The *Establishing* phase finalizes the strategy and supporting plans for the software process improvement program. It sets the direction and guidance for the next three to five years, including strategic and tactical plans for software process improvement.

The *Acting* phase takes action to effect changes in organizational systems that result in improvements in these systems. These improvements are made in an orderly manner and in ways that will cause them to be sustained over time. Techniques used to support and institutionalize change include defining software processes and measurements, pilot testing, and installing new processes and measurements throughout the organization. The SW-CMM provides guidance for the improvement actions in its informative components: the key practices and their subpractices and examples.

The *Leveraging* phase completes the process improvement cycle. Lessons learned from the pilot projects and improvement efforts are documented and analyzed in order to improve the process improvement program for the future. The business needs that were determined at the beginning of the cycle are revisited to see if they have been met. Sponsorship for the program is revisited and renewed for the next software process improvement cycle.

Effective CMM-Based Process Improvement

The SW-CMM identifies practices for a mature software process and provides examples of the state-of-the-practice (and in some cases, the state-of-the-art), but it is not meant to be either exhaustive or dictatorial. The SW-CMM identifies the characteristics of an effective software process, but the mature organization addresses all issues essential to a successful project, including people and technology, as well as process.

Using the SW-CMM correctly means balancing conflicting objectives. CMM-based appraisals require the use of professional judgment. Although the SW-CMM provides a significant amount of guidance in making these judgments, removing subjectivity implies a deterministic, repetitive process that is not characteristic of engineering design work. Although the SW-CMM is sometimes referred to as a set of process requirements, it does not contain any "shall" statements.

Judgment is required to use the SW-CMM properly. The goals, key practices, and subpractices of its key process areas are arranged in a hierarchy useful for interpreting or tailoring the SW-CMM. As Charlie Weber has pointed out, "To tailor a goal, you should sweat blood (and then be conservative in mapping the relationship to "the" CMM). To tailor a key practice, you should just sweat. To tailor a subpractice, your conscience should hardly bother you."

The following recommendations reflect ways to implement the SW-CMM that I consider especially effective (Paulk 1996). To a knowledgeable software professional, many of these recommendations may seem obvious. I fear, however, that the consistent implementation of even well-known good engineering and management practices is far from common.

Management Sponsorship

Trite though it may seem, obtaining senior management sponsorship is a crucial component of building organizational capability. As individuals, we can exercise professionalism and discipline within our sphere of control, but if an organization as a whole is to change its performance, then its senior management must actively support the change. Bottom-up software process improvement, without sponsorship and coordination, leads to islands of excellence rather than predictably improved organizational capability. Effective sponsorship will only occur, however, when there is significant dissatisfaction with the status quo.

Sponsorship involves such activities as setting policies and providing resources for process work, but the most crucial factor is that senior management believe in the strategic importance of process improvement, demonstrate their support, and *pay attention*. If, on the other hand, management provides annual speeches on the importance of quality and then returns to the "really important" issues of cost and schedule, then that message will be received also.

"Paying attention" may lead to the Hawthorne Effect: people increase their efforts as a result of the attention, and productivity and quality could improve even if the work environment becomes worse. Sustaining improvement over an extended period, however, implies systematic change aligned with the business objectives of the organization.

If process improvement is a true priority, then management will monitor it closely. Management should set aggressive improvement goals, while at the same time recognizing that achieving higher levels of software process maturity is incremental and requires a long-term commitment to continual process improvement. It may be useful to view software process improvement as a project, with goals, plans for achieving those goals, and management of progress. Do not forget, however, that changing the culture requires the long view. CMM-based improvement is most effective within a systematic approach to software process improvement, such as the SEI's IDEAL approach.

Commitments and Management by Fact

Management by fact is a paradigm shift for most organizations, which must be based on a measurement foundation. To make data analysis useful, you need to understand what the data means and how to analyze it meaningfully. Begin by collecting a simple set of *useful* data – I recommend the Goal/Question/Metric approach (Basili, 1992). Then deploy standard definitions for typical measures that support aggregating and comparing data from different environments.

A corollary to collecting data is to believe what it tells you. If you have historically developed 100 lines of code per month, "betting the company" that you can bring in a critical project at 1000 lines of code per month is certainly unwise. And changing the size estimates without changing commitments and functionality is not likely to help matters.

You also have to be sensitive to the potential for causing dysfunctional behavior by what you measure (Austin 1996). People will focus their efforts where management is paying attention. If management only pays attention to schedule, then quality is likely to suffer. The act of measuring identifies what is important, but some things are difficult to measure. Management needs to ensure that attention is visibly paid to all critical aspects of the project, including those difficult to measure, not just those it is easy to measure and track. The corollary to believing the data is to recognize that no limited set of data can completely capture a complex operation.

Establish an internal commitment process based on what you know you can do. If the workers buy in to a set of commitments as being realistic, even if aggressive, then they have made a personal commitment. If they consider the stated commitments as being a management pipe dream, they are likely to consider the resulting problems to be management's. This implies that management should seek worker feedback when establishing commitments.

Process Focus

In most organizations, a software engineering process group (SEPG) or some equivalent should be formed to coordinate process definition, improvement, and deployment activities. The SEPG should be formed early, e.g., in the Initiating phase of the IDEAL approach, and be staffed with competent and respected individuals possessing both managerial and technical skills. It is crucial that these individuals have good interpersonal skills. Success of the SEPG depends on their ability to communicate, teach, negotiate, and consult.

One of the reasons for dedicating resources to an SEPG is to ensure follow-through on appraisal findings. Many improvement programs have foundered simply because no action resulted from the appraisal. Improvement comes from action planning, assigning individuals to do the work, piloting and deploying improved processes, and management oversight throughout.

It is desirable to have part-time participants on process improvement teams. In saying that SEPGs should coordinate process activities, the word "coordinate" was deliberately chosen. The skills and knowledge of the organization's best people should be brought to bear on addressing its problems. In a world-class organization, I would expect everyone to participate in process and quality improvement activities. Worker participation and empowerment enable successful process improvement.

Begin with the "as is" process, not the "should be" process, to leverage effective practices and co-opt resisters. Mandating top-down that everyone will follow the new "should be" process, particularly if not developed by empowered workers, is a common recipe for failure. The "as is" process evolved because the people doing the work needed to get the job done – even if that meant going around the system. The "should be" process may, or may not, be feasible in the given culture and environment. With an organizational focus on process management and improvement, the "as is" and "should be" processes will converge, resulting in organizational learning.

It is important to remember that software process improvement occurs in a business context. There may be many other crucial business issues being worked at the same time; there may even be a Total Quality Management (TQM) initiative under way. Since CMM-based improvement is an application of TQM principles to software, the synergy of aligning these initiatives seems obvious. Some of the most effective software process improvement programs I have seen operated under the umbrella of a TQM initiative. In the instances where I have observed disjoint TQM and software process improvement programs, both were ineffective.

Useful Processes

Document your processes. The reasons for documenting a process (or product) are 1) to communicate – to others now and perhaps to yourself later; 2) to understand – if you can't write it down, you don't really understand; and 3) to encourage consistency – take advantage of repeatability.

Building software is a design-intensive, creative activity. While the discipline of process is a crucial enabler of success, the objective is to solve a problem, and this requires creativity. Software processes should be repeatable, even if they are not repetitive. The balance between discipline and creativity can be challenging (Glass, 1995). Losing sight of the creative, design-intense nature of software work leads to stifling rigidity. Losing sight of the need for discipline leads to chaos.

Processes need to be tailored to the needs of the project (Ginsberg, 1995). Although standard processes provide a foundation, each project will also have unique needs. Unreasonable constraints on tailoring can lead to significant resistance to following the process.

The following list contains some process attributes that are important for useful and effective processes:

- documented – but keep it simple
- trained – impart the skill to use it
- practiced – used even in a crisis
- measured – simple measures for insight into critical questions
- owned – by a responsible party
- maintained – updated as needed
- supported – in plans and by management expectations and rewards
- controlled – process changes are disciplined
- verified and validated – checked for correct execution

Keep the process simple because we live in a rapidly changing world. Processes do not need to be lengthy or complex. The SW-CMM is about *doing* things, not *having* things. A 1-2 page process description may suffice (Humphrey 1995), and subprocesses and procedures can be invoked as needed and useful. Order is not created by complex controls, but by the presence of a few guiding formulae or principles (Wheatley, 1992, page 11).

Training

Documented processes are of little value if they are not effectively deployed. Training, via a wide variety of mechanisms, is critical to consistent and effective software engineering and management. Training is an investment in our greatest asset – our people.

The reason for training is to develop skills. There are many "training mechanisms" other than formal classroom training that can be effective in building skills. One that should be seriously considered is a formal mentoring program. In this case, formality means going beyond assigning a mentor and hoping that experience will rub off. Formality implies training people on how to mentor and monitoring the effectiveness of the mentoring. For example, the Onboard Shuttle project considers mentoring in doing causal analysis of defects (Paulk, 1995).

Management training is particularly important because ineffective management can cripple a good team. People who are promoted to management because of their technical skills have to acquire a new set of skills, including interpersonal skills (Weinberg, 1994).

Having taken the Personal Software Process (PSP) course, I can highly recommend it for building self-discipline (Humphrey, 1995). Note that the effect of reading the book is not the same as taking the course and doing the work! Where the SW-CMM addresses the organizational side of process improvement, PSP addresses building the capability of individual practitioners. The PSP course convinces the individual, based on his or her own data, of the value of a disciplined, engineering approach to building software. Even if the organization is not interested in PSP, I would recommend it for professional development; applying PSP can be a survival skill in a Level 1 organization.

Risk Management

Some argue that software project management is really risk management. In one sense, the SW-CMM is about managing risk. We attempt to establish stable requirements so that we can plan and manage effectively, but the business environment changes rapidly, perhaps chaotically. We try to establish an island of order in the sea of software chaos, but both order and chaos have a place. As Wheatley suggests, "To stay viable, open systems maintain a state of non-equilibrium, keeping the system in balance so that it can change and grow." (Wheatley, 1992, p. 78) Although we can establish processes that help us manage the risks of a chaotic world, we also need to change and grow.

This implies that you should use an incremental or evolutionary life cycle. If you want to focus on risk management, the spiral model may be the preferred life cycle model. If you want to focus on involving the customer, perhaps rapid prototyping or joint application design would be preferable. Few long-term projects have the luxury of the stable environment necessary for the waterfall life cycle to be the preferred choice – yet it is probably the most common life cycle. (Note that for small projects, the waterfall life cycle may be an excellent choice.)

Customer-Supplier Relationship

Establish a good, working relationship with the customer and end user based on open communication and integrity. This requires cooperation from the customer, and the danger of customer irrationality is difficult to overstate. Building good customer-supplier relationships is a long-term effort, but if the organization has (or wants to have) a stable customer base, the customer must appreciate the complexities of software engineering just as the supplier must understand the application domain and business environment for the software product. Even those organizations that operate in an environment where customer "churn" is the norm can profit from a (deserved) reputation for quality and integrity.

Talk to the customer. In the exercises for our *Introduction to the CMM* training, one of our most common observations is that the teams do not talk to their customer. This is crucial in making good cost, schedule, functionality, and quality tradeoff decisions. In a study done for one multinational company, which focused on leading-edge technology as its primary competitive advantage, the #1 and #2 issues for satisfying customer expectations were "quality" and "meeting commitments." "Technology" was #7.

One of the drivers behind software capability evaluations (SCE) is the customer's desire for predictable costs and schedules. Communication is critical for setting (and changing) customer expectations. Our contact at one of the early SCE pilot organizations commented later that he considered the improved customer-supplier communication to have been worth the (nontrivial) cost of doing SCEs, even if there had been no other improvement.

And, of course, be a good customer yourself when the tables are turned.

Peer Reviews

Although you can argue over the best kind of peer review, the simple fact is that the benefits of peer reviews far outweigh their costs. The data suggests some form of inspection should be used, but any form of collegial or disciplined review, such as structured walkthroughs, adds significant value.

Some time ago, one of my colleagues had an interesting question about peer reviews. He had been consulting with a company that had its supervisors review code. This had been successful for them and was firmly entrenched in their culture. His question was whether this satisfied the Peer Reviews key process area. My answer was that it did not. Peers are not supervisors, and having only one person do the review seems inadequate.

My colleague later reported that the company had performed an experiment comparing their supervisor reviews with peer reviews. They discovered that peer reviews found significantly more errors than supervisor reviews. They also discovered that supervisor reviews tended to identify "form errors" while peer reviews tended to identify "content errors." They also performed reviews with designers as well as "code-level peers" and found that these reviews identified more (and more design-oriented) errors than the code-level peer reviews.

The significant point to me was that they experimented to find a data-based answer to an open issue. They did not take a consultant's advice on faith. I consider this one of the hallmarks of a learning organization, one that can take advantage of the SW-CMM effectively.

Conclusion

To quote George Box, "All models are wrong; some models are useful." The SW-CMM represents a "common sense engineering" approach to software process improvement. The maturity levels, key process areas, common features, and key practices have been extensively discussed and reviewed within the software community. While the SW-CMM is not perfect, it does represent a broad consensus of the software community and is a useful tool for guiding software process improvement efforts.

The SW-CMM provides a conceptual structure for improving the management and development of software products in a disciplined and consistent way. It does not guarantee that software products will be successfully built or that all problems in software engineering will be adequately resolved. However, current reports from CMM-based improvement programs indicate that it can improve the likelihood with which a software organization can achieve its cost, quality, and productivity goals.

Never forget why process improvement is important. Standards and models such as the SW-CMM can help organizations improve their software process, but focusing on achieving a maturity level without really improving the underlying process is a danger. Maturity levels should be a measure of improvement, not the goal of improvement. What are the business needs and business goals of the improvement effort? What is the impact on cost, cycle time, productivity, quality, and — most importantly — customer satisfaction?

For further information regarding the SW-CMM and its associated products, including training on the SW-CMM and how to perform software process assessments and software capability evaluations, contact: SEI Customer Relations, Software Engineering Institute, Carnegie Mellon University, Pittsburgh, PA 15213-3890, (412) 268-5800, Internet: customer-relations@sei.cmu.edu. The SEI Web page is *http://www.sei.cmu.edu/*. For information specifically on the SW-CMM, the Web page for the CMM Correspondence Group is *http://www.sei.cmu.edu/technology/cmm/*.

Endnotes

This work was sponsored by the U.S. Department of Defense. CMM, Capability Maturity Model, and IDEAL are service marks of Carnegie Mellon University.

[1] A software process assessment is an appraisal by a trained team of software professionals to determine the state of an organization's current software process, to determine the high-priority software process-related issues facing an organization, and to obtain the organizational support for software process improvement.

[2] A software capability evaluation is an appraisal by a trained team of professionals to identify contractors who are qualified to perform the software work or to monitor the state of the software process used on an existing software effort.

[3] A number of CMMs inspired by the CMM for Software have now been developed. To minimize confusion, we use SW-CMM to identify the original CMM for Software.

[4] SW-CMM.

References

Austin, Robert D. (1996). *Measuring and Managing Performance in Organizations*, Dorset House Publishing, New York, NY, 1996.

Basili, V.R. (1992). "Software Modeling and Measurement: The Goal/Question/Metric Paradigm," University of Maryland, CS-TR-2956, UMIACS-TR-92-96, 1992.

Ginsberg, Mark and Lauren Quinn (1995). *Process Tailoring and the Software Capability Maturity Model*,Software Engineering Institute, CMU/SEI-94-TR-024. November 1995.

Glass, Robert (1995). *Software Creativity*, Prentice Hall, Englewood Cliffs, NJ, 1995.

Herbsleb, James, Anita Carleton, et al.(1994). *Benefits of CMM-Based Software Process Improvement: Initial Results*,"Software Engineering Institute, CMU/SEI-94-TR-13. August 1994.

Humphrey, Watts S. (1989). *Managing the Software Process*, Addison-Wesley, Reading, MA, 1989.

Humphrey, Watts S. (1995), *A Discipline for Software Engineering*, Addison-Wesley Publishing Company, Reading, MA, 1995.

Lawlis, Patricia K., Robert M. Flowe, and James B. Thordahl (1995). "A Correlational Study of the CMM and Software Development Performance," *Crosstalk: The Journal of Defense Software Engineering*, Vol. 8, No. 9, , pp. 21-25.

Paulk, Marck C. (1995). Carnegie Mellon University, Software Engineering Institute (Principal Contributors and Editors: Mark C. Paulk, Charles V. Weber, Bill Curtis, and Mary Beth Chrissis), *The Capability Maturity Model: Guidelines for Improving the Software Process*, Addison-Wesley Publishing Company, Reading, MA, 1995.

Paulk, Mark C. (1996). "Effective CMM-Based Process Improvement," *Proceedings of the 6th International Conference on Software Quality*, Ottawa, Canada, 28-31 October 1996, pp. 226-237.

Weinberg, Gerald M. (1994). *Quality Software Management*, Volume 3: Congruent Action , Dorset House, New York, NY, 1994.

Wheatley, Margaret (1992). *Leadership and the New Science*, Berrett-Koehler Publishers, San Francisco, CA, 1992.

CHAPTER

11

Inter-Organizational Networks and Diffusion of Information Technology: Developing a Framework

Jacky Swan
Sue Newell
University of Warwick

Maxine Robertson
Coventry University

To introduce innovative solutions, either in products or processes, a firm needs to have access to new ideas. This chapter presents a framework which depicts the importance of inter-organizational networks for diffusing new ideas. It is argued that the links between a firm and a variety of other organiszations established by individual employees involved in inter-organizational networks, are crucial to understanding the innovation process. While a variety of inter-organizational networks are seen to be important for the diffusion of new ideas, this chapter focuses on the role of professional associations in the diffusion process. The chapter presents data from two firms introducing a new information technology (IT) system, one of which was relatively more successful in implementing this system. Those involved in the decision-making process in these two firms made rather different use of inter-organizational networks. The different outcomes from the innovation process can at least be partially explained by these differences in inter-organizational networking. The importance of professional associations for the diffusion of ideas was evident in both firms, but the more successful firm had more active involvement in a particular professional association, as well as access to a broader range of alternative ideas via other professional groups. A more detailed analysis of the activities of the professional association that both companies highlighted as central demonstrates that while professional associations can stimulate the diffusion of ideas, the actual ideas diffused through their formal activities tends to be partial and restricted. More active and diverse involvement in networks by firm members is important in order to understand the difficulties as well as the advantages associated with particular new ideas and the broader range of potential solutions.

This chapter focuses on understanding the diffusion of Information Technology (IT) and emphasizes the role of inter-organizational networks in this process. A framework is developed to depict this process and is used to analyze the diffusion into manufacturing firms of IT to support operations management. Van de Ven (1986) defines the process of innovation as "the development and implementation of new ideas by people who over time engage in transactions with others in an institutional context" (p. 591). This definition is useful as it highlights the importance of interpersonal networks in diffusing new ideas ('transactions with others') needed for innovation. It also emphasizes the importance of the wider organizational and societal ('institutional') contexts into which new ideas are introduced. This is in keeping with other researchers on innovation who argue that the innovation process is socially and societally embedded (e.g. Clark, 1987; Bijker, Hughes and Pinch, 1987). An important aspect of the context for firms in industry is the firms' links, both formal and informal, with a variety of other types of organizations (e.g., other firms, technology suppliers, professional associations and so forth). Inter-organizational networks refer here to the links between a focal organization and a variety of other organizations (Hauschlidt, 1992; Swan & Newell, 1995). It is clear that an organization could never contain within its boundary all the information that it would ever benefit from using. While it may be possible to reconstruct existing internal information to create new ideas, the "fundamental information contribution to change within the firm must come from outside" (Macdonald & Williams, 1993, p. 418). It is through these inter-organizational networks then, that ideas new to the focal firm may be generated, communicated and diffused (Alter & Hage, 1993; Axelsson & Easton, 1992).

Special emphasis is given in this chapter to the networks provided by professional associations. These professional associations have been found to be key networks in the innovation diffusion process but have been under-researched in the past (Newell and Clark, 1990; Swan and Newell, 1995). It will be argued that professional associations representing the domain of operations management are important because they disseminate information about IT relevant to operations management directly to members who work in manufacturing firms. This occurs through the formal activities organized by the professional association, such as meetings, conferences, journal publications. Also importantly, professional associations may foster weak or loose interpersonal links among members from across a variety of different industry sectors and occupations. In keeping with other research, this chapter will argue that the knowledge acquired through these links have the potential to influence the development and diffusion of IT solutions (Granovetter, 1973; Rogers, 1983).

Two case study examples of the introduction of IT for operations management into manufacturing firms will be explored. These illustrate the inter-organizational networks which were found to be relevant to the innovation process with regard to the adoption of operations management information systems. Different outcomes from the innovation process are at least be partially explained by differences in inter-organizational networking. In both firms, professional associations are seen to be important for the diffusion of ideas. Data will then be presented on a number of professional associations relevant to this domain of operations management in order to provide a more detailed understanding of the role they play in the innovation diffusion process. The chapter focuses on data collected in the UK, but the discussion will also consider the relative importance of different inter-organizational networks for the diffusion process in other European countries, based on research being conducted currently in Sweden, France and the Netherlands.

Research Position and Theory

To develop an understanding of the diffusion of IT it is necessary to consider two separate issues. Firstly, the communication of new ideas across a community of organizations, through inter-organizational networks. Secondly, the translation of these new ideas within particular organizations as firm-specific solutions.

Taking the organization as the unit of analysis, organizations manage to forge links with a variety of other organizations. This can occur, for example, through formal contractual arrangements such as joint ventures, collaborative agreements, mergers, acquisitions and so forth. Such links, once established, are superordinate to particular individuals within organizations: that is, they will persevere beyond the involvement of particular individual employees. However, such links will inevitably be limited - an organization could never extend its control over all other organizations that might be useful (Macdonald and Williams, 1993). Thus, much of the information that permeates an organization's boundaries is brought in by individual members who are able to formulate new ideas through contacts made with individuals in other organizations i.e., utilizing interpersonal networks (Allen, Tushman & Lee, 1979). It is these interpersonal networks that link individuals across organizations that are the focus of this paper.

Individuals who are active in inter-organizational networks are referred to in the literature as boundary spanners (Tushman & Scanlan, 1981). Through their networking activity outside the organizational boundary, these individuals are involved in the production, articulation and dissemination of knowledge which potentially can be translated into innovation within their own organization. Boundary spanners can thus act as technological gatekeepers (Allen, 1967; Aldrich, 1979). Gatekeepers "act as mediators, transforming external information so that it can be digested within the organization" (Macdonald and Williams, 1993; p. 425). Boundary spanning is depicted as a two step process (Allen, Tushman & Lee, 1979; Tushman & Katz, 1980; Hage, 1980). Firstly, boundary spanners need to have access to new ideas through being involved in a variety of inter-organizational networks. Secondly, they need to be able to communicate these new ideas into terms that are meaningful and useful to their internal colleagues so that they influence decisions relating to innovation. However, boundary spanning is not simply a question of individuals receiving knowledge broadcast from somewhere 'out there.' Rather, boundary spanners may also play a key role in articulating or constructing the knowledge base that is diffused via the networks in which they are involved. All organizational members are to some extent boundary spanners, but some individuals are clearly in a more central gatekeeping position in terms of penetrating a variety of external networks ***and*** also being influential internally within their organization. This may be either because of their formal position or function in the organization or because of their informal personal contacts.

In this chapter we concentrate mainly on the first issue, i.e., the communication of ideas through inter-organizational networks. The issue of the translation of these ideas into IT solutions has been dealt with elsewhere (Swan & Clark, 1992; Swan & Newell, 1994). Suffice to say here that it is mediated by cognitive and political processes of social actors within organizations. This section therefore reviews literature on inter-organizational networks as it pertains to the spread of new ideas and then focuses on a particular inter-organizational network - that created by a professional association.

Knowledge Communication

The Van de Ven (1986) definition given above emphasises that the innovation process involves using knowledge which already exists within a particular institutional context to create a new solution for an organization ('new' being a relative concept - the solution may be well-established in other firms). Much of the diffusion of innovations literature has thus focused on understanding the transmission of knowledge and its adoption or acquisition by potential users, i.e., on the process of communication (Rogers, 1962; 1983).

Information can be transferred among individuals by a process of communication. But communication which results in the furthering or development of knowledge is not simply about transferring information; it is a social process in which two or more people share information in order to come to a mutual understanding of reality. Communication is derived from the Latin word 'communis' which means common. When we communicate with someone we are trying to establish a 'commonness' with them, sharing information, an idea or an attribute. While complete mutuality of understanding may never be achievable, the continuous interaction among people leads over time to the development of shared meanings, underpinned by the convergence of their cognitive frameworks and articulation in the form of common languages and symbols (Schwenk, 1988). Communication without at least partially overlapping cognitive frameworks among those involved in the social exchange will not facilitate the development of knowledge. For example, if you listen to a lecture or read an article in an unfamiliar language, information is transferred, but this will not result in furthering your knowledge about the subject.

The development of knowledge through communication thus depends upon the prior knowledge of those involved built through past experience and reflected in their cognitive frameworks (Cohen & Levinthal, 1990). For example, Rogers (1983) notes that a decision about whether or not to implement a particular new idea depends on the individuals' evaluations about the relative advantages of the new idea over existing solutions: its complexity, trialability and so forth. These evaluations are subjective; based on past experience as well as on what has been learnt about the new idea in the process of communicating with others. The same information may, therefore, be perceived, understood and utilised very differently by two people.

Scarbrough (1996) suggests that there are basically three modes of knowledge communication which he refers to as professionalism, objectification and organizational sedimentation. Briefly, professionalism refers to 'the communication of knowledge through its embodiment in the learning and experience of individuals and groups'. Thus, through professionalism new ideas can be diffused through particular occupational groups or 'communities in practice' (Spender, 1992). Objectification refers to the process of creating standardised solutions which embody the knowledge and make it portable. Finally, organizational sedimentation refers to 'the knowledge communicated via rules, standards, routines and structures' (p. 999).

The earliest sociological models describing the innovation diffusion process emphasised the objectification mode of knowledge communication. The most well-known model was developed by Everett Rogers (1963, 1983). The model describes the diffusion process as an attempt by a central agency to create the universal application of a new idea or technology through standardisation. It is premised on the notion of a 'best practice' technology. New, improved technologies are generated, based on 'scientific' evidence, by some central agency. This central agency then engages in a process of communication with potential users who must be persuaded of the superiority of the new technology over existing practices. The knowledge that is communicated during this process is objectified in the form of technological artefacts which are, as far as possible, standardised and removed from any dependency on the organizational or social relations that exist locally in the context of the user. This can be termed 'blackboxing' (Scarbrough, 1996). An example would be the diffusion of a new software package or upgrade sold as a new form of 'best practice' with universal applicability. Communication focuses on the ability of the technology to improve existing operations rather than on unpacking the features of the technology per se.

There have been a number of criticisms of these models (e.g. Clark & Staunton, 1989). They underplay the importance of context and assumes that it is possible to identify a universal best practice. This objectification or 'blackbox' strategy can be seen to be one of the major causes of IT adoption failure. This is because in complex, dynamic organizational or technological contexts, knowledge relevant to the use of IT is organizationally or socially embedded. This knowledge is not easily surfaced, codified and made portable as a 'black box' solution. Technologies diffused by IT suppliers therefore require appropriation: that is the 'blackbox' needs to be opened, investigated, unpacked, and realigned alongside changes in organizational structures and routines in order to be implemented. Users persuaded by 'off the shelf', quick fix solutions may purchase technologies which they only partially understand, which are not necessarily suited to their particular organizational context, and which they find impossible to appropriate (Cooper & Zmud, 1990; Swan & Clark, 1992).

Objectified package IT solutions ('blackboxes') also distance the user from their design and development: the knowledge needed to further design such solutions remains with IT suppliers. Thus users who have bought particular packages can become locked into a dependent relationship with the IT suppliers. Importantly diffusion models underpinned by a notion of best practice and objectification depict a one-way communication process (from broadcaster - the central agency developing new ideas; to the receiver - the potential adopter). More recent research suggests that this is an oversimplification. Rather, as noted above, the communication processes involved in diffusion provide a means for the creation, social construction and articulation of new forms of knowledge and expertise (Alter and Hage,

1993; Scarbrough, 1996).

The professionalism mode of knowledge communication emphasises the ways in which members of a particular occupational group develop shared understanding of knowledge through their education and training. In a sense professionalism also operates on the basis of a broadcaster (teacher) - receiver (student) model. However, most established occupational groups will also develop forums (professional associations) in which communication continues among members once the formal education period is ended. Spender (1992) refers to this as a 'community in practice.' As long as individuals continue to identify with this group, they will form part of a network in which there is more or less regular communication, involving the sharing of ideas among members. Moreover, the professional identity can help to generate mutual trust, credibility and shared cognitive schema from which knowledge exchange might develop (Scarbrough, 1996). While the formal activities of such networks (e.g., journal articles, conference presentations, taught courses, qualifications) may follow a broadcaster-receiver mode of knowledge communication, members can also be involved in a variety of social informal communication exchanges. These communication channels are fluid networks rather than uni-directional informational flows. The framework presented in this chapter highlights professionalism as a mode of knowledge communication because the innovation diffusion literature has been dominated by objectification modes, and the professionalism mode has been relatively neglected.

The final mode of knowledge communication described by Scarborough is organizational sedimentation whereby knowledge is transferred by rules, standards, routines and structures. This mode refers to knowledge which is communicated internally and embodied in the organizational scripts and routines which influence day-to-day behavior. All innovation processes involve some sedimentation as the new IT solutions become embedded within a specific organization. The degree to which this mode of knowledge communication is activated is perhaps a function of the degree to which IT suppliers have attempted to create objectified solutions and the extent of professionalization within the firm. Where solutions are highly objectified there will be less scope for organizational sedimentation. For example, it would be extremely difficult for a firm to decide not to use the QWERTY keyboard design in a new computer system. Where solutions are less heavily objectified there is perhaps more scope for sedimentation. Further, where there are a diverse range of professionals within a firm there will be greater scope for organizational sedimentation as these individuals will bring with them knowledge from a variety of arenas that can be used to redesign and appropriate IT. An important point, however, is that some sedimentation will almost always occur. This is the reason why broadcaster-receiver analogies are limited—they underplay the extent to which users are active in the process of innovation, developing, articulating and implementing new ideas (Scarbrough & Corbett, 1992).

Professionalism and objectification both involve knowledge communicated via channels from outside the organizational boundary being diffused internally, albeit in very different guises: professionalism as ideas 'in the heads' of individuals; objectification in the form of specific technological artefacts. This knowledge diffusion process is therefore predicated on the existence of inter-organizational networks.

Inter-Organizational Networks

The links between individuals in inter-organizational networks can be very varied, from highly formal, business partnership type arrangements, to informal personal or friendship ties. In this respect, Grandori and Soda (1995) identify different forms of inter-organizational network, differentiating these by the extent to which the links are formalised and centralised and their characteristic mix of coordination mechanisms. Networks can be described as either 'bureaucratic', where there is a formal, contractual agreement, or social, where the links are informal and personal, or proprietary, where the link concerns ownership of intellectual property.

Providing another useful categorization, Aldrich and Glinow (1992) make a distinction between personal and social networks. Social networks refer to the population of people that are interconnected within a particular boundary, for example a professional or occupational group. Personal networks refer to the particular interpersonal ties which are actually activated within this wider social network. Thus, within any social boundary only some of the potential links between people will actually be activated. Some people within the social network will be 'isolates,' unconnected to anyone else in the network. There will also be 'cliques,' small groups of people who all link regularly to each other, usually because they share some common characteristic. For example, members of a professional group who work in the same industries or regions may form their own clique in the form of trade associations, or regional interest groups. Social networks may also have 'bridges,' or individuals who link two or more cliques by having links with both groups. These individuals are potentially very important within a social network as they can create 'weak ties' (Granovetter, 1973) which stimulate the diffusion of more radical ideas across cliques.

Aldrich and Glinow (1992) argue that some networks play brokerage roles, linking people with complimentary interests, transferring information or resources and generally stimulating links among people who would otherwise remain unconnected. Professional associations provide a good example of this brokerage role, linking people with particular professional interests who may work across a range of different types of industry and sector. For example, in the UK the Institute of Operations Management (IOM, formerly known as the British Production and Inventory Control Association - BPICS) links practitioners in a wide variety of industries, IT suppliers, consultants, academics, and so forth. Such broker-type relationships have parallels with the symbiotic networks described by Alter and Hage (1993). They argue that symbiotic networks, which link firms not in direct competition, are more likely to facilitate the joint production of knowledge and ideas than competitive networks, because trust can more easily be built where there is no direct competition. Similarly, professional associations link individuals who belong to firms that are not in direct competition with one another.

There are two important distinctions which these different typologies reveal. Firstly, the extent to which the network is formalised. This can range from a highly formal, contractual linkage between two or more firms to a social, informal personal tie between two or more individuals. Secondly, the extent to which the network links organizations or individuals with common interests and backgrounds, as opposed to linking people with very different interests and backgrounds. The professional association networks which are the focus of this chapter could be described as networks which link individuals (rather than organizations) on a largely informal (rather than formal) basis and which have the potential to create links between people from diverse types of organizations with different (rather than common) interests.

Implementation of New Ideas

New ideas that are diffused and then actually adopted by a particular organization do not necessarily translate into improved organizational performance. This can be seen in the many examples of organizations that adopt but fail to appropriate IT (e.g., Wilson et al, 1994). DiMaggio and Powell (1983) note that, where organizations are highly interconnected, they may adopt new ideas in order to gain legitimacy and power within their field rather that to improve efficiency directly i.e., the new idea is seen as 'the thing to do' among a particular community of organizations. In this way organizations that face the same environmental conditions can become more similar to each other. DiMaggio and Powell (1983) refer to this as the process of 'isomorphism' and suggest this is more likely to occur where strong links among organizations exist. As DiMaggio & Powell suggest "after a certain point in the structuration of an organizational field, the aggregate effect of individual change is to lessen the extent of diversity within the field" (p.65). The concept of isomorphism helps to explain why inefficient technologies can also be diffused (Abrahamson, 1991). Thus inter-organizational networks, whilst enabling faster diffusion of ideas, may paradoxically constrain innovation by creating social and operational norms that serve to restrict the types and range of new ideas that are generated and diffused. In support, Swan and Newell (1995), for example, found that professional association networks in the UK and North America encouraged the diffusion of information technologies for operations management among manufacturing firms but only promoted specific types of information systems that were legitimised as 'best practice' at the time. In fact, other types of information technology existed that may have been more appropriate for improving efficiency in particular firms.

This discussion suggests that there is not a direct or simple relationship between the development of networks, diffusion of IT, and outcomes of firms' attempts to implement IT. Rather, an analysis of inter-organizational networks in the diffusion process needs to consider the different types of relationships among partners in the network, different modes of knowledge communication, and the organizational, institutional and societal contexts in which the networks operate.

The Role of Professional Associations

As seen, professional associations are particularly relevant to innovation diffusion because they can facilitate the emergence of symbiotic relationships as they bring together individuals from different industries and sectors, i.e. playing a brokerage role (Alter & Hage, 1993; Aldrich & Glinow, 1992). Professional associations also play an important role in the professionalism mode of knowledge communication, acting as a forum through which common understandings can be established though education and training for a particular professional group. The 'weak ties' possible through membership of a professional association have long been seen to be beneficial to the innovation process (Granovetter, 1973), providing people with ideas which are not bounded by their normal rationality. As Macdonald and Williams (1993) state: "..obvious sources of external information may well yield only obvious information, predictable information that fits readily within the organization because it is so similar to that already in use there" (p. 419). For more radical change, information from a variety of diverse sources is likely to be needed.

Individuals who engage in personal networking are more likely to be able to tap into a diverse range of sources, and professional associations create the arena in which such relationships can develop. However, they do not create the relationships per se. This will depend on the individual members' propensity to be actively involved with others in the association, outside their own network of industry-specific acquaintances. Thus individual members can take greater or lesser advantage of the brokering role of their association and, arguably, get more or less value from association membership. Some members are relatively passive in associations, paying their subscription and getting access to the formal knowledge base of the association as communicated through journal articles. Others are more active, perhaps attending seminars and conferences. Research has demonstrated that more active involvement by an individual in a professional association's activities is associated with greater technological development within that individual's firm, in those functions most clearly related to the knowledge domain of that particular association (Newell & Clark, 1990; Swan & Newell, 1995). The relative activity of the individual in a network is likely to depend in part on the extent to which they perceive themselves as getting any kind of personal benefit from this networking activity. A paradox is that firms may discourage or attempt to control individual networking because they fear that company secrets may be given away. Ultimately though, individuals judge what is useful to reveal or not to reveal through a network. This will depend in part on the degree of trust that exists between network partners.

The framework developed in this chapter presents professional associations as a central agency within the set of inter-organizational networks that potentially influence the diffusion of new ideas (King et al., 1994). Clearly other networks exist and play an important role in the innovation process but professional associations are examined in detail here because, as noted earlier, the professionalism mode of knowledge communication has received relatively little attention in the innovation diffusion literature. Professional associations are often established with the explicit aim of providing a network for encouraging the development of knowledge and expertise in particular problem domains (Lynch, 1989). They may also facilitate trust among members because their objective is to further the interests of an entire professional domain, not a specific company or sector. However, the roles of professional associations as enablers of innovation will depend on the body of knowledge that they promote and how relevant this knowledge is for helping firms to design, implement and use appropriate innovative solutions.

Professional associations in the Operations Management Knowledge Domain

Taking the example of the UK, operations management is an area with a large number of relevant professional associations. These include the IOM (the UK association most central to the operations management domain) but also the Institute of Mechanical Engineering

(IMechE), the Institute of Electrical Engineers (IEE - Manufacturing Division), the Institute of Logistics (IOL), the Institute of Purchasing and Supply (IPS), the Operations Management Association (OMA). Each association focuses on a somewhat different aspect of the entire logistics chain, which includes a number of processes - bringing in needed parts from suppliers (the focus of IPS), designing the manufacturing process (the focus of IMechE & IEE), planning how and when to manufacture the different parts (the focus of the IOM), and distributing the finished goods to customers (the focus of IOL). Each association also disseminates information about the latest technological developments in their area to members. However, given that new information systems increasingly necessitate integration of these different parts of the logistics process, there is now considerable overlap between the knowledge disseminated by these different associations. Thus, the knowledge domain of each of these associations has broadened and the boundaries between different associations have become blurred. For example, at the 1994 IOM annual conference there were several papers on supply-chain-integration - a topic which the IPS would also deem very much as part of its core.

The overlap in the knowledge domains addressed by these different logistics associations suggests a need for co-operative alliances to form between these associations. However, to date such alliances have not been very successful. Rather, each of these associations attempt to maintain their own 'market niche' and compete for both members and recognition. Where mergers among associations have occurred, these have tended in reality to result in one association being taken over by another (Drazin, 1990). There are a number of examples of this. For example, the Institute of Production Engineers, with a membership of approximately 18,000, 'merged' with the much larger IEE, but in doing so lost identity, power and membership. These examples suggest that these professional associations are continuing to encourage the production of a disciplinary based and explicit type of knowledge (described as 'Mode 1' knowledge - Gibbons, 1994). In contrast, the current pressure, in part from development in IT, is to integrate across functions within organizations. This requires a multi-disciplinary and tacit type of knowledge production (or 'Mode 2' knowledge - Gibbons, 1994).

Another issue is that professional associations, sometimes explicitly, attempt to establish rules and norms of operating (i.e., via the professionalism mode of knowledge communication). As a result, the likelihood of isomorphism is increased and some firms may adopt technologies that, for them, are not the most appropriate (DiMaggio and Powell, 1983). For example, during the 1980s, the American Production and Inventory Control Society (APICS) was very active in promoting a particular information system for operations management (known as Materials Requirements Planning - MRP, and later

Manufacturing Resources Planning - MRP2). In the US, the APICS network was used as a vehicle by software vendors and consultants to sell this particular version of 'best practice' to practitioners in manufacturing firms. Since other IT solutions for managing operations were available at that time, it could be argued that the association actually distorted the diffusion process by presenting a limited and partial account of 'best practice' that may have led firms to adopt MRP/MRP2 without considering the alternatives (DiMaggio & Powell, 1983).

Research Framework: Inter-organizational Networks and Innovation Diffusion

Figure 1 presents the framework, premised on the literature above, which has been developed to illustrate the importance of inter-organizational networks, and professional associations in particular, for understanding the diffusion of IT. The framework, emphasises the personal links created by individual firm members (acting as boundary spanners) and other organizations, rather than contractual links between a firm and another organization. The framework depicts inter-organizational networks (top of Figure 1) as central to the innovation process. In this case, the focal organization is a manufacturing firm (represented in the bottom half of Figure 1) and the focal technology is IT for operations management. The networks are shown in the top half of Figure 1, with the circles representing the various types of organizations with which a boundary spanner in a manufacturing firm may have contact. Ideas relevant to the development of IT for operations management can potentially be diffused through any one of these links. A central tenet of this framework is that firms are more likely to adopt new IT solutions where their members have developed a more extensive and diverse set of inter-organizational relationships (Kelly & Brooks, 1991) and that professional associations play an important role in facilitating such networking. Through these networks ideas needed to develop and implement information technology for operations management within firms can be created and diffused. However, the particular IT solutions communicated via these networks may be more or less appropriate for particular organizational contexts.

The innovation process within firms is depicted in the bottom of Figure 1 as a decisionmaking process comprising a number of decision episodes. These are adapted from Newell et al., (1993) and Clark et al, (1992) and include: agenda formation, where problems and potential solutions are defined; selection, where a particular technology and corresponding information system is chosen; implementation, where the information system is introduced to the firm and developed; and usage, where the information system is incorporated into new routines of organizational and technical practise. This decisionmaking process will be influenced by the objectified and professional knowledge communicated via the boundary spanners from outside the organizational boundary. This knowledge will be interpreted and used in the light of current organizational knowledge (sedimented knowledge), as found articulated in organizational scripts, routines, equipment, languages and so forth (Clark and Staunton, 1989). Through the process of innovation, these scripts may be modified, so affecting subsequent decisions. For example, at Time 1 'organizational wisdom' may say that it is better to develop a bespoke information system. A particularly problematic experience of this method of technology adoption, may modify this organizational 'recipe' for a subsequent system development, such that the firm 'learns' that buying and using unmodified packages is a preferable solution.

There are a number of general issues related to this framework which need to be borne in mind. First, the decision episodes are rarely, if ever, linear or sequential (there is very little evidence of this in the literature - Wolfe, 1994) but are likely to be iterative, circuitous, tortuous and have both intended and unintended outcomes (Koopman and Poole, 1994). Second, innovation entails both cognitive and political processes, particularly where decisions concern the adoption of IT that crosses specialist or disciplinary-based knowledge domains within firms. Different individuals with a variety of cognitive perspectives and political interests will be active in the decision arena at different times (Swan and Clark, 1992; Scarbrough and Corbett, 1992). Third, 'boundary spanners' play a key role in the innovation diffusion process (Tushman and Scanlan, 1981).

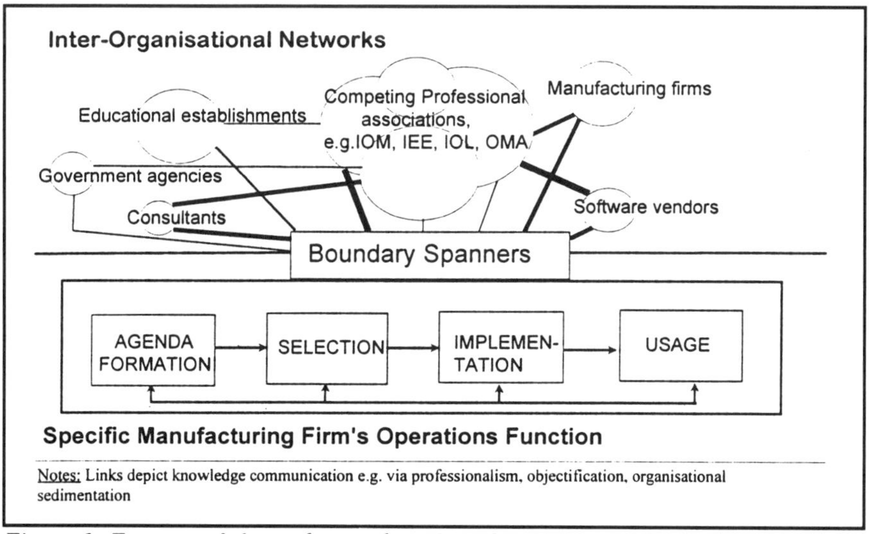

Figure 1: Framework for understanding the role of inter-organizational networks in the innovation diffusion process

The discussion that follows illustrates the framework and focuses upon networks that have been found to be influential in diffusing knowledge about IT for operations management. However, it is hoped that the framework will have more general applicability for analyzing inter-organizational networks in other contexts.

Methodology

In order to illustrate the operation of this framework data from three main sources will be presented. Firstly, data from two UK case studies will be presented. These case studies focused on identifying the importance of various inter-organizational networks in the knowledge diffusion process. The two case companies were similar in terms of manufacturing profile, establishment size and approximate timing of the adoption of a production control system. In both cases, an MRP2 system had been implemented in the late 1980s. Company M attempted to implement MRP2 for central planning and control of shopfloor operations. Company L attempted to implement MRP2 rather differently, with MRP2 being used for central planning alongside Just-In-Time (JIT) for control of shopfloor operations. Both companies operated in the vehicle manufacturing sector in the UK and both operated on a make-to-order basis. Both companies also operated in the same geographical region so that they had similar opportunities available for networking, at least externally.

Semi-structured interviews were conducted with individuals in the two companies who had been involved in the decisionmaking process leading to the implementation of the new operations management IT system. In both cases this involved interviewing individuals who had been involved in the project team which had been set-up to oversee the choice and implementation of the system. The purpose of the interviews was to explore the nature of relationships among social actors within inter-organizational networks and to consider how these relationships had influenced the diffusion process. Each interview lasted approximately one and a half hours and focused on agenda formation and selection and on the networks that informed decisions at that time. In addition secondary sources of information were used, such as minutes of meetings and company reports, to support interviewees' retrospective perceptions of the decisionmaking process. These case studies reinforced the importance of professional association networks in the innovation diffusion process (Robertson et al, 1996).

The second source of data explicitly considers the activities of professional associations and the knowledge communicated via this route. The most central operations management association (the IOM) is described in detail, focusing on the constitution of its membership and how involved different types of members were in association activities. Finally, the content of the journals of all of the major professional associations in the UK associated more broadly with the area of operations management are analysed covering the period from the mid-1980s to the mid-1990s. This analysis is conducted in order to establish what ideas regarding the adoption of information systems for operations management were being formally diffused by the relevant professional associations while the two case companies were making their decisions. The analysis also illustrates differences in the types of IT being promoted as best practice by these associations at different times.

Analysis

This analysis is divided into two parts. In the first part, empirical data is presented from two companies both involved in the implementation of a new IT system, one of which was relatively more successful than the other. This data shows that those involved in the two companies made use of the same kinds of inter-organizational networks, but that those from one of the companies tended to exploit these more actively. In particular, both companies made use of the same professional association as a focus for developing their ideas, but in the less successful company this involvement was much more limited. The importance of a more active involvement is demonstrated in the second part of this section where data on this particular professional association is presented. Here it is shown that through the formal channels of the association, the information disseminated can be somewhat restricted. More active nvolvement with a greater range of participants using more informal kinds of contacts, can help to counter this bias by opening access to a wider range of ideas and information. In addition, this second section looks at the ideas presented by different, but related, professional associations. This illustrates the partiality of what is presented by a particular professional association. Companies may therefore need to integrate the ideas from a range of professional groups. The data demonstrate that one of the case companies was better placed to do this because it involved a wider range of experts in its project team.

Company Case Studies

Individuals in the two case companies who had been involved in the decision to adopt MRP2 (i.e., in agenda formation and selection episodes) were interviewed about the source of the idea. The results are presented in a summary form in Table 1. This depicts the two

	Company M	Company L
Inter-organisational networks		
• Professional associations	• IOM	• IOM
• Educational establishments	• Local University	• Local University
• Manufacturing Consultants	• IT Ltd.	• IT Ltd.
• Other manufacturing firms (reference sites)	• Sites selected by software vendors	• Informal visits
• Software vendors	• Single	• Multiple
Internal networks		
• Project team	• Logistics manager; Special projects manager	• Logistics Director; 4 senior managers; senior external management consultant.
Outcomes		
• Level of integration achieved	• Partial integration	• Full integration

Table 1: Inter-organizational networks used in the agenda formation and selection episodes of an MRP2 adoption in two UK companies

companies' involvement in the inter-organizational networks outlined in Figure 1 above (names are not the original to protect anonymity).

Inter-Organizational Networks. Table 1 suggests that the same kinds of external networks were accessed by the two companies. However, this masks subtle but important differences between the two companies in terms of their use of networks. These different patterns of network usage are explored next.

Professional Associations. Both companies referred to professional association networks as being an important stimulus for learning about MRP2 technology, in particular the IOM which at that time were involved in training and education for MRP/MRP2. Members of both project teams had involvement with the IOM. The Logistics Director, who was the project leader at Company L, had been an active member of the IOM and for many years. He was involved in different branch committees at various times, regularly attended monthly branch meetings, and had presented papers about logistics at the annual IOM conference. The Logistics Manager at Company M was not a member of the IOM himself, but had access to the journals and was made aware of IOM events by one of his staff who was an IOM member. As a guest of IOM, he was able to attend free seminars about operations management. These were given by software vendors and manufacturing consultants who at that time tended to objectify MRP2 as a universally applicable best practice standardised solution (e.g., Wight, 1984). Arguably Company L benefited more from involvement in the professional association network, as there was the chance for social and informal links to develop between the project team leader and other IOM members at a variety of IOM events. Knowledge from the association was derived over a longer period of time and was provided via both formal and informal routes. Company M derived knowledge from a more limited source within the association represented by the formal presentations given by IT suppliers.

Educational Establishments. Both companies also had established links with the same Local University and this promoted a more detailed understanding of MRP2. The Special Projects manager at Company M had taken an MBA course which had included a ten week course focusing on the use of MRP2 information systems for achieving optimal control. The course was actually delivered by a visiting lecturer from the US, who was also a manufacturing consultant. This member of the project team derived most of his knowledge about MRP2 from this University course. Company L had formal research arrangements with the University and these had promoted the development of a number of informal links with faculty members across the University. Thus members of the project team at Company L could also discuss informally logistics/operations management issues and problems with a range of academics. Again, it is clear that Company L had access to a variety of informal information sources at the University whereas Company M derived information from a single, more formalised source.

Manufacturing Consultants. Manufacturing consultants were an obvious network for both companies to use in the decision making process. Members of the project teams in both companies became aware of 'IT Ltd.', an international consultancy specialising in the provision of MRP2 education and training, through the consultant's own publicity material, mailed direct, and through extensive advertising in IOM journals. Both project teams believed that information from this organization could contribute to the adoption decision and arranged to meet IT Ltd. consultants. IT Ltd. emphasized the need for an in-depth understanding of MRP2 concepts at all levels within the company, commencing with the Board. Following presentations to the Board in both companies, the decision was taken to invest in education and training with IT Ltd.

IT Ltd. promoted a standardized methodology for implementing MRP2 including a checklist to determine that all of the modules comprising MRP2 were implemented as prescribed. Thus, this is another example of the objectification of knowledge, in this case about MRP2 implementation, as standard packaged solutions. However, this objectification of the process under-emphasized the importance of the organizational context in the appropriation of MRP2 systems, discussed earlier. Whereas Company M fully committed to the package solution offered by IT Ltd., Company L recognised that they were being presented with an objectified solution that did not neatly fit their manufacturing environment. Company L therefore used the information and training provided by IT Ltd. selectively, blending it with their own in-house training and education and deciding to use MRP2 alongside JIT (an approach that was not being recommended by IT suppliers at that time).

Manufacturing Firms. Both firms were aware that other firms' practical experience of using MRP2 technology would be an important source of information. Company M had purchased an information system three years previously, which they were given to believe (see software vendors), would support MRP2. The team therefore developed informal links with one other manufacturing firm which was using the same software package, in order to assess suitability. This firm operated in a different industrial sector and its manufacturing profile was mass production rather than make-to-order which led to problems of comparability when trying to assess suitability. The Logistics Director at Company L was aware that a range of technologies existed other than just MRP2 to support operations management. He actively encouraged all internal members of the project team to use their informal links with other firms and arrange site visits to look at and discuss their operations management function and the technologies they were using. The external member of the team, working for a leading UK consultancy, was also asked to arrange site visits to firms who were operating with a make-to-order profile. Through this activity, Company L developed informal links with a number of firms operating in a variety of industrial sectors and acquired knowledge about a range of technologies for operations management and their applicability. This was, in part, what led to their decision to use MRP2 with JIT.

Software Vendors. As mentioned in the previous section, Company M already had a software package that was being used in a limited capacity for one part of the production process. Both IT Ltd and the vendor that had sold the original software suggested that this package would quite easily support MRP2 technology in an integrated way across the whole site. Following this advice, Company M became largely dependent on the software vendor who had the knowledge needed to develop the existing IT package. This provides a good example of problems associated with the 'blackboxing' strategy discussed earlier. In contrast, Company L drew up a list of six software vendors and asked them to tender for the supply of software to support MRP2. The external consultant who was a member of the project team provided advice as to which vendors should be approached based on his knowledge of the current platform Company L were using and his knowledge of the software industry. Following submission of tenders and presentations by all six vendors, the team visited one reference site provided by each one of a shortlist of three vendors, before making a decision on the final choice of package.

Internal Networks: The Sedimentation of Ideas. It is important to consider the constitution of the project team in each company and its links with internal networks in order to more fully understand the innovation process. How information and knowledge acquired by the project team was received and diffused internally ultimately affected the implementation of the technology in these firms.

The team at Company M represented a rather narrow range of professionalisms and relied on knowledge and information mainly acquired from suppliers of technology (vendors and consultants) in order to reach a decision regarding the adoption of MRP2. Both members of the team discussed with managers from other functions, the implications of the introduction of MRP2, but they had no direct professional experience or knowledge of other functional areas. The team reported directly to the Board who, apart from the MD, were all new external appointees. Their knowledge of MRP2 and company processes was limited and they were reliant on the project team for information and recommendations. Conversely, the team at Company L represented a range of professional specialisms relevant to the logistics process. Each member brought their own expert knowledge to the decision forum and were able to assess the impact of the innovation adoption decision on their own functional area, their assessments being based on both their own direct experience of organizational processes, and on information and knowledge derived from a number of inter-organizational networks including other varied manufacturing firms. This team also benefited from the inclusion of an external member who provided both expert IT and industry-specific knowledge. It can be argued then, that Company L benefited more from knowledge communicated via the professionalism mode than Company M.

Outcomes

While success of innovation with IT is difficult to measure, it was clear that Company L experienced relatively more positive outcomes than Company M in terms of the degree to which the IT was appropriated (Clark & Staunton, 1989). Company L appropriated MRP2 technology using a software package which interfaced with a PC-based bespoke system which was being used for shop floor data collection. Detailed shop floor control was achieved using JIT technology which interfaced with the information systems thus achieving full integration. Company M experienced significantly more problems. Despite recommendations to the contrary, the software did not support a fully integrated MRP2 system, and significant enhancements and modifications to the package were required. The software vendor was unwilling to accommodate these extensive revisions, because they were developing a new package at that time to replace the existing one. Not having the knowledge needed to develop the system in house, Company M were locked into a dependent relationship with an uncooperative partner. Ultimately individual software modules were installed which provided some of the functionality of MRP2 and some interfaces were made with existing manufacturing systems. However, only a partially integrated system was achieved.

This differential success can at least partially be explained by the different usage made of the inter-organizational networks tapped by members of the two project teams. Two factors appear to be crucial in this respect. Firstly, size and make-up of the project team may well have been a significant factor. By involving a more diverse range of professional experts in the project team, Company L had access to information from a greater range of professional sources and exhibited more diverse boundary spanning activity than Company M. The more detailed analysis below, will suggest that this is important because the different, professional associations, related to this knowledge domain each present only a limited range of ideas.

Secondly, the nature of members' involvement in inter-organizational networks appeared to be a major difference. The networks identified by those involved in the two companies were remarkably similar, supporting the relationships depicted in Figure 1. However, Company L appeared to appreciate the importance of more active involvement in these networks than Company M. The team at Company L were more likely to tap informally into these networks, as well as formally, thus sharing ideas via a much more fluid process which was likely to provide a means for the creation, construction and articulation of new forms of knowledge and expertise (Scarbrough, 1996). In contrast, Company M was more passive, 'receiving' information that was being 'broadcast' from these different sources, especially consultants, software vendors and professional associations. Whereas Company L questioned, unpacked and redefined the solution presented to them by the vendors, Company M were willing to accept the 'blackbox' solution. These differences are summarized in Table 2.

Table 2 shows that both companies identified contact with the IOM as an important source of information about new technologies in operations management. But Company L also used a broader range of professional expertise and more informal networking activities. The reasons as to why this was important can be more clearly understood by studying the activities of professional association which both these companies identified as being an important source of their initial ideas about the particular IT implemented. The second part of this section therefore considers the activities of the IOM in more detail and compares the ideas disseminated by this association with other professional associations in the UK, more broadly relevant to the domain of operations management.

Professional Association Networks

Both companies identified a particular professional association - the IOM - as being crucial to the early awareness and understanding of the MRP2 concepts subsequently implemented. However, as seen,

	Company M	Company L
Inter-organisational networks • Professional associations	• IOM contact - but not member and only attended formal events	• IOM contact - long-term member benefiting from wide range of social and informal links, as well as attending formal events
• Educational establishments	• Local University - attended MBA as student so restricted to formal presentation of ideas in lectures	• Local University - research partnership between company and university providing range of opportunity for formal and informal links
• Manufacturing Consultants	• IT Ltd. - accepted standard package solution offered	• IT Ltd. - used only parts of offered solution integrated with own in-house education and training
• Other manufacturing firms (reference sites)	• One firm using same package - visited and informal links developed, but not really comparable in terms of its operation	• Informal visits to range of firms - operating in variety of sectors to acquire broad view of options available
• Software vendors	• Single source - as 'locked' into relationship from previous IT implementation	• Multiple sources - 6 vendors asked to tender and present and provide reference sites
Internal networks • Project team	• 2 members with limited functional & professional representation	• Broad range of functional & professional representation
Outcomes • Level of integration	• Limited success - Partial integration	• High success - Full integration
• Implementation	• Adopted but did not fully implement MRP2 packaged solution	• Adopted MRP2 solution and appropriated this with JIT

Table 2: Comparison of the inter-organizational networks used during agenda formation and selection episodes of MRP2 adoption in two UK companies

the nature of their involvement with this particular association was rather different. No one in the project team at Company M was actually a member of the IOM and involvement was restricted to formal presentations as a guest. A member of the project team in Company L was a long-standing member who, as well as attending formal events, had a wider range of informal links with other individuals within the association. The importance of this difference is highlighted by looking more specifically at the information diffused through the networking activities (formal and informal) of the IOM. This section therefore analyzes in more detail the activities used by this association to disseminate information, the boundary spanners involved in such activities, and the nature of the information about IT for operations management that was disseminated.

The Institute of Operations Management (IOM). The IOM currently has about 4500 members, this number having increased marginally over the last ten years. The association uses a number of formal activities to disseminate information. Each member receives two membership journals, 'Control' and the 'Journal of Production and Inventory Management'. In addition, the IOM organize a range of events, including courses, seminars and meetings. The major event is the annual conference which attracts around 500 participants. Table 3 illustrates clearly that although this professional association is dominated by practitioner members, the bulk of the information that is disseminated by these formal routes is communicated via the much smaller group of consultant and vendor members (software and hardware vendors) and academic members. Practitioner members do get involved in disseminating new ideas, but they are very under-represented in this role. This raises potential problems as far as practitioners (i.e., users of IT) are concerned. It is, understandably, in the interests of IT consultants and vendors to present solutions as best practice packages that can easily be slotted into any context because this will help them to increase sales. However, as seen in the case of Company M, firms often fail to appropriate adopted IT solutions. This suggests that the role of professional associations in this domain is complex. They serve to establish a professional mode of knowledge communication which might benefit user firms, but the knowledge communicated through their formal activities is also subject to objectification and simplification by IT suppliers, which might prove problematic.

A related point here is that stories of failure of IT tend to be rarely discussed in the formal presentations organized by professional associations (for example at conferences). Again, this may reflect the rosy pictures of best practice painted by the supply side or it may result from a reluctance on behalf of practitioner members to publicly announce incidents, which they themselves may have been involved in, where technology has failed. An individual is much more likely to gain access to these stories through informal contacts with other

Percent %	Membership	Conference participation	Conference presentation	Journal article contribution
Practitioners	70%	60%	26%	20%
Consultants & vendors	23%	31%	55%	63%
Academics	5%	7%	16%	14%

Table 3: Percentage of different types of IOM members involved in various IOM activities

members with whom they have developed some degree of trust and who have themselves been involved in unsuccessful implementation projects (of which there are many - Wilson et al., 1994). A crucial implication for the two companies reported above is, therefore, that the individual from Company M who used formal presentations as his main source of information would have heard an objectified and simplified account of the IT system presented by IT suppliers and consultants. In contrast, the individual from Company L had more informal and long standing contacts and is likely to have heard, in addition, stories from other practitioners who had been through the implementation process themselves and who understood, and were prepared to discuss, the real problems of trying to introduce new ideas into different organizations. It could be argued then, that the project team in Company L was exposed to a richer and more realistic understanding of the IT, not just the idealized 'blackboxed' solution that Company M were presented with.

Analysis of Professional Association Journals. Another way of considering the role of a professional association is to analyze the types and diversity of ideas being presented through its main dissemination medium, its journals. The focus of the IOM has been concentrated on the production and inventory control (PIC) aspect of the logistics chain. As technologies in this area have integrated PIC with other parts of the logistics chain, so the focus of the IOM has broadened over time. The society finally decided to change its name from BPICS to reflect this. Other professional associations covering different parts of the logistics chain in the UK have done the same thing. For example, the current Institute of Logistics was the result of a merger between the former Institute of Physical Distribution Management and the Institute of Materials Management. Nevertheless, each association, including the IOM. continues to specialise in a fairly narrow knowledge domain, despite the growing need to integrate this knowledge at the practical level (Gibbons, 1994). This is reflected in the coverage of articles in their respective journals.

The journals of the four major professional associations in the Logistics knowledge domain in the UK were content-analysed, looking at articles disseminated over a ten-year period from 1984 to 1994. A record was kept of articles whose content was related to the following categories (articles in other categories were not recorded):

MRP2: if the article was discussing Manufacturing Resource Planning technologies.
JIT: if the article was discussing Just-In-Time technologies.
MRP2+JIT: If the article was discussing combining MRP2 and JIT technologies.
SCI: If the article was discussing the concept of Supply Chain Integration.

PHIL: If the article was discussing more general operations management philosophies such as Business Process Reengineering, Total Quality Management.

Table 4 shows that the four different professional associations have disseminated information about IT for operations management very differently, in terms of what they have promoted at different times. For example, the IOM and the IEE have been much more active in promoting MRP2 than either the IPS or the IOL. However, the attention of the IOM shifted somewhat from MRP2 to JIT across the two time periods. In contrast, the IPS has focused more heavily on JIT and SCI throughout both time periods. These findings suggest that professional associations disseminate partial information about IT that enhances the status or centrality of their members. For example, the IPS (with its focus on Purchasing) and the IOL (with its focus on distribution) are more likely to promote concepts such as JIT and SCI in which purchasing and materials management figure prominently. This analysis suggests that different professional groups develop different shared understandings about 'best practice' IT solutions. The challenge for the firms then is to harness the knowledge base of each group in order to inform the innovation decision episodes.

In terms of the two case companies, the differences between the functional representation and professional experience of members of the two project teams is, therefore, significant. In Company M there were only two members of the project team and they had a more restricted range of professional experience and knowledge. While they did talk to people in other functional areas they did not tap directly into these other bodies of knowledge, nor access information from the professional associations which represent them. On the other hand, the project team in Company L had representatives from different functional areas with a much broader range of professional backgrounds. While only the IOM was specifically mentioned as an important source of ideas for the MRP2 implementation, it is likely that knowledge of alternative solutions, being promoted by other profes-

	MRP2	MRP2	JIT	JIT	MRP2 + JIT	MRP2 + JIT	SCI	SCI	PHIL.	PHIL
	84-89	*90-94*	*84-89*	*90-94*	*84-89*	*90-94*	*84-89*	*90-94*	*84-89*	*90-94*
IOM	26	17	50	63	5	6	3	11	0	9
IOL	0	0	13	4	2	0	26	21	18	10
IPS	1	2	5	14	1	1	7	19	0	6
IEE	43	11	13	3	3	3	2	1	3	3

Table 4: Number of articles focusing on operations management technologies in professional association journals between 1984 and 1994

sional groups, like that represented by the IPS, indirectly influenced the team by alerting them to the possibility that MRP2 was not the only solution. This has been found elsewhere (Swan and Clark, 1992; Robertson et al., 1996). This larger project team, with their greater range of professional expertise, were better able to integrate the ideas which were being offered by different professional groups at that time.

Discussion and Conclusions

The two case studies demonstrated the importance of inter-organizational networks for the development of new ideas that can lead to new IT solutions (in these cases the introduction of MRP2). It appeared from the comparison of the two cases that it was important to develop a number of links with various other types of organizations and to develop informal as well as formal links. Company M which mentioned fewer inter-organizational links and which tended to rely more on formal types of network relationships appeared to be less successful in implementing the new technology because those involved had a less comprehensive understanding of both MRP2 and the alternative solutions open to them. Clearly, it is not possible to generalize from two case studies. However, other research, using larger sample sizes, supports this conclusion (e.g. Newell & Clark, 1990; Swan & Newell, 1995; Axelsson & Easton, 1992; Kelly & Brooks, 1991).

The two case studies also demonstrated the important role that a professional association played in diffusing knowledge about the particular technology of MRP2. Professionalism and objectification modes featured in the knowledge communication process. An important feature of professional associations is that they can provide links among members from different industrial product markets (e.g., the IOM has members from the chemical, textile, car industries, etc.). This provides the opportunity for individuals to discuss ideas with people who might have very different perspectives. This is more likely to lead to radical solutions than is discussing ideas with someone from the same industrial sector (Macdonald & Williams, 1993). There may also be more open exchange of ideas among individuals in these networks because they see themselves as coming from firms which are not in direct competition with each other and therefore, trading secrets is permissible (Alter and Hage, 1993).

Another important feature of professional associations is that they can bring together individuals from different occupational sectors (manufacturing industry, academia, consultancy, etc.). This may again stimulate the development and diffusion of more radical ideas. The analysis of the IOM membership, however, highlighted a potential problem. The majority of IOM members worked in manufacturing industry and a minority worked as consultants or software suppliers but this minority was more active in the association than the practitioner majority. They tended to dominate in terms of the presentations at formal events and writing the material published in journals. They also had, and have, a strong presence on national and regional organizing committees. The knowledge communicated through the formal activities of professional associations is more likely to be objectified by IT suppliers in whose commercial interests it is to use 'blackboxing' strategies to encourage firms to adopt particular standardized IT solutions (Scarbrough, 1996).

Finally, the analysis of the various professional associations which relate to the knowledge domain of operations management demonstrated the potential problems of a contested knowledge domain (Abbott, 1988). Professional association networks, while acting as a major vehicle for the promotion of new ideas, may purvey knowledge which is bounded both by the desire to defend a particular knowledge domain (so perpetuating a Mode 1, discipline-based type of knowledge production - Gibbons, 1994) and by the political interests of particular members. The problem is that members often perceive the knowledge diffused via professional association networks as open, objective and 'factual' rather than as knowledge that is socially constructed by individuals from different occupational groups with vested interests in the promotion of selected IT solutions (Bijker, Pinch & Hughes, 1987). Perhaps, then, it is the informal networks created through joint membership of professional associations which are more helpful in encouraging the diffusion of innovation. The formal activities create a forum in which individuals from different industry sectors and occupational groups can meet informally and share ideas and experiences about IT successes but also, importantly, about IT failures. In this way, knowledge can be shared, jointly constructed and articulated (Scarbrough, 1996), rather than merely transferred or exchanged.

The roles of professional associations across countries can be rather different (Child & Fores, Glover, & Lawrence, 1983). Interviews by the authors of members of professional operations management associations (comparable with the IOM) in Sweden, France and the Netherlands indicate that, despite focusing on a similar knowledge domain, each professional association operates somewhat differently. In Sweden, for example, where there is a strongly developed egalitarian ideal, formal qualifications were considered to be somewhat elitist. The Swedish association has not run a formal qualification programme until very recently. This is very different to the UK, where the IOM has for the past 20 years run courses which lead directly to formal qualifications - certificates or diplomas - where MRP2 technology has figured prominently. In contrast, Sweden, the activities have generally been on a more informal basis with the association providing a forum for individuals with similar professional interests to meet and discuss issues (it was described by one of the interviewees as "a professional club"). Interestingly, standardized MRP2 packages have been slower to diffuse in Sweden and have also been less heavily emphasized in the activities of the professional association (Swan, Newell and Robertson, 1996). More recently, the Swedish association has introduced a certification programme imported from the association in the United States (APICS) - the Certificate of Production and Inventory Management (CPIM). However, only a handful of members have this qualification. The majority who took it initially were software vendors and consultants who gained legitimacy through having such credentials. The Swedish association does run an annual conference, but unlike the UK association, this has only recently involved a formal 'call for papers.' Previously, consultants, software vendors and academics had been invited to give papers. A related feature of the Swedish association is that the membership itself is even more heavily dominated by consultants and software vendors than its counterpart in Britain.

Conversely, in France, there appears to be a strong emphasis on professionalism through formal qualifications as a means for establishing status and position in the organizational hierarchy. This is reflected, for example, in the elite status attributed to qualifications obtained from the Ecoles Supérieurs and the Ecoles de Commerce over other university degrees. In fact, some interviewees commented that they utilised their exclusive 'alumni' network formed by their respective Ecoles when seeking out new ideas (alumni networking was never mentioned in the UK). The professional CPIM qualification has also proved very popular, with around 600 people having taken or taking the CPIM since it first became available three years ago (a very different situation to the one in Sweden described above). The majority who take this qualification are relatively senior managers in industry and may therefore be in good positions as gatekeepers. Interestingly, this qualification was attributed status because it was recognized as filling an educational niche specializing in operations management not covered by existing traditional engineering programs, and also because it was taken in English! Unlike the UK, French members taking the CPIM also meet regularly (monthly) in study groups where there are possibilities to develop more informal types of relationships for sharing and exchanging information.

Finally, in the Netherlands the association operates in a highly formalized way with an emphasis on education and training and the diffusion of knowledge through formal channels. Education and training is carried out by a separate organization which administers the APICS qualification, the CPIM. This certification has been offered in the Netherlands since 1979 and is now nationally recognized. Consequently the largest group of CPIM qualified practitioners in Europe are found in the Netherlands. The association also arranges formal activities across five regions in the Netherlands and a monthly journal is produced in conjunction with the Netherlands Purchasing Association, with which it has a close working relationship. A Research Council exists to perpetuate the acquisition and dissemination of knowledge. Ad hoc special interest groups are created from volunteers who will meet regularly over a period of months to write reports on new areas of interest and relevant logistics problems. These reports are published and are made generally available to anyone working in the logistics field. Informal networking is encouraged through an annual directory of members is produced and sent to all members for their personal use. It contains member profiles, details of any special interests and contact telephone numbers. The general pattern in the Netherlands, then, is one of an association which is active both in encouraging formal and informal networking among members working in operations management, and in forming broad alliances with other professional groups relevant to the wider domain of logistics.

This discussion, while focusing on the network provided by professional associations, suggests that wider national and societal contexts the nature and importance of inter-organizational networks in the diffusion of information technologies. Application of the kind of framework presented here to different countries may reveal interesting differences in the types of inter-organizational networks and the nature of networking relationships that shape the diffusion of IT.

Acknowledgements

The authors would like to thanks the Economic and Social Research Council, Swindon, UK, for providing the funding needed to carry out this research.

References

Abbott,A. (1988). *The system of professions: An essay on the division of expert labour.* Chicago: University of Chicago Press.

Abrahamson,E. (1991). Managerial fads and fashions: The diffusion and rejection of innovations. *Academy of Management Review, 16*(3), 586-612.

Aldrich,H. (1979). *Organizations and environments.* Englewood Cliffs,NJ: Prentice-Hall.

Adlrich,H. & Glinow,M.A. (1992). Personal networks and infrastructure development. In: *The Technopolis Phenomenon: Smart cities, fast systems, global networks.* D.V. Gibbon, G. Kozmetsky, & R.W. Smilor (Eds.), pp. 125-145. New York: Rowman & Littlefield Publishers.

Allen,T. (1967). Communication in the research and development laboratory. *Technology Review, 70*(1), 31-37.

Allen,T; Tushman,M. & Lee,D. (1979). Technology transfer as a function of position in the spectrum from research through development to technical services. *Academy of Management Journal, 22*(4), 694-708.

Alter,C. & Hage,J. (1993). *Organizations working together.* Newbury Park, CA: Sage.

Axelsson,B. & Easton,G. (1992). *Industrial networks: A new reality.* London: Routledge.

Bijker,W.E., Pinch,T.J. & Hughes,T.P. (1987). *The social construction of technology.* Cambridge, MA: MIT Press.

Child,J; Fores,M; Glover,I; & Lawrence,P. (1983). A price to pay? Professionalism and work organization in Britain and West Germany. *Sociology, 17*(1), 63-78.

Clark,P.A. (1987). *Anglo-American Innovation.* New York: De Gruyter.

Clark,P., Bennett,P., Burcher,P., Newell,S., Swan,J., & Sharifi,S. (1992). The decision-episode framework and computer aided production management. *International Studies of Management and Organization, 22*(4), 69-80.

Clark,P. & Staunton,N. (1989). *Innovation in technology and organization.* London: Routledge.

CohenW.M. & Leventhal,D.A. (1990). Absorptive capacity: A new perspective on learning and innovation. *Administrative Science Quarterly, 35,* 128-152.

Cooper,R.B. & Zmud,R.W. (1990). Material requirements planning systems infusion. *OMEGA International Journal of Management Science, 17,* 471-481.

DiMaggio & Powell (1983). The iron cage revisited: Institutional isomorphism and collective rationality in organizational fields. *American Sociological Review, 48,* 147-160.

Drazin,R. (1990). Professionals and innovation: structural-functional vs. radical-structural perspectives. *Journal of Management Studies, 27*(3), 245-263.

Gibbons,M. (1994). *The new production of knowledge. The dynamics of science and research in contemporary societies.* London: Sage.

Grandori,A. & Soda,G. (1995). Inter-firm networks: Antecedents, mechanisms and forms. *Organization Studies, 16,* 183-214.

Granovetter, M.S. (1973). The strength of weak ties. *American Journal of Sociology, 78,* 1360-1380.

Hage,J. (1980). *Theories of organization: Form, process and transformation.* New York: Wiley.

Hauschlidt,J. (1992). External acquisition of knowledge for innovations - A research agenda. *R & D Management, 22,* 105-110.

Kelly,M. & Brooks,H. (1991). External learning opportunities and the diffusion of process innovations to small firms. *Technological Forecasting and Social Change, 39,* 103-125.

King,J., Gurbaxani,V., Kraemer,K., McFarlan,F.W., Raman,K.S. & Yap,C.S. (1994). Institutional factors in information technology innovation. *Information Systems Research, 5*(2), 136-169.

Koopman,P.L. & Poole,J. (1994). Decision-making in organizations. In: *Key Reviews in Managerial Psychology: Concepts and research for practice.* C.Cooper and I.Robertson (Eds.). Chichester: John Wiley and sons.

Lynch,J. (1989). Looking overseas for new members. *Association Management,* March, 110-115.

Macdonald,S. & Williams,C. (1993). Beyond the boundary: An information perspective on the role of the gatekeeper in the organization. *Journal of Production and Inventory Management, 10,* 417-427.

Newell S & Clark P (1990). The importance of extra-organizational networks in the diffusion and appropriation of new technologies: the role of professional associations in the USA and UK. *Knowledge: creation, diffusion, utilisation, 12*(2), 199-212.

Newell,S. & Swan,J. & Clark,P. (1993). The importance of user design in the appropriation of new information technologies: The example of PICS. International *Journal of Operations and Production Management, 13*(2), 4-22.

Robertson,M., Swan,J. & Newell,S. (1996). The role of networks in the diffusion of technological innovation. *Journal of Management Studies,* .

Rogers,E.M. (1962, 1983, 3rd Edition). *Diffusion of Innovations.* New York: Free Press.

Scarbrough,H. & Corbett,J.M. (1992). *Technology and Organization: Power, meaning and design.* London: Routledge.

Scarbrough,H. (1996). *The management of expertise.* London: Macmillan Business Press.

Schwenk,R.C. (1988). The cognitive perspective on strategic decision making. *Journal of Management Studies, 25*(1), 41-55.

Spender,J. (1992). Knowledge management: Putting your technology strategy on track. *Management of Technology III.* New York: Institute of Industrial Engineers.

Swan,J. & Clark,P.A. (1992). Organizational decision-making in the appropriation of technological innovation: Cognitive and political dimensions. *European Work and Organizational Psychologist, 2,* 103-127.

Swan,J. & Newell,S. (1994). Managers' beliefs about factors affecting the adoption of technological innovation: A study using cognitive maps. *Journal of Managerial Psychology, 9*(2), 3-11.

Swan,J. & Newell,S. (1995). The role of professional associations in technology diffusion. *Organization Studies, 61*(5), 847-874.

Swan, J.A., Newell, S. and Robertson, M (1996). The illusion of 'best practice' in information systems for production management. *Proceedings of the European Conference on Information Systems.* Lisbon, July.

Tushman,M. & Katz,R. (1980). External communication and project performance: An investigation into the role of gatekeepers. *Management Science, 26*(11), 1971-1085.

Tushman,M. & Scanlan,T. (1981). Boundary spanning individuals: their role in information transfer and their antecedents. *Academy of Management Journal, 24*(2), 289-305.

Van de Van,A.H. (1986). Central problems in the management of innovation. *Management Science, 32*(5).

Wight,O. (1984). *Manufacturing Resource Planning: MRP2.* Vermont, Oliver Wight Publications.

Wilson,F., Desmond,J. & Roberts,H. (1994). Success and failure of MRP2 implementation. *British Journal of Management, 5,* 221-240.

Wolfe,R.A.(1994). Organizational innovation: review, critique and suggested research directions. *Journal of Management Studies, 31,* 405-431.

CHAPTER
12

Diffusion of Technological Innovations: Bringing Businesses onto the Internet

The Case Study of Connect

Colin Charlton, Chris Gittings, Paul Leng, Janet Little, Irene Neilson
University of Liverpool, England

This chapter explores the extent to which the exploitation of the Internet for Electronic Commerce can be adequately described by traditional models of the diffusion of an innovation.

Two levels of analysis are employed. First, the generalized mathematical model offered by Mahajan and Peterson (1985) to describe the diffusion of an innovation is examined. The impact of the Internet globally on Electronic Commerce is shown to be best described by the "internal influence" model. The key role the Internet plays in its own diffusion is emphasized. Mathematical models of diffusion are however shown to be limited. Within these models, diffusion is viewed primarily as a function of time. The specific social and organisational context in which a diffusion takes place is neglected. This restricts the value of such models when applied to the micro-level of analysis. In the latter context, the research framework that is predominantly used is that of descriptive analysis (Rogers, 1983).

We thus adopt a descriptive/analytic framework for our second level of analysis: a micro-analysis of the key factors that have influenced the adoption of Internet related technology at a local community level, that of the Merseyside region of the UK. Within this region, an Internet Centre for Merseyside Businesses, Connect, has been created to act as an instrument to accelerate the diffusion of Internet awareness and skills among the local community. The impact of Connect's activities on the process of diffusion is reviewed. The myriad of interacting factors which influence the diffusion process and the synergetic nature of their interactions are emphasised.

The chapter concludes by emphasizing the need for models of diffusion to become process oriented, to allow for evolution in both the characteristics of an innovation and the mechanisms through which it is diffused.

The Internet is widely recognised as an innovation of outstanding importance for the global world of business. The emerging "Information Superhighway," for which the Internet may be seen as a prototype, offers the prospect of a transformation in business practices comparable with that produced by the other major technological developments of the twentieth century: telecommunications, the television and the computer itself. It now appears likely that, in a relatively short span of years, most significant businesses in the developed world will not only be linked to the Internet but will be using it for a substantial proportion of their communications, promotional and marketing activities. Conversely, a widely accepted corollary to this is that businesses and societies which fail to take advantage of this development will be significantly disadvantaged. In this context, a study of the process by which this transformation will take place is of particular interest and importance.

Twenty years have passed since the U.S. Defense Advanced Research Projects Agency (DARPA) initiated the research programme[1] that gave rise to the system of networks we know as the Internet. Yet it is only now, in the 1990's that its potential as a vehicle for the support of commerce is being realised. Why has it taken so long? Can the traditional models of innovation diffusion explain this time lag? Is the Internet comparable to any other innovation in terms of the factors which have influenced its diffusion or does it possess unique features that challenge some fundamental assumptions of traditional models of Innovation and Diffusion?

In this chapter, we aim to explore the extent to which exploitation of the Internet for Electronic Commerce can be adequately described by the features traditionally identified in the literature as critical to the diffusion of an innovation. Our analysis will be performed at two levels: that of the impact of the Internet *globally* on Electronic Commerce, and that of the *micro-level* of a case study of the adoption of the Internet in the service of Electronic Commerce within a particular business community.

In the former context, the generalized mathematical model of the diffusion of an innovation as developed by Mahajan and Peterson (Mahajan and Peterson, 1985) will be explored. The fact that the

Internet is a computer network, means that *relatively* accurate statistics are available in terms of its growth. This circumvents a major problem with mathematical models of the diffusion process; namely, the reliability of the data on which they are based. Most statistical studies of the diffusion of an innovation have been retrospective, based on people's ability to recall when they adopted the innovation.

In the latter context, the case study of an initiative, Connect, undertaken by the Department of Computer Science at the University of Liverpool to promote the use of Internet related technologies by small businesses will be examined. Connect, the Internet Centre for Merseyside Businesses, is part of a regional economic regeneration program for the Merseyside area which was established to assist businesses in Merseyside, the region surrounding Liverpool, to take advantage of the opportunities that will arise through exploitation of the Internet and the emergence of the Information Superhighway. Connect is an example of a diffusion in progress, the features of which will be examined as to the extent to which they reflect those characteristics of the diffusion process identified in the meta-analysis of Rogers (Rogers, 1983).

Finally, the extent that the mathematical model of the diffusion of the Internet, based on *global* statistics, can be applied at a *local* level will be examined. Within this chapter this cross-referencing of the mathematical model is done at two levels. The first is in terms of its applicability to the development of a Public Domain Archive on the Internet, for which accurate statistics have been kept. The second is in terms of the consistency of the mathematical model with the factors known at the micro-analysis level to have influenced the diffusion of the innovation of the Internet to the business community in Merseyside. In making these comparisons, some general weaknesses in models of diffusion are highlighted.

Research Position and Theory

In this section we consider models of the diffusion process and why the Internet may be considered a special case of innovation. The particular analytic perspective adopted in this chapter is outlined.

Models of the diffusion of innovations

The process of diffusion of an innovation has been studied from a wide variety of perspectives. Research on the diffusion process has been reported in at least two dozen academic disciplines, including: sociology, economics, anthropology, geography and education, with each discipline applying its own characteristic methodology.

There does, however, seem to be a general agreement that there are four principal elements of the process by which diffusion of an innovation takes place (Rogers, 1983). These are:

- the innovation itself
- the channels of communication through which it is propagated
- the time over which this takes place
- the social system within which it takes place.

This agreement is reflected in the commonly quoted definition of the diffusion process, from the meta-analysis of Rogers (Rogers, 1983) p. 5 as:

"The process through which an innovation is communicated through certain channels over time among members of a social system."

Diffusion is frequently referred to as a special form of communication in that it is centerd upon the reduction of the uncertainty associated with the communication of a new idea (the innovation). Diffusion as a process is argued to centrally involve the provision of information.

Why the Internet is a Special Case of Innovation

The Internet is an innovation which has certain unique attributes. These raise interesting questions for models of its diffusion. Four attributes are described below.

The innovation itself provides a medium for its own diffusion

The first point of interest relates to the channels of communication by which an innovation is disseminated. In the case of the Internet, the innovation is itself a medium of communication which may indeed become of importance equal to, or greater than, all other such media. The Internet as an innovation is capable of acting as a medium for its own diffusion, including diffusion into organizations which have no direct access to it.

The innovation can be conceived as part of a global, worldwide social system

The second point is that the social system within which information about this innovation is being diffused is *global* and almost unconstrained: in one interpretation, it may include, eventually, every individual and organisation in the developed and developing world.

For the purpose of this study, we will take a narrower view of the social system as being the totality of businesses throughout the world. Within this definition, it is reasonable to assume that adoption of the innovation will be in the interest, eventually, of almost all members of the social system.

The time lag in adoption of the innovation

Thirdly, the Internet has been around for a considerable length of time and yet it is only recently, in the 1990s, that its potential as a vehicle for Electronic Commerce has been attended to.

This has been partly ascribed to the development of the World Wide Web as only the hypermedia environment realized by this interface permits the realization of one major facet of Electronic Commerce, namely, advertising and access by a large population base of customers.

The innovation is constantly changing

In the case of the Internet, we have an innovation which is itself evolving. While studies of diffusion have acknowledged that an innovation may be customized locally through a process termed "re-invention," few have attended to the problem of studying how change in the features of the innovation itself over time should be treated in studying its diffusion.

Analytic Perspective Adopted in this Chapter

Each of the aforementioned features will be considered in our analysis of the diffusion of the Internet. Our analysis will proceed at two levels.

Diffusion of the Internet at a global level

At a global level, the generalized mathematical model of the diffusion of an innovation as developed by Mahajan and Peterson will be explored. The available statistics on rates of adoption of Internet related technologies will be examined and the extent to which the process of adoption of the technology can be more appropriately described by an "internal influence" rather than "external influence" diffusion model will be discussed. Finally, the limitations of the Mahajan and Peterson model will be reviewed.

The micro-level analysis: a case study of Connect

At a micro-level of analysis, we will explore the adequacy of models of the diffusion process as applied to the activities of a change agent. This analysis will focus on the case study of Connect, an initiative to accelerate the diffusion of knowledge about the Internet to a local

	Framework of first level of analysis	**Framework of second level of analysis**
Level of Analysis	Macro-level	Micro-level
Social System to which the analysis is applied	World Business Community	Local Business Community in Merseyside, UK
Type of analysis	Mathematical modelling of the process of diffusion over time	Descriptive analysis of factors known to influence the diffusion process
Theoretical framework employed	Mahajan & Peters (1985) Generalised Mathematical Model	Rogers (1983) Meta-Analysis across Social Science disciplines
Empirical data	Statistics on the growth of the Internet	Participant Analysis Interviews Questionnaires

Table 1 : Analytic framework employed to study the diffusion of Internet related technologies among the business community.

business community. This analysis will be shown to highlight the need for models of the diffusion process to attend to the specific technical, organizational or social context in which a diffusion takes place and to accommodate the complexity of the interactions between factors that characterizes diffusion in practice. The analysis presented here is largely qualitative; monitoring procedures for the gathering of quantitative data are currently being established. The dimensions of the analysis center on key features of the diffusion process identified by Rogers (Rogers, 1983) meta-analysis.

Table 1 summarizes the main features of each level of analyses.

We begin our analysis by considering diffusion of the innovation at a global level.

Diffusion of the Internet at a Global Level

The Mahajan and Peterson model

In promoting the adoption of a new technology, the rate of adoption within the target system is of central importance. A generalized model to describe this is offered by Mahajan and Peterson (Mahajan and Peterson, 1985) as a differential equation:

$$\frac{dN(t)}{dt} = g(t)\,[\overline{N} - N(t)]$$

where N(t) expresses the cumulative number of adopters at time t; $\overline{N}$ is the total number of potential adopters in the target system; and g(t) is a coefficient of diffusion. Different manifestations of g(t) may describe characteristics of this diffusion process in different contexts. At one extreme, the coefficient of diffusion may be a constant. In this model, referred to as *the 'external influence' diffusion model*, the rate of adoption is influenced only by factors external to the community of adopters, such as a promotional campaign. The characteristic consequence is that rate of adoption is highest in the early stages of diffusion, declining as saturation of the target approaches.

In a different case, the coefficient of diffusion is expressed as bN(t), that is, it is linearly proportional to the current number of adopters. This case, referred to as *the 'internal influence' model*, is characteristic of a process in which diffusion is effected wholly by communication within the social system. The rate of adoption in these circumstances is accelerated by increase in the numbers of existing adopters who act as agents for further diffusion. Clearly, in other cases, a mixture of external and internal influences may apply, and the internal influence may be non-linear.

Evaluation of Mahajan and Peterson's model

Is this external or internal influence model more valid as a description of the diffusion of the Internet as an innovation amongst the business community?

Adoption of the Internet can be measured in several ways. For our purposes, we will take the number of Internet "host" computers as a convenient measure which is more stable than measurements which may be related to personal usage. The number of hosts is conventionally measured by the number of entries in the Domain Name System (DNS) and pre-DNS host tables. DNS entries are collected by a program called ZONE, which searches the Internet and retrieves data from all known domains. Pre-DNS host table data are retrieved from system archive tapes. Full details of this methodology are available in RFC 1296 (Lotter, 1996). The predominant function of a host computer, whether educational, commercial, military, etc. may be identified from its suffix i.e. .edu (ac), .com (co) and .mil respectively.

Our source data is from Network Wizards (http://www.nw.com) and Genmagic (http://www.genmagic.com/).

Figure 1, taken from Rutkowski (Rutkowski, 1996), illustrates the growth in the number of recorded Internet hosts over the period 1989-1996. It is clear from this graph that the growth pattern is closer to that of the *'internal influence'* model, and, indeed, that the coefficient of diffusion in this case is greater than linear in N(t).

Table 2, using figures provided by the Internet Society (ftp://ftp.isoc.org/isoc/charts), tabulates growth in the number of Internet hosts in North America, Western Europe and the Pacific region, taken separately, in 1994.

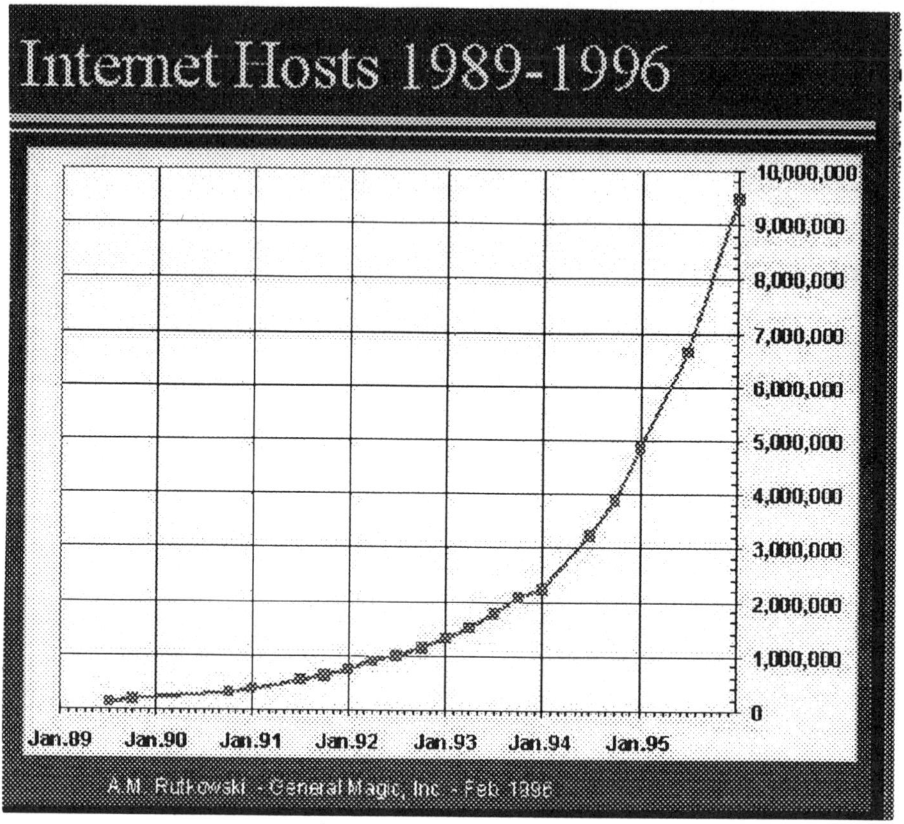

Figure 1: Growth in Internet Hosts 1989-1996 (Rutkowski, 1996)

If we make the simplifying assumption that the number of outstanding potential adopters remains approximately constant, then in each case the coefficient appears to be of order N^2 suggesting an 'internal influence' model which is very highly geared.

Applicability of the model to the local context

Does an "internal influence" model of adoption of the Internet also apply in a local context? From the data we hold the answer is "yes." The above characterization of the adoption of the Internet as a vehicle of communication is consistent with what we have observed in our local context of using the Internet to facilitate the distribution of public domain software for Hewlett Packard computers.

In 1992, the Department of Computer Science at Liverpool University established a Porting and Archive Centre for Public Domain software on Hewlett-Packard computers. This Centre has built up a library of over 1000 software packages which are made available over

Geographic Region	Number of hosts (in thousands)			Growth	
	Jan '94	July '94	Jan '95	Jan-July '94	July'94 - Jan '95
North America	1686	2177	3372	29%	55%
West Europe	551	730	1039	32%	42%
Pacific	113	142	192	26%	35%

Table 2: Regional Growth of the Internet during 1994[2]

the Internet from Liverpool and a number of associated partner centers throughout the world.

Figure 2 illustrates the growth in the number of packages withdrawn from the Archive throughout its existence until mid-1996. Bearing in mind that an Internet user will not normally need to withdraw a package more than once, we may see this graph as a representation of the diffusion of the software archive throughout the Internet-using community.

Obviously, any interpretation of this graph is complex, as the growth illustrated compounds global Internet growth with growth in

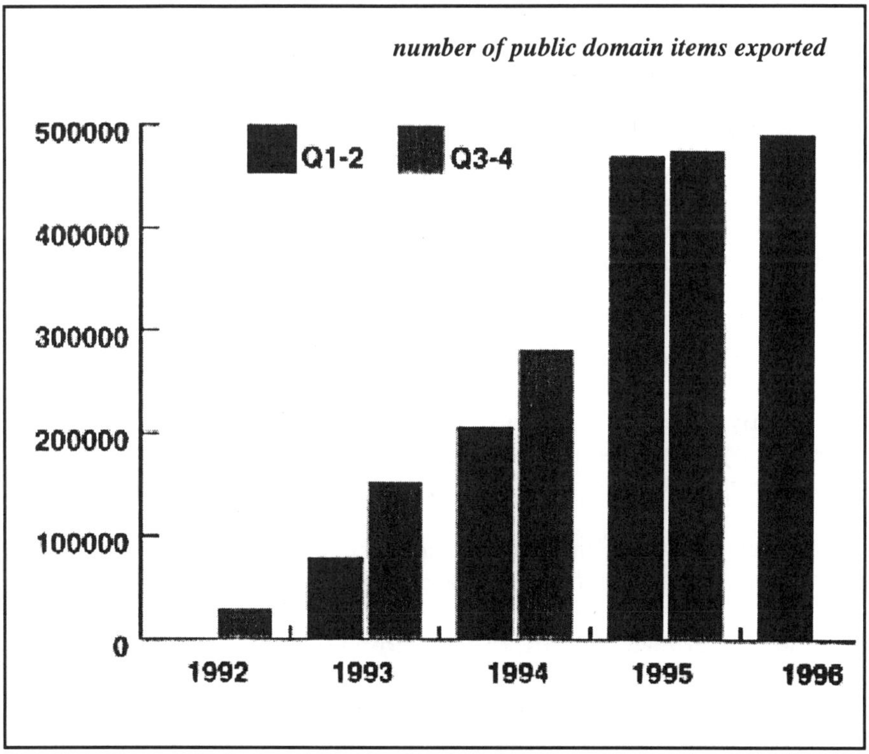

Figure 2: Growth in the number of public domain items exported from the HP Archive, 1992-96

awareness of the Archive, and may also be influenced by the emergence of rival archives and other factors. However, the accelerating growth curve displayed is consistent with an "internal influence model" in the early stages of diffusion (i.e., growth is accelerating and has not yet approached saturation). The advertising of the Public Domain Archive has been carried out almost exclusively via the Internet, and it has not been promoted actively even in this medium. It thus appears reasonable to see this as an instance of an innovation (the Archive) whose diffusion is brought about entirely by internal communication within the community of Internet users.

The "internal influence" model as applied to the Internet

It appears clear that in both the overall diffusion of Internet use and in the diffusion of an innovation relevant to Internet users, internal factors predominate.

In the latter case, it is not hard to see why this is so, but at first sight the growth in use of the Internet does not appear to be a typical instance of "internal" diffusion. In this case, the "social system" is a very diverse and widely distributed set of businesses which operate in a competitive rather than a cooperative context, and are for this reason unlikely to be active in encouraging imitation. Competitive pressures can, of course, stimulate imitative behaviour, but it is relevant to note that at this stage in the growth of the Internet, there appear to be very few instances of *direct* competitive advantages deriving from its adoption (*The Economist*, 1995).

What appears to be happening is a very rapid growth in Internet adoption by businesses driven by an expectation of *future* economic advantage and necessity. Why should this correspond to an 'internal influence' model of the diffusion of the Internet as an innovation within the business community?

Why does an "internal influence" model apply?

We hypothesize that this pattern of growth in adoption of the Internet by the business community has an "internal diffusion" pattern for three reasons:

1. *The growth in the number of Internet users, both business and private, creates greater opportunities for others to become aware not only of the innovation, but of the extent of its possibilities.* This creates a social system wider than, but including, business representatives, in which imitative behaviour is stimulated.
2. *The innovation and the spread of its adoption has created a public phenomenon which has become the subject for very widespread media attention.* Thus, even without any direct promotional cam-

paign, what we observe is a self-generating and accelerating program of publicity for the medium. A particular characteristic of the innovation is having an influence on its diffusion.

3. *The growth in Internet usage has the effect of increasing the value of participation in the process.* In the latter respect, Internet adoption is similar to the adoption of other communication devices such as the telephone or fax: adoption becomes hugely more valuable in the later stages of the diffusion process.

It also has similarities to the growth in television adoption: as Internet adoption grows, the 'program content' of the medium also becomes more diverse and substantial. In this respect, *the emergence of the World Wide Web* has been a *key development*, which we might compare to the emergence of public broadcasting services in the television system.

Our assumption is that these factors are combining to create a very powerful pressure for growth which, at this stage, is increasing with the increase in adoption. We also conclude, from our experience with the Public Domain Software Archive, that within this process of growth the Internet is a very powerful and effective agent for disseminating information about itself to the community of users.

Criticism of the generalized diffusion model

The generalized diffusion model is based around several key assumptions, each of which may be challenged in the case of the Internet.

Firstly, it is assumed that the diffusion process is binary (adoption or non-adoption of the innovation). Stages in the diffusion process such as awareness, acquisition of know-how knowledge, trialling and evaluation (Rogers, 1983) are ignored. Secondly, the original model of Mahajan and Peterson assumes that there is a distinct and constant ceiling on the number of potential adoptees in the social system. It is difficult to define how the social system should be referenced in the case of the Internet. If it is referenced with respect to host connections, then both the geographic boundaries of the system and the number of potential adoptees are changing over time as the number of countries connected to the Internet are changing[3]. In our preceding analysis, our underlying assumption is that the number of potential adopters is very much greater than the number of current adopters, so the rate of growth is proportional to the coefficient of diffusion[4]. Thirdly, the basic diffusion model assumes that there is a complete mixing of members in the social system. This is obviously not the case in terms of the adoption of the Internet. Fourthly, the value of the Internet as an innovation is changing as more businesses go on-line and as access by the general public to the network changes. Thus, the effect of interaction between prior and potential adoptees of the Internet is unlikely to be independent of time. Finally, increase in on-line numbers is also affecting the development of associated innovations such as software for on-line connectivity, software to facilitate HTML authoring, etc. that in turn affects business interest in the original innovation. The Internet as an innovation is not independent of other innovations.

The major criticism of the basic diffusion model discussed above, however, is that it is of little use to agencies interested in diffusing an innovation because diffusion is only considered as a function of time. The model is constructed without reference to a specific technical, organizational or social context. While various extensions and refinements of this fundamental diffusion model[5] have been made, all suffer from the fact that they involve the relaxation of only one or two of its assumptions and consequently fail to capture the complexity of the interactions between factors that characterises diffusion in practice.

The micro-analysis of Connect's activities, described below, clearly illustrates the need for developing new models of the diffusion of an innovation that are more attentive to the feedback loops that exist between the factors which influence a diffusion.

The Micro-Level Analysis

Background

Our interest in the process of diffusion of interest in the Internet within the business community arises from the wish to encourage adoption of the Internet within a local business community, in this case, businesses active in the City of Liverpool and its surrounding region. Our aim in establishing Connect, the Internet Centre for Merseyside Businesses, has been to stimulate early adoption of the Internet within the local business community in order to assist in the economic regeneration of the region. Our reflections on Connect are reflections on a diffusion in process. Connect as an agent for the adoption of the Internet within the local business community is itself continually evolving both in response to changes in the technology associated with the innovation and in response to the interests of the potential adopters of the innovation, the local business community. Mechanisms for statistical monitoring of the diffusion process as it occurs are only now being formally established. Much of what we have to say at this micro-level of analysis of the diffusion process will be qualitative rather than quantitative and centred around those key features of the diffusion process identified by Rogers' (Rogers, 1983) meta-analysis.

Connect as a change agent

Within the framework described at the start of this chapter, Connect may be viewed as an instrument *(change agent)* for the process of diffusion of the Internet. The principal purpose of Connect is to accelerate the process of diffusion within a localized subset of the global social system, namely, the Merseyside area. The aim is to give Merseyside a competitive edge in the world of Electronic Commerce by having its business community well versed and skilled in the use of the Internet.

In this micro-level analysis, we identify the means by which such localized acceleration can in principle, be effected according to the literature on the diffusion of an innovation. We then describe the implementation of such principles in Connect's program of activities. Finally, we review the results of our experience arising from the first year of Connect's operation, and discuss how these may be used to widen the scope of the processes described.

But first: the need for a historical perspective

The creation of Connect as an organisation for the diffusion of information about the Internet to the local business community *is itself an innovation.*

Before proceeding to a micro-level analysis of the diffusion of the Internet as an innovation, we must first pay attention to the factors that influenced the creation of Connect. These are critical to understanding the particular strategies for diffusion adopted by Connect and emphasize the interwoven nature of Connect's activities. In doing so, they emphasize a critical limitation in many discussions of diffusion: models of diffusion rarely allow for the effect that feedback on diffusion activities and change in the innovation that is being diffused have on the diffusion process itself. This point will be returned to in the final section of the chapter.

The Origins of Connect as a Change Agent

Overview

To quote from Rogers (Rogers, 1983), pxv1:

"One important type of change (in studies of diffusion) has been to view the diffusion process in a wider scope and understand that diffusion is part of a larger process which begins with a perceived problem or need, through research and development on a possible solution, the decision by a change agency that this innovation should be diffused, and then its diffusion (leading to certain consequences).

Such a broader view of the innovation-development process recognises that many decisions and activities must happen before the beginning of the diffusion of an innovation; often diffusion cannot be very completely understood if these previous phases of the total process are ignored."

This broadening of the context in which research on diffusion takes place, while welcome, still makes the implicit assumption that a change agent exists prior to the innovation rather than being created as a consequence of a recognition that the innovation is worth diffusing.

Connect, as a change agent, was created as a consequence of a recognition by members of the academic staff at the Department of Computer Science, the University of Liverpool. It was recognized that the Internet had potential as a "Shop Window on the World" for Merseyside and that some mechanism was required through which the business community could become aware of this perspective on the Internet and develop the necessary skills required to exploit it.

In the above quote Rogers also makes the assumption of a linearity to the process of diffusion that was certainly not observed in the case of Connect. In the latter's case, awareness of the potential of the Internet as a means of distribution of goods and services arose in the specific context of the creation of the Public Domain Archive for Software Distribution. The need for an archive of this type was recognised from observation *of the amount of time academics* in the Department of Computer Science spent searching for suitable UNIX based software for use in teaching. The success of this Archive in turn gave rise to an *awareness of the potential commercial value* of the Internet for the distribution of goods and the establishment of a commercial development from the Software Archive, in the form of an Internet-based distribution source for conventional software, called Soft&Net. The value of the Internet as a distribution network for *all types of goods*, a Shop Window on the World, was *only subsequently realised.*

Even with this realization, the funding necessary to create Connect as an agent for the realization of this vision was not initially available. The concept of the Internet as a tool for Electronic Commerce won initial acceptance only by the Department emphasising its established links with Hewlett-Packard and the success of the Public Domain Archive for software distribution. The initial funding for Connect came from the idea of the Internet as a means of distribution of goods in general being hidden behind the argument for its importance in the specific case of distributing software. Parallel developments in technology, such as the World Wide Web (CERN, 1990) and the creation of special European Funds to promote economic regeneration of regions such as Merseyside (Objective 1 status)[6] through technology were necessary in order to create a situation in which the idea of the Internet being exploited by businesses in the region as a

Liverpool as European Cup Winners	1976-77 season
Liverpool as European Cup Winners	1977-78 season
Graphical User Interfaces	1979
Liverpool as European Cup Winners	1980-81 season
Liverpool as European Cup Winners	1983-84 season
WWW Server at Cern	November 1990
Beatles	1962
XFTP	May 1992
XARCHIE	June 1992
Software Technology Retraining Programme	June 1992
Public Domain Software Archive (PDSA)	August 1992
Mosaic	1993
Soft & Net	August 1993
Soft & Net (UK branch)	March 1995
European Objective 1 call for funding	June 1994
WS ARCHIE	September 1994
Funding of Connect	December 1994
Everton as European FA Cup Winners	1994-95 season
WS FTP	January 1995

Figure 3: Historical Influences on Connect

"Shop Window on the World" became politically acceptable. Only then could Connect become established as a change agent for the adoption of Internet related technology by the local business community in general.

Connect as a change agent was established through the interaction of a considerable number of factors. These are illustrated in Figure 3. This graphic also displays other, apparently incidental events, that are of relevance to the success of Connect as a change agent. Each of these factors is discussed in more detail in the text that follows. We begin our discussion with a consideration of the importance of the HP Public Domain Archive and Soft&Net to the establishment of Connect.

The importance of the HP Public Domain Archive and Soft&Net

These have had three types of influence on the creation of Connect:

- The concept of using the Internet as a "Shop Window on the World" arose from the success of using the Internet to distribute software.
- The rapid growth of the archive led to a need for more trained staff than could be funded. This led to the idea of a Work Experience Programme which, as described later, plays an important part in the diffusion activities of Connect.
- The Archive created a local pool of expertise that could immediately profit by the opportunities realised by the development of the World Wide Web. The work on Soft&Net created a paradigm for Electronic

Commerce that could readily be adopted.

The concept of the Internet as a means of distribution in Electronic Commerce

The Department of Computer Science had had a successful liaison for several years with Hewlett-Packard through the HP Public Domain Software Archive (PDSA) Project. As indicated above, the origins of this contact lay in *perceived academic needs* rather than business community needs. Academic members of staff were observed to spend considerable time searching for good, public domain software for use with students. A good source of public domain software tested for use on Hewlett-Packard computers was not readily available. The idea of running a PDSA based in Liverpool was born (Charlton *et al.*, 1994b).

The rapidity with which interest in the Archive grew, through the mechanism of the Internet itself, gave rise to the realization that the same distribution mechanism could be used to support Electronic Commerce, a realization that gave rise to Soft&Net and subsequently to Connect. It also gave clear support to the ability of off-centre locations such as Merseyside to profit and participate effectively in European or world industry through exploitation of the geographic insensitivity of the Internet. Partner sites to duplicate the Liverpool Archive are being, or have been established, at Toulouse, Karlsruhe, Wisconsin, Delft, Milan and Tokyo. Liverpool has been made a leader of a major international project through the Internet.

Soft&Net is a small commercial software archive and marketing service which enables Internet-connected customers to: browse a catalogue of software products; obtain trial versions for examination; and finally purchase (by credit-card transaction) and download the required software to their own computer. Much of this is a *paradigm for Electronic Commerce,* and the experience and expertise established in these developments have underpinned the formation of Connect.

The need for trained personnel to support Electronic Commerce

The rapid development of the Public Domain Archive gave rise to a need for more and more skilled personnel to resource it. As it was a Public Domain Archive, however, money was not available to resource it to the extent required by its rapid growth. In response to this perceived need, a Work Experience Scheme was created within the Department whereby unemployed computing professionals could continue to develop their skills in a useful context while seeking employment. The Work Experience Scheme was able to draw on trained personnel on the Department's Software Technology Retraining Programme (described later) who did not immediately find employ-

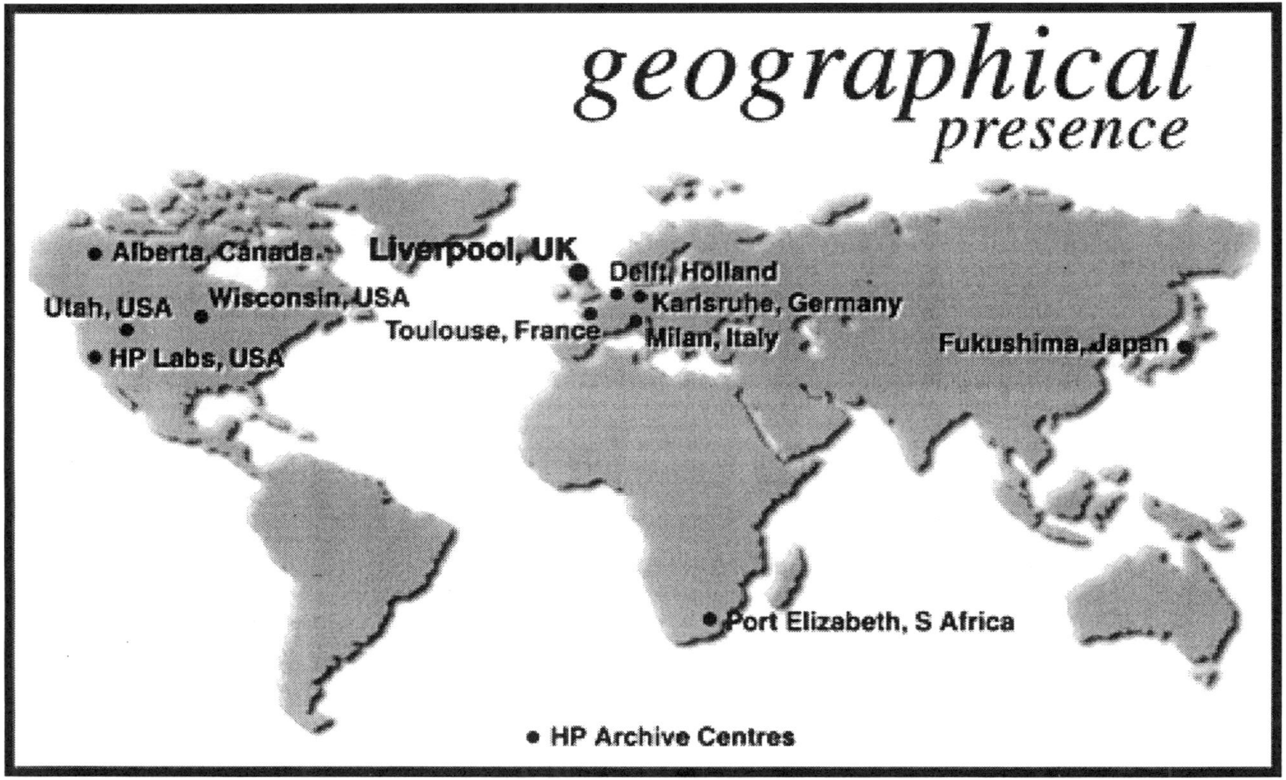

Figure 4: Partner sites to the Liverpool Public Domain Software Archive around the World

ment. The Work Experience Scheme has since become a key component in the development of Connect's activities. As is described in detail later, when Work Experience people who have been attached to Connect find employment within the local business community, they themselves act as change agents for the diffusion of information about the Internet to that community.

Development of specialist technical skills through the Public Domain Archive and Soft&Net

We have already spoken on the importance of Departmental staff's experience with the Software Archive and Soft&Net in creating an insight into the potential of the Internet for Electronic Commerce at a time when this relevance was not generally realised. This allowed Connect to be a leader in the field, a position that in turn reaped rewards which further facilitated the role of Connect as change agent in the local community. New telematics initiatives in the Merseyside area now actively seek Connect's involvement in their activities, which in turn facilitates Connect's role as a diffusion agent.

The Public Domain Software Archive and the experience of academic staff with Soft&Net was also important for the level of expertise that it built within the Department, which allowed new developments in Internet related technology such as the World Wide Web to be readily exploited by staff.

Perceived need for a new strategy for economic regeneration

The Merseyside region has suffered particularly severely from the decline of traditional industries, particularly shipping, and has in consequence had to come to terms with problems of high unemployment and inner-city decay. A compounding factor in Liverpool's problems is that its geographical location, on the North West coast of England, leaves it poorly situated for conventional, non-electronic trade with the European Union.

These problems have given the region a priority status for regional development within the EU (The European Community, 1994). This made available special funding, known as "Objective 1 funding" from the European Community for which local organizations with projects focused on regeneration through telematics could bid. The University of Liverpool thus was able to obtain funding from both the European Regional Development Fund and the European Social Fund for the establishment of Connect and related activities. The ability of the Department to call on private sector backing through its established contacts with Hewlett-Packard was an important determinant in obtaining this funding.

Avoidance of a disincentive to adoption

The key agents in the growth of Electronic Commerce are recognized as being small businesses; only they have the flexibility to restructure rapidly in response to this rapidly changing activity (US Small Business Administration, Office of Advocacy, 1995) (The European Commission, 1994) (The Bangemann Report, 1994). However, this requires an investment not only in the technology, but in the training of staff. Such an investment is problematic for many small businesses who are going through a process of business re-engineering and whose staffing complements in the future cannot be guaranteed. It is unsurprising that small businesses are unwilling to bear the costs of such an investment in anticipation of a payoff which is still unlikely to arrive for some time to come. The potential of Electronic Commerce has yet to be realized. Access to the Internet is still not widely publicly available. Forrester Research (Forrester Research, 1995) estimates that it will take more than four years for consumer users to pass corporate users even in the United States. Many areas of the world, including areas of Europe, currently have few connections to the Internet (Internet Society, 1996). The exponential growth of the Internet (Lotter, 1996), however, indicates that the potential will be realised (*The Economist*, 1995).

This problem was circumvented by the availability of the European Funding to pump prime the adoption of the Internet for Electronic Commerce in the Merseyside region. It has been vital to the success of Connect's mission that the cost of establishing a critical mass for Electronic Commerce locally was borne by the region.

Associated technologies and experiences: the impact of graphical interfaces to Internet tools and the advent of the WWW

From the 1970s, the Internet has been used for communicating research ideas. However, interaction with the Internet up until the 1980s was usually through the command line interface offered by UNIX. This type of interface is well recognized as off putting to new or infrequent users (Nielson, 1993). In the 1980s, however, the role of interface design as a determiner of the frequency and willingness with which software was used was increasingly recognised. Graphical interfaces to tools for transferring files, such as xftp, and to tools for searching for files, such as xarchie, made accessing resources on the Internet easier. Within the Department of Computer Science in Liverpool, a graphical interface to the Public Domain Archive, xbrowser, was developed.

This awareness of the importance of interface design in distributed computing led the Department to monitor with interest the developments that were taking place at CERN and other locations in respect of a new protocol, HTTP, that would support the exchange of hypermedia in a distributed computer environment. Indeed, the Department was granted permission to embed the HTML widget developed at Illinois into its own xbrowser to the PDSA. Eventually, however, that interface has been superseded by WWW browsers such as Netscape.

In 1989, Tim Berners-Lee suggested the HTTP protocol as a means through which scientists anywhere in the world could view each other's research papers and make links from sections of these papers to other sections or other papers/resources of interest. The idea was raised by Vannevar Bush (Bush, 1945) as long ago as 1945. The concept of HyperText was coined by Ted Nelson in 1965 (Nelson, 1965). In 1993 the first graphical browser to the World Wide Web that supported other Internet access protocols was invented by Marc Andreessen,[7] at that time at NCSA, and public/commercial interest in the Internet took off. The World Wide Web as a distributed hypermedia environment for the *presentation* and *communication* of information was born. Commercial interest in the Internet was taking off just at the time as Connect was becoming established.

Thus there was a fortuitous combination of circumstances acting to make Connect an effective agent of change within the community. Just at the time that Connect received funding to promote the use of the Internet to the business community, the innovation-the Internet-underwent a transformation[8], that both self promoted the innovation and radically altered the ease with which it could be used by members of the public.

This change in the innovation allowed Connect to alter accordingly. The original idea underlying the development of Connect, that of promoting the Internet as a "Shop Window on the World" to the local business community gained critical acceptability and publicity through media attention given to the World Wide Web. Connect, as an organization, accordingly adapted to its new role. The World Wide Web increased the scale on which Connect could achieve its mission.

The effect of the World Wide Web on the growth of easy-to-use Internet connection software

Connect provides the business community with the knowledge about the technology necessary to exploit the Internet and the characteristics to look for in an Internet Service Provider. Tools for connecting to the Internet have to be easy to use, and local points of presence or equivalent have to be available if that knowledge is to be translated into practice.

As commercial interest in the WWW has grown, so too has the ease with which Internet connections can be made by small businesses. The availability of Internet Service Providers and good software packages for connecting to the Internet have increased. In March 1995 there were only three Internet Service Providers with local Points of Presence (POPs) in the Merseyside area. Now, in May 1996, there are approximately 25 with POPs or 0345 numbers (which in the UK means that a call is charged at a local area rate). In a very competitive market, Internet Service Providers vie with each other to develop a competitive edge through better connection software packages, more economical charges, etc; all of which reduce the barriers to going on-line and therefore facilitate the realization of Connect's mission.

The decline in the cost of personal computers obviously also means that the cost of exploring the potential of the Internet in business transactions is not prohibitive. Similarly, the development of user friendly software packages for exploiting the Internet and the WWW such as HTML authoring tools and PC equivalents of xftp and xarchie all influence the probability that a business will see Electronic Commerce as a reality in the making, and become connected to the Internet.

Availability of skilled personnel through a co-existing programme of Software Technology Retraining

Of importance to the origins of Connect as a change agent is the fact that the Department of Computer Science at the University of Liverpool has had a long history of providing courses for industry. Of particular relevance to Connect was the prior establishment within the Department of a Software Technology Retraining Programme for unemployed graduate or graduate-equivalent computing professionals. This course arose as a response to the depth of the recession in the UK in the early 1990s, which, combined with fundamental organizational changes brought about by the desire of companies to "downsize" their computing systems, resulted in large scale unemployment in the professional, computing workforce. Many of those made redundant are skilled in aspects of computing which are no longer the preferred skills for industry, either large or small (Entickmap, 1992, Vail, 1993). The 13-week intensive course offered by the Department, leading to a postgraduate Certificate in Software Technology, focuses the skills of these unemployed professionals into areas of national skill shortage. The course programme is further described by Charlton *et al.* (1994a).

Of importance to the realization of Connect's mission is the fact that this course is providing the Merseyside region with a pool of skilled personnel trained in skills necessary to exploit fully the capabilities of the Internet at an advanced level. Within Connect itself, at least 50% of the current staff complement have come through the Software Technology Programme.

The program is supported by the European Social Fund. Thus once more, the priority status for regional development that Liverpool has within the European Union (The European Community, 1994) is acting for the benefit of Connect.

Local attitudes: Merseyside has been a World Leader

Merseyside has been a world leader in commerce. Historically, it was a major trading port. In the field of entertainment, it received international acclaim as the home of the Beatles during the 1960s and 1970s. It is home to internationally famous football teams, Everton and Liverpool. Liverpudlians are aware of this heritage. Thus, despite the current economic problems that plague the region and the urban poverty that gives rise to the region's Objective 1 status, Liverpudlians have a belief in themselves so that the message "Merseyside could be a World Centre of Electronic Commerce through the Internet" is not immediately dismissed as absurd. This self belief may be a factor enabling the adoption of new technology into businesses within the region. "If men perceive situations as real, they are real in their consequences" (Thomas and Znaniecki, 1927).

Personalities involved

The "people" factor certainly operated in both the establishment of Connect and in the drawing together of a variety of activities, each of which proved subsequently critical to the success of Connect. The people involved in the establishment of Connect could see clearly what they wanted to achieve, had the technical skills necessary to reach the desired goals, had good connections internationally in respect of the electronic distribution of software and in respect of private sector backers such as Hewlett-Packard for the project. Their personalities were complementary: one being strong on vision, ideas and ability to persuade and motivate people; another being strong on detail and the ground work necessary to persuade the University authorities to support a scheme of this kind. In turn, they drew into Connect a team of people who were equally motivated and to whom the challenge of transforming an economically depressed region into a centre for Electronic Commerce was not simply a job.

It is interesting to note, that at the inception of the project, Connect *did not have a strong connection to local* commercial organizations. These have developed as Connect's reputation for technical expertise has been recognised within the business community. Interestingly, parallel projects which had a stronger political base but which were weaker in technical know-how have foundered.

Summary of historical factors

The above discussion has emphasised the number of interweaving factors that have contributed to Connect and which have been

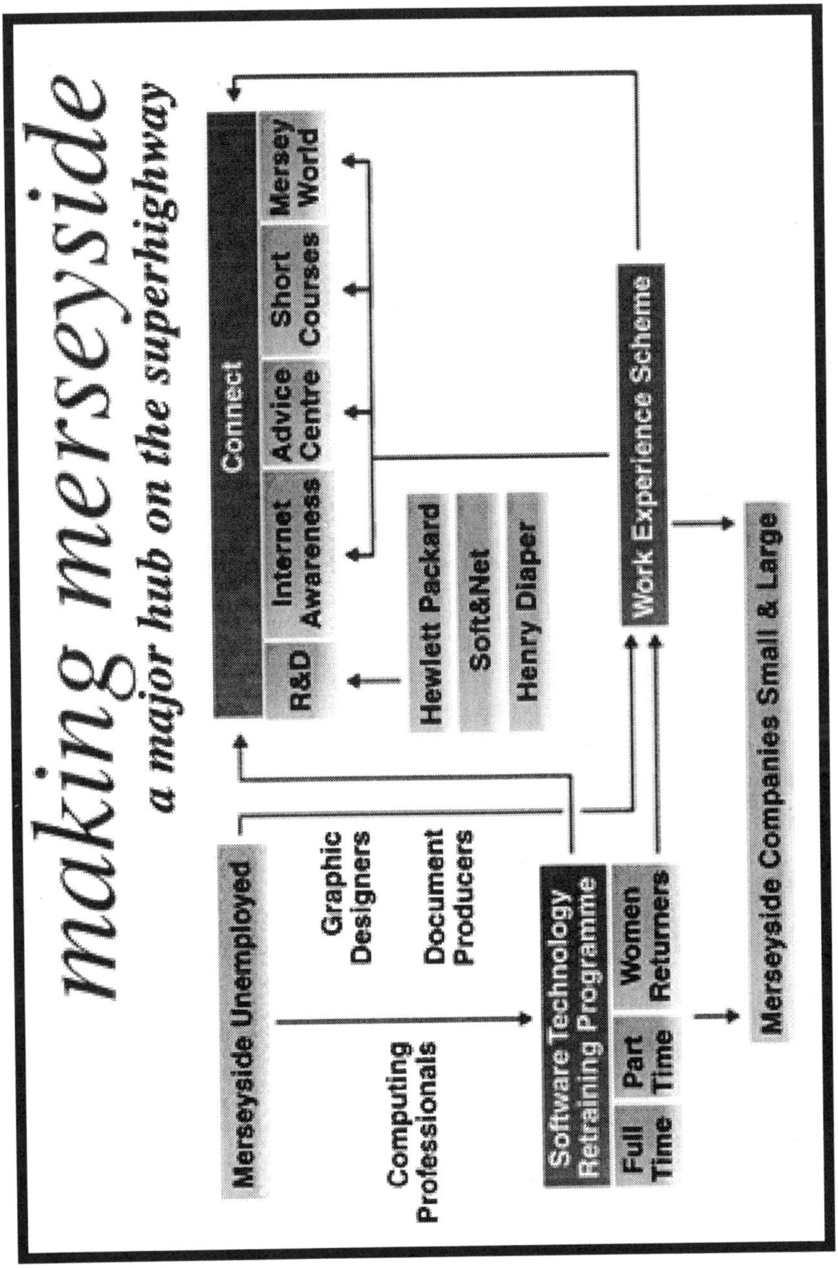

Figure 5: Strategies employed by Connect to promote use of the Internet

instrumental to its ongoing development and success as a change agent. Of particular current relevance to Connect's role as a change agent are the Software Technology Retraining Programme and the Work Experience Scheme, both of which are providing the skilled personnel that businesses will require as the move to Electronic Commerce gains momentum.

Connect's Strategy as a Change Agent

Connect's strategy is complex, far more complex that the characterisation of change agents taken from Rogers (1983), p. 168 below:

"Most change agents seem to concentrate their efforts in creating awareness knowledge, although this goal often can be achieved most efficiently in many client systems by mass media channels. Change agents could perhaps play their most distinctive and important role in the innovation-decision process if they concentrated on how-to knowledge, which is probably most essential to clients at the trial and decision stage in the process."

Connect avoids the above criticism of the activities of change agents. In promoting the adoption of the Internet as a vehicle for Electronic Commerce within the Merseyside region, it has sought both to raise the level of small business awareness of the commercial possibilities and to provide the opportunity through a Short Course Program for businesses to develop the necessary know-how knowledge to do so. Such activities have been undertaken as part of a large program of interwoven activities, all of which are viewed as critical to the realization of Connect's mission. The strategies employed by Connect are shown in Figure 5.

As Figure 5 illustrates, the Awareness Raising and Short Course Programme developed by Connect are interwoven with the Software Technology Retraining Programme and the Work Experience Scheme, both of which are seen to provide the necessary skill base within the region for full exploitation of the innovation by companies. An infrastructure for the expression of know-how knowledge, the MerseyWorld Web Site has also been created. Direct advice and consultancy to clients embarking upon Internet-related activities is also available. Finally, clients are able to try out an Internet connection for themselves for a limited trial period. Connect's activities thus embed all the key ingredients for adoption of an innovation: *awareness, know-how, trial and evaluation.* It also does not cost a business a lot to adopt the innovation for themselves.

The creation of Awareness Knowledge

While mass media is now playing an ever increasing role in the promotion of the Internet in the public consciousness, media interest in the Internet and in Electronic Commerce in general was not substantial at the birth of Connect. General media hype in any case does not make individuals aware of local opportunities for investigating an innovation. Thus at the inception of Connect, it was necessary to develop a mechanism whereby the awareness of the local business community in the potential of the Internet could be raised. This was done through running an ongoing programme of Internet Awareness Days, offered at no cost to clients. The program for each seminar includes both talks and hands-on practical work to introduce participants to aspects of the Internet, especially those relevant to business activities, including the World Wide Web, electronic mail, Electronic Commerce, and security issues.

The Internet Awareness Days were stimulated by an initial mailing. This mailing was based on a list that was purchased from the organizers[9] of a meeting for local businesses that the initiators of Connect were invited to attend because of local interest in the Public Domain Software Archive. This occurred just around the time the Connect initiative had received funding. The list proved an excellent base from upon which to advertise Connect's activities. There has been very little need for subsequent mailings. The demand for Internet Awareness Days has been consistent: the take-up has been achieved with little advertising, constrained by the limits on the capacity of the program. During 1995, participants in the Internet Awareness Day program have predominantly stated 'word of mouth' as their source of information about this activity.

Effect of the Awareness program

Establishment of a client base interested in know-how knowledge. As well as raising awareness in the business community, the effect of this seminar program has been to provide a client base of interested companies who can be targeted in mail shots about further training opportunities offered by Connect.

Defining the training needs of the client base. The seminar program has also given Connect the opportunity to find out more about the needs and interests of local businesses through the responses of participants to a questionnaire distributed at the end of the seminars. Analysis of the questionnaires is being used to help focus an extended program of activities on the particular needs of local businesses.

Responsiveness to perceived needs of clients. Responsiveness to feedback from clients as to the extent to which the information and training provided by Connect is meeting their perceived needs has been a key aspect of Connect's strategy. Connect has thus sought to avoid the criticism that is often made of change agents, of being overly focused on the innovation in their diffusion activities as opposed to the needs of the client population the innovation was meant to serve. Both

	Course Title	% respondees
1.	How to Connect to the Internet	62
2.	Using e-mail	66
3.	Using FTP	68
4.	Using Usenet News	49
5.	Using the World Wide Web	83
6.	Authoring documents for the World Wide Web	N/A
7.	Interacting with WWW documents via the Common Gateway Interface	N/A
8.	Business on the Internet	79
9.	Connecting a small LAN to the Internet	N/A
10.	Running a small LAN	54
11.	Security	56

Table 3: Business Interest in Short Courses Proposal

the training courses offered and the format in which they are offered (half day and at no cost) is determined by perceived client need. Managers of small businesses prefer not to be away from the office for a full day.

Anticipation of the consequences of technological changes. Connect has also acted to employ its superior technical knowledge in the interest of clients by being pro-active in drawing clients' attention to new technological developments, such as the invention of Java, that are liable to have wide reaching implications for use of the Internet.

The how-to knowledge transfer program

The main elements of this programme are: further seminars and short courses aimed at developing skills and understanding within local businesses; direct advice and consultancy to clients embarking upon Internet related activities; and activities aimed at creating a skilled workforce, capable of meeting the staffing needs of clients.

Short course program

The short course/seminar program is intended to meet the needs of businesses who are embarking upon or contemplating some use of the Internet in their commercial activities. Eleven distinct courses have been developed, each addressing a different issue, summarised in Table 3. The tabulation also identifies for each course the number of respondees, from the first 256 attendees on the introductory seminars, who expressed themselves to be "*interested/very interested*" in attending a course on the topic named (not all the courses developed were identified explicitly in the questionnaire). These figures give some indication of the current perceived priorities of small businesses contemplating these developments in 1995. The program is evolving both in response to clients' needs (as the knowledge base of the clients increases the type of training required changes) and in response to new developments in Internet-related technology.

Further technical assistance to business

Alongside the scheduled course program, Connect offers direct advice and assistance to local businesses on any aspects of Internet use. In this respect, it is able to draw on the expertise available not only within the Centre, which has twelve full-time technical staff, but also in the parent Department of Computer Science. Most of this advice is available at no charge, especially to small businesses who are simply seeking help in beginning to make use of the Internet. For more ambitious ventures which may require extensive technical assistance from Centre staff, our procedure is to agree a Research and Development contract between Connect and the client organization. A number of such contracts are now in progress, the first of which is identified in Figure 5 above.

Training and Work Experience

The Short Course Programme has to date been principally aimed at the needs of businesses who are only beginning to consider possible uses of the Internet. Our expectation, however, is that this will lead in the relatively near future to developments which will require specialist staff within these organizations or in support organizations emerging to meet a new demand for Internet Consultancy and programming tasks. This is where the Software Technology Retraining and the Work Experience Programme have a particular role to play in Connect's diffusion activities. At this stage in the progress of the Internet, however, there are still very few small businesses that are ready to employ specialist staff for these developments. The Work Experience Scheme provides a critical bridge to full employment. This scheme is fed partly by trainees emerging from the Software Technology course, and partly by direct entry from unemployed professionals. The latter category includes, in particular, staff with skills in graphic design and multimedia that are also very relevant to applications of the Internet. Both the Software Technology Retraining and the Work Experience Programmes are a critical part of Connect's overall strategy for fostering regional development and job creation through Electronic Commerce centred on Internet related technology. As many of Connect's activities as a change agent are sustained by the Work Experience trainees, it is worth describing this component of Connect's overall strategy in more detail.

Figure 6: MerseyWorld Home Page (http://www.connect.org.uk/merseyworld/)

Work Experience Scheme

The Work Experience Scheme takes place in two stages. In the first stage, work experience trainees gain important skills and experience while providing additional effort for the program of work being carried out by Connect. For trainees who have been without work for a substantial period prior to entering the scheme, it provides an important reorientation to the rhythm of a working day. In the second stage, trainees are attached to businesses identified by Connect that are beginning to make further use of the Internet. This work placement serves both *as a tool for technology transfer*, in which trainees pass on skills gained from their work with Connect to businesses outside, and as a further stage in their training for work. Ideally, of course, we hope that the activities initiated and progressed by work experience trainees in their host organizations will lead directly to their subsequent employment.

Utilisation of know-how: MerseyWorld

It was anticipated that most small businesses in the region would wish to use the Internet for marketing their products on the World Wide Web but might be deterred from doing so by the cost of maintaining their own Web server. It was also recognised that while a single business in isolation might not draw much attention on the Web, a large, attractive, multi-faceted Web Site, like a well-populated shopping center, would be more likely to do so. A comprehensive regional World Wide Web Site was accordingly created by Connect. This brought together in a single site a wide range of information and services relevant to the region, as well as information about businesses operating from within Merseyside. It created the necessary infrastructure for the expression of the know-how knowledge clients were developing in the training programme. Clients were provided with a personalised illustration of the potential of the technology for Electronic Commerce. The Home Page of MerseyWorld is shown in Figure 6.

Subsidiary pages contain information about local government, business organisations and education providers, as well as the Connect organization itself. All of this provides not only an information service for the community, but also attractions to draw browsers to the site.

Participation by individual small businesses in MerseyWorld is collected together within a subsidiary section, "MerseyMall." The aim of MerseyMall is to provide an electronic shop window, displaying the products and services offered by regional businesses. An option, taken up by many, is to include in their MerseyMall entry not just promotional material, but the facility to place an order directly over the

Internet, possibly paying by credit card. The service provided by Connect includes translating enquiries and orders received in this way into e-mail or fax messages which are passed on to the client businesses. The communications facilities of the Internet are thus integrated seamlessly with those forms of communication that businesses are more familiar with.

The multifaceted nature of the MerseyWorld Site is shown in Figure 7.

Current New Developments

Transferring management of the innovation to the client

As businesses have become familiar with the Internet and developed the skills required to exploit it, the responsibility for managing the clients' Web pages has been passed from Connect to the clients themselves. This currently poses a new technological challenge in respect of ensuring a consistent level of maintenance across the whole of MerseyWorld in order that this does not lose its collective identity.

Maintaining a profile in Electronic Commerce: new initiatives

As other regions and organizations start to explore the potential of the Internet, it has become increasingly important to provide MerseyWorld with *added value* in order to maintain its potential as a vehicle for stimulating Electronic Commerce within the region. One move in this direction, has been the maintenance at the MerseyWorld site of a 'technology watch,' in the form of information pages intended to allow readers to keep up to date in new Internet related technology developments. Topics monitored include: Internet providers; Archive sites; Web authoring tools; security; anti-virus measures; browser and server developments; and electronic cash. This kind of technical information, apart from its intrinsic value, will help maintain the reputation of the site within the Internet community.

Ensuring a consumer base for Electronic Commerce

It will also be increasingly important to attract the attention and interest of the wider public, the consumer base within the region. Connect is thus in the process of broadening its programme of awareness raising to reach the general public to promote not only Electronic Commerce, but also the educational and cultural value of Internet related technologies. This is being done through a community action program centered around Internet cafes based in local libraries and promoted through existing action groups in the community such as the Pathways projects. In order that the interest which is stimulated

Figure 7: The multifaceted nature of the MerseyWorld Site.

by such activity is maintained, free or relatively cheap Internet access for the casual user will need to become the norm. This is, however, an issue outside Connect's direct control though as an organization it seeks to influence the opinion of the local authority which has the budgetary power to establish kiosk based Internet Access points for the general public.

Wider, public interest in MerseyWorld is also being stimulated by the creation within the site of pages of topical interest, such as a Lottery page. While the original purpose of this page was to demonstrate to clients how aspects of information processing could be automated through use of CGI-scripts, the topic itself has proved of such interest that it has drawn a large number of viewers to the site who subsequently browse other pages.

To ensure the necessary infrastructure within the region for exploitation of the Internet across a wider sector of the community, Connect has been involved in the establishment of strategic alliances with companies such as Cable North West who are the franchised providers of ISDN lines in the region.

Reflections on Connect's Activities

The above is a formal description of Connect's activities as a change agent. We now consider these activities in the context of issues raised in the diffusion literature.

Is Connect a case of centralised diffusion activity?

Research on diffusion has in the past distinguished between diffusion and dissemination (planned) activities. In recent years, this distinction has been dropped on the grounds that it has not proved particularly useful and the terms *centralized* and *decentralized diffusion* have been used instead. In many ways, the activities of Connect could be seen as promoting a centralised model of diffusion as described by Rogers:

"Decisions about such matters as when to begin diffusing an innovation, who should evaluate it, and through what channels will it be diffused, are made by a small number of experts and/or technical experts at the head of a change agency."

Such a characterization would only partially apply to Connect. As described below, word of mouth has played an important role in the dissemination of information about Connect's activities and thus about the Internet, particularly recently. Informal human networks now play an important role in the local diffusion process. Further, as emphasized earlier in discussion of the global impact of the Internet on commercial activities, once people have access to the Internet, *the Internet itself begins to act as an agent in its own diffusion.* Businesses do not have to wait to hear from other local businesses as to the success or otherwise of going on-line. The Internet itself can be used as a vehicle through which to explore what others are doing and to access reviews of the profitability or otherwise of such initiatives. Even if a business does not initially choose to adopt the Internet in the service of Electronic Commerce, it nonetheless may continue to use it to learn further about its potential for Electronic Commerce. In some cases known to Connect, the innovation has not been adopted for the purpose aimed at by the diffusion activities, i.e., Electronic Commerce, but has been adopted for another purpose (child education).

Connect itself has changed in response to the client community it seeks to serve: its role as a diffusion agent is continually changing. Classical perspectives of the diffusion of an innovation tend to neglect the time varying nature of the activity (Rogers,1983):

> *"Social science methods better at getting snapshots of behavior rather than moving pictures......*
>
> *Survey research on the diffusion process is a convenient methodology of the researcher, but it is intellectually destructive of the process aspect of diffusion."*

Currently we are trying to establish monitoring methods that will capture the changing nature of the diffusion process in which Connect is involved. Indeed, the design of new monitoring methods sensitive to the changing requirements of business is an research area of concern to Connect.

The importance of informal networks in the dissemination of Connect's activities

Traditional research on diffusion has been criticized for conceiving diffusion too narrowly, as a linear, unidirectional communication activity. Within such a framework, Connect would simply be seen as a change agent, developing courses on the Internet (the message) in order to influence the attitudes and/or behavior of passive recipients (the business community). In our experience, such a criticism would be more than justified. Our experience fully concurs with the sentiments of this quote, taken from Rogers:

> *"Information about innovation is sought from near-peers, especially information about their subjective evaluations of the innovation. This information exchange about a new idea occurs through a convergence process involving interpersonal networks. The diffusion of innovation is essentially a social process in which subjectively perceived information about a new idea is communicated.*

This dependence on the communicated experience of near-peers suggests that the heart of the diffusion process is the modelling and imitation by potential adopters of their network partners who have adopted previously."

Of critical experience to the diffusion of information about the Internet and the possibilities it holds for Electronic Commerce has been informal networks within the community as a whole. As indicated earlier, the predominant source of information for attendees at Connect's Internet Awareness Days is 'word of mouth.' The local media have also shown considerable interests in Connect's activities without any prompting.

In the case of the Internet as an innovation, it is particularly in the interests of the clients themselves to act as change agents, to publicise the innovation. Exploitation of the Internet in Electronic Commerce requires that both potential customers and that suppliers of goods are on-line. A business can only exploit electronic communications if the desired partner in the communications is also connected.

MerseyWorld has also played a role in providing a showcase through which local businesses can readily view the outcome of colleagues' contacts with Connect. It has undoubtedly facilitated the modelling and imitation processes that are well recognized in the diffusion literature. It has also resulted in businesses become interested in the Internet through curious associations. For example, a local hotel learned about the Internet and MerseyWorld through a football fan from Norway who happened to be staying at the hotel and who had visited MerseyWorld's Sports Pages. Since going on the Web, the hotel itself has received bookings from around the world from prospective visitors to Liverpool.

The issue of heterophily in communication

Within the literature on the diffusion of an innovation, a considerable amount of attention has been paid to the effect of asymmetry of knowledge/background between the change agent and the client base that is being persuaded to adopt an innovation. Change agent/client communication is termed an instance of heterophilous communication. Generally, it is argued we relate best to those with whom we share common knowledge/interests (homophilous communication) (Rogers, 1983), p19:

"Individuals tend to be linked to others who are close to them in physical distance and who are relatively homophilous in social characteristics."

Thus, the asymmetry in the knowledge about an innovation that by definition exists between the change agent and the client is seen to be potentially problematic, a factor acting against the successful

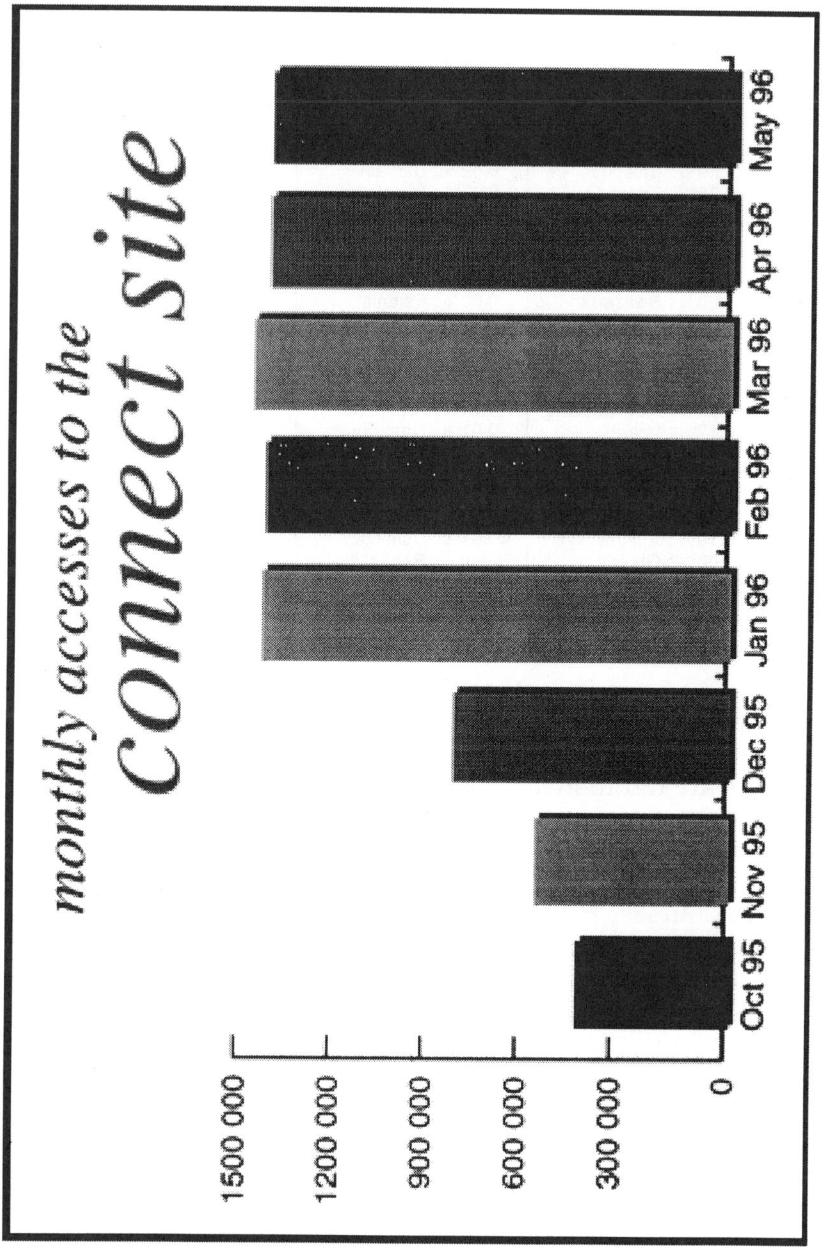

Figure 8: Monthly accesses to the Connect Site since the launch of MerseyWorld

statistics
for access by country

44.04%	United Kingdom	0.07%	Brazil	<0.01%	Trinidad and Tobago
23.02%	USA	0.05%	Bermuda	<0.01%	Turkey
16.44%	unresolved	0.04%	Malta	<0.01%	China
3.66%	Norway	0.04%	Mexico	<0.01%	Latvia
1.61%	Canada	0.03%	Portugal	<0.01%	Slovak Republic
1.52%	Singapore	0.03%	UAE	<0.01%	Cayman Islands
1.32%	Sweden	0.02%	Taiwan	<0.01%	Romania
1.05%	Australia	0.02%	Czech Republic	<0.01%	Nepal
0.80%	Iceland	0.02%	South Korea	<0.01%	Cyprus
0.79%	Ireland	0.02%	Bahrain	<0.01%	Ukraine
0.78%	Finland	0.02%	Indonesia	<0.01%	Saint Lucia
0.53%	Germany	0.02%	Former USSR	<0.01%	Faroe Islands
0.46%	Japan	0.02%	Slovenia	<0.01%	Croatia
0.43%	Netherlands	0.02%	Poland	<0.01%	Costa Rica
0.37%	Hong Kong	0.02%	Estonia	<0.01%	Colombia
0.30%	Switzerland	0.01%	Macau	<0.01%	Egypt
0.28%	Thailand	0.01%	Chile	<0.01%	Ecuador
0.28%	Denmark	0.01%	Liechtenstein	<0.01%	Philippines
0.27%	South Africa	0.01%	Hungary	<0.01%	Zambia
0.24%	Malaysia	0.01%	Argentina	<0.01%	Fiji
0.23%	New Zealand	0.01%	Uruguay	<0.01%	Panama
0.17%	Israel	<0.01%	Luxembourg	<0.01%	Antigua and Barbuda
0.16%	Spain	<0.01%	Russian Federation	<0.01%	Peru
0.14%	France	<0.01%	Kuwait	<0.01%	Jamaica
0.14%	Italy	<0.01%	Dominican Republic	<0.01%	Venezuela
0.12%	Belgium				
0.10%	Austria				
0.08%	Greece				

Figure 9: Accesses to MerseyWorld from around the world

diffusion of an innovation. In the case of Connect, the change agent is also located in an environment, that of a University, that has been traditionally regarded with suspicion by the business community. Academics are traditionally salaried and therefore free of the need to 'earn a living.'

In Connect's case, there has been no evidence of either of these factors influencing interest in either the Internet Awareness Days or in the Short Courses. Businesses have been only too willing to avail themselves of free technical advice. Perhaps here *the fact the advice was free* and coming from within a University setting acted positively: a business could readily obtain the desired information and develop the necessary skills to exploit the innovation without being required to adopt the innovation or buy anything. The fact that the change agent is situated in a very different environment from that of business, an environment that is traditionally associated with *impartiality* may have had a positive influence. Certainly, the fact that Connect is staffed by academics used to a rapidly changing discipline requiring the adaptation of old skills and the learning of new skills has helped Connect as a change agent to be both responsive to the emergence of new client needs and to the changing nature of the Internet itself as a vehicle for Electronic Commerce.

The characteristics of businesses involved with Connect

We cannot comment in detail on the type of businesses that have become involved in Connect's activities or whether this has changed over time. The businesses that have attended our Internet Awareness days and the Short Courses are a very heterogeneous group as judged from our records of the markets they serve. This heterogeneity has remained constant. Similarly, those businesses with which Connect has contractual R&D arrangements do not differ in any significant dimension from those who to date have simply come on the training courses. They are an eclectic group, the one salient dimension which they have in common, according to the comments of Connect staff, is a particularly entrepreneurial leadership. Certainly, we do not have data that would support the claim that the early adopters of an innovation tend to have a high degree of mass media exposure/wide interpersonal networks. As we indicated at the beginning of this micro-level of analysis of Connect, the mass media was not publicizing the Internet heavily at the inception of Connect. Once connected to the Internet, everyone's potential interpersonal network is as big as everyone else's.

The consequences of adoption of the Innovation to SMEs in the region

Currently, a systematic study of the impact of Connect's programme of activities is being implemented. Therefore, we can therefore say very little about the innovation-decision process as applied to our client base. However, we can comment upon the success of MerseyWorld as a strategy for promoting Electronic Commerce in the region.

Success of MerseyWorld

MerseyWorld as a showcase for businesses in the region is proving successful. Figure 8 indicates a steady rise in the number of hits recorded by our server since the launch of MerseyWorld in October 1995, up to a steady peak of about 1.4 million accesses per month.

Most of these hits are from browsers located within the UK.

For example, Figure 9 displays the frequency with which MerseyWorld was accessed from around the world during the week from January 1, 1996 to August 1, 1996. During this week, an average of 53,276 hits per day were recorded by the MerseyWorld server. The majority of these accesses (44.04%) were from browsers based within the UK. The USA was second with 23.02% of accesses. Access from any

Type of analysis	**Mathematical modelling of the process of diffusion over time**	**Descriptive analysis of factors known to influence the diffusion process.**
Theoretical framework employed	Mahajan and Peters (1985) Generalised Mathematical Model	Rogers (1983) Meta-Analysis across Social Science disciplines
Question(s) asked	Which coefficient of diffusion, g(t), best describes the diffusion of the Internet among the global business community?	Have the factors described in the literature on diffusion been observed to be important in our case study of Connect?
Empirical data	Statistics on the growth of the Internet (Rutkowski, 1996)	Participant Analysis Interviews Questionnaires
Result	The coefficient of diffusion expressed as bN(t) This corresponds to the 'internal influence' model	• Evolving nature of the technology, the change agent and the processes of diffusion within the local community • Significance of historical and co-incidental factors • Awareness, know-how, trial and evaluation important in the adoption of an innovation • Heterophily can be of benefit
Supporting Analyses	Hewlett-Packard Public Domain Archive Exports, Liverpool University	Importance of Internet as mechanism in its own diffusion
Issues raised by analysis	• Non-dynamic model • Assumption of constancy in both the nature of the innovation and the social context in which the diffusion is taking place	• Traditional models of diffusion neglect the synergistic nature of the interaction between factors in the diffusion process • Innovations are rarely diffused in isolation from other innovations.

Table 4: Summary of Analysis

other country then dropped below 4%. Altogether accesses to MerseyWorld were observed from 78 countries, from 16,344 distinct hosts.

Anecdotal evidence from companies on MerseyWorld also indicates a positive impact of the adoption of the Internet on local businesses. There are no negative tales to tell to date[10].

Two case studies may be cited as examples; that of Henry Diaper Co. Ltd., an established small business in Merseyside and that of Glaxo

Neurological Centre, a registered charity for neurological disorders.

The examples of Henry Diaper Co Ltd and Glaxo Neurological Centre

Henry Diaper Co. Ltd. is an established small business in Merseyside. They are a distribution company who distribute goods to small retail outlets such as wine merchants. Currently, the retail shops order goods from Diaper's using a paper-based system involving orders via the post, telephone or fax. Connect has been developing with Diaper's an on-line electronic ordering method, by which retail outlets can order over the Web. This will bring to Diaper's all the advantages currently reserved for large distribution chains having their own private networks. A prototype is currently being tested.

Glaxo Neurological Centre, a registered charity for neurological disorders, reported a stream of enquiries within days of going on-line with a Web Site. Of particular interest here was their report of enquiries from all round the world as to help available for certain rare conditions. This is a clear illustration of how the Internet can overcome geographic barriers and equalize access to facilities. Those who are in a minority need no longer be disadvantaged in respect of access to information. Glaxo Neurological Centre now plan to utilize the Web to get information about disabilities into doctors' surgeries. Surgeries currently only have a limited number of leaflets available, many of which are out of date. Glaxo views the Web as a way of overcoming this problem so that doctors could access the latest information pack at the site, download it to the surgery and print it for the patient. Utilisation of information filtering techniques could allow the package to be tailored to the particular needs of the individual patient.

Evaluation of Connect as a change agency

At the time of writing, 763 individuals have attended the Connect programme of Internet Awareness days. Four hundred twelve local SMEs have made use of the Short Course Program and services offered by Connect, including 34 software suppliers[11]. In addition, another 111 Merseyside organisations have been advised. The fact that interest in the Connect program exceeds the provisions we can make on the basis of our current resources clearly indicates the success of the Centre's activities promoting awareness and information about the Internet.

Since the launch of MerseyWorld in October 1995, 69 companies have been incorporated. Interest in MerseyWorld has been high as was indicated in Figure 9 above. Accesses to MerseyWorld are beginning to stabilize at around 1.2 million hits per month. Approximately 100,000 of these are to companies listed in MerseyMall. At present, a high proportion of these potential trading accesses are to software-related businesses, such as NA Software, reflecting the current dominance of technology-linked usage of the Web. We expect this pattern to change rapidly as Internet penetration extends to home and non-technical business use.

At a European level, Connect has been approached by a wide variety of regional authorities/organizations in locations as diverse as Southern Ireland, Portugal, Sweden, Italy as well as by other authorities in the UK interested in developing a similar initiative within their own country. Bids are being made to the European Regional Development Fund for funding to extend and develop the strategies and ideas that underpin Connect at a trans-European level. At a local level, Connect's activities are now well integrated with other initiatives that service the business community, both traditional such as Business Link, Chamber of Commerce, Merseyside Partnership, City Council and telematics based such as GPT who manufacture the ISDX telephone switches used by many of the larger organizations in Merseyside[12] and Cable North West.

Summary of Analysis

In this chapter, we have engaged in two levels of analysis. At the global level, we have evaluated mathematical descriptions of the diffusion process among the world business community. At the micro level of analysis, we have studied the factors influencing the diffusion of the Internet among the business community at a local regional level through the activities of a change agent, Connect.

A summary of the results of these analyses is given in Table 4.

Discussion and Review of our Analysis

We began this chapter by highlighting some key features of the Internet as an innovation.

First, we raised the paradox that the Internet, as a global network of computers which could act as a medium for the distribution of any form of digitized information, has been in existence since the 1970s, yet it is only now, in the mid-1990s, that it is being adopted by commercial organizations as a vehicle for Electronic Commerce. Does our preceding analysis of the factors influencing the diffusion of the innovation at a global and/or local level allow us to resolve this paradox?

Secondly, we posed the question of whether the Internet is comparable to other innovations in terms of the factors which have influenced its diffusion. Does our analysis of the factors influencing the diffusion of the innovation at a global and/or local level suggest that the diffusion of the Internet has unique features which challenge some fundamental assumptions of traditional models of Innovation and Diffusion?

The paradox surrounding business interest in the Internet

The first point we wish to make is that the Internet is not unique in being an innovation in which the interest of different communities has varied over time. The literature on the diffusion of innovations contains many examples of innovations whose adoption has been restricted to particular contexts, with cross context generalisation taking considerable time. Rogers (1983) quotes the particularly salient example of scurvy control. Scurvy was the worst killer of the world's sailors. The ability of citrus fruit to control the disease was experimentally demonstrated in 1601 by Lancaster and again in 1747 by Lind. It was not until 1795, 48 years later that the British Navy adopted the policy of giving all ships' crews citrus fruits! One might choose to argue that this example is not relevant as it occurred in the days before easy world wide communications. However, the lack of world wide communications cannot be used to explain the fact that it took another 70 years before the British Board of Trade adopted this innovation for the merchant navy. This example is not so dissimilar from the academic and military's exploitation of the Internet as a means of communication over a 25 year period before the commercial world became interested.

Why has business interest in the Internet grown in the 1990s?

What, however, explains the increase in the commercial world's interest in the Internet in the 1990s? An explanation appears to be found in the change which has taken place in the tools for interfacing to and exploiting the Internet. As we indicated in our discussion of the origins of Connect, tools for interfacing to the Internet changed considerably during the late 1980s. Graphical user interfaces replaced hard to use command line interactions. This gave a greater range of people ready access to the Internet. It was however the advent of the World Wide Web and the embodiment of protocols such as Ftp and Gopher within easy-to-use Web browsers that transformed access to the technology on a world-wide scale.

The growth of the WWW has been remarkable even compared with the Internet at large, as shown by the number of hosts per Web server. In January 1995, one in 94 machines on the Internet was reported to be a Web server (Gray, 1996). It is estimated at least 50% of all Web sites are now .com sites (Genobject, 1996; Gray, 1996). To the person in the street and to many businesses, the World Wide Web is the Internet.

The increase of commercial interest in the Internet thus appears rooted in the interface to the network that the World Wide Web provides. The World Wide Web allows the Internet to be readily used as both a medium of *communication* and a mode of *presentation*. It is this *transformation in the accessibility of the technology* that has made it attractive to the business community. As we pointed out in our discussion of the historical precursors of Connect, it was technically feasible to realize the functionality of using the Internet as a "Shop Window on the World" by other means[13] prior to the advent of the World Wide Web. However, the accessibility of the technology and hence, the *perceived* advantage of the Internet to the business community has been substantially altered by the advent of the WWW.

A redefinition of the Internet

The above analysis highlights the difficult in determining what exactly is being referenced when one discusses the Internet as an innovation. If the Internet is defined simply as a network of computers, then as that network has existed since the early 1970s. However, when one attempts to analyze change in its exploitation over time, the importance not only of the network itself, but its associated protocols, such as HTTP[14], and tools for viewing information being transmitted over the network, such as Web browsers, become important. This suggests that the Internet as an innovation should therefore be redefined as *a technology cluster* of the network, the protocols that can exploit it, the type of information transferable by the protocols and the tools for viewing that information. Viewed from this perspective, since its inception in the 1970's the Internet has undergone a series of transformations; it has reinvented[15] itself as the technologies associated with its use have changed.

The paradox disappears

If we accept that the Internet has been reinvented through change in associated technologies, the paradox we started this chapter with disappears. The academic and military communities' exploitation of the Internet focused on its *communications* aspect. In our experience at Connect, any knowledge about the Internet that businesses bring to the Internet Awareness Days is predominantly centred on the *advertising* potential of the World Wide Web. The strength of the Internet as a communications medium is often not realized.

The Internet only became of interest to the business community when the World Wide Web provided an easy interface to use the Internet as a publishing agent. The availability of this type of functionality was not necessary for academic and military interest in the technology.

The Uniqueness of the Internet as an innovation

What of our second question? Is the Internet comparable to any other innovation in terms of the factors which have influenced its diffusion or does it have unique features that challenge some fundamental assumptions of traditional models of Innovation and Diffusion?

Our analysis supports the latter position. The Internet has several features that make it unique. First, throughout our analysis. the fact that the Internet can act as a medium for its own diffusion has been a consistently important factor in its adoption. Little to no formal advertising surrounded the success of the Public Domain Archive. MerseyWorld advertises the power of the Internet as a presentational medium. The need for the small business to get customers or suppliers on-line before the Internet can fully benefit their own interests results in the interested adopter also becoming an agent in the diffusion process. Once on-line, the potential client base of the adoptee changes dramatically but without the need for the local structure of the business to change. Unlike innovations that require adaptation of the client's normal way of working, the Internet allows the old and the new method to operate in parallel. This is very clearly seen in the case study of Henry Diaper that we described above. The Internet is an innovation that is value added in the sense that, while it may be adopted for one purpose, that of Electronic Commerce, it can readily be made to serve others, such as on-line training of staff, children's education, etc.

The Internet as an innovation is also unique in that in reality, as discussed above, it defies being defined. We suggest that it is best thought of as a technology cluster that is constantly evolving, features of which are liable to be of greater interest to one community as opposed to another. Our initial analysis of growth of the Internet simply treated the Internet as the number of host computers that were connected. If we redefine the Internet as a technology cluster, that growth pattern would need to be analyzed relative to changing features of the technology cluster that were deemed relevant. Such an analysis is difficult to conduct as features of the technology cluster are not independent in their effects.

What the term *social system* means in the context of the Internet poses a challenge for diffusion theory. The fact that this is a medium that can be used for a wide variety of purposes, in a wide variety of business and other situations, that transcend national boundaries, that create alliances as well as reinforce existing ones, poses major conceptual problems. The traditional S-shaped curve observed in studies of diffusion is conventionally explained with reference to an underlying normal distribution of number of adopters at a given moment in time in the social system referenced by the innovation.

There is, however, no boundary to the social system within which the Internet is being adopted. The Internet as an innovation certainly challenges the adequacy of traditional models of the diffusion process.

Conclusion

Within this chapter, we have examined the diffusion of the innovation we refer to as the Internet at a global and at a micro-level of analysis. Our analysis has been partly mathematical, partly descriptive. Throughout our analysis, we have sought to emphasize the time varying nature both of the innovation and of the diffusion process itself. Classical models of innovation and diffusion were found to be overly static for our purpose. This we suggest may reflect their predominant base in retrospective analysis. The process of diffusion we sought to describe is ongoing and one in which we are active participants. It involves an innovation, the Internet, which we ourselves argue is really an evolving technology cluster, acting as a medium for its own diffusion. These factors combine to suggest to us that a new model of diffusion is needed to deal with phenomena of this type, one rooted in methodologies that can handle synergy and interaction between contributing factors in the diffusion process. Perhaps, such a tool may be found in Cybernetics.

Acknowledgments

The establishment of Connect and the other work described here has been funded by the European Union through the European Regional Development Fund and the European Social Fund, and supported also by Hewlett-Packard and Soft&Net Distribution SA of Geneva. The success of the Center depends, of course, on the abilities and efforts of many staff who have worked for/been associated with it, and also on the involvement of many organisations in the Merseyside region and elsewhere. Our thanks are here expressed to them all.

Endnotes

[1] The research goal was to develop communications protocols that would allow network computers to communicate transparently across multiple, linked packet networks.

[2]Equivalent figures for 1995 were not available at press time.

[3]This assumption has been queried in other contexts such as Mahajan and Peterson's (1978) study of membership of the United Nations from 1945-1974. They developed a dynamic diffusion model to accommodate the fact that because the number of countries of the world nearly doubled during this period, the ceiling on the number of potentially joining countries (adoptees) was not static.

[4] To date, there is no clear evidence that this assumption is not a reasonable one.

[5]Examples of extensions are dynamic diffusion models, multi-innovation diffusion models, multistage diffusion models, multiadoption diffusion mod-

els, space and time diffusion models and models that incorporate influencing or change agents.

[6]Objective 1 status means that a region is recognized as particularly deprived and in need of economic regeneration. Special funds are set aside by the European Commission for regional development proposals.

[7] Andreessen left NCSA in 1994 to co-found Netscape Communications Corporation with a former member of Silicon Graphics, Jim Clark.

[8] Exactly how this change should be conceived is problematic. It is possible to view the invention of the HTTP protocol and the development of Web browsers as a case of 'reinvention' of an innovation (the Internet as a distribution mechanism) by its users (the academic community at CERN).

[9] The organizers were the Economic Development Group of the Mersey Council and Partnership for Enterprise.

[10] We acknowledge that this may reflect reporting bias. Perhaps those clients with bad experiences don't contact us again, even to tell us of these experiences.

[11] There have been more people interested in the Short Course Programme that we have been able to accommodate to date. The Internet Awareness Programme was established before the former. There is therefore a lag between going on the latter and being able to attend a short course.

[12]This resource offers the potential of ISDN access between any points of the private networks of the large organizations in Merseyside. ISDN lines would permit the use of teleconferencing tools and allow Connect to develop distance learning courses which businesses could access as required from the workplace.

[13] In our case through the use of dedicated clients for Ftp and Gopher and the local development of a software tool, xbrowser, for perusing a catalog on-line.

[14] HyperText Transfer Protocol: the protocol on which the World Wide Web is based.

[15] Reinvention is a term that is normally used within the diffusion literature to refer to adaptation of the innovation to a different purpose by an adoptee. Here we suggest there is also a need to recognise that an innovation may reinvent itself through change in associated technologies. This suggests that a systems or cybernetics perspective might usefully be adopted in some forms of diffusion research.

References

Bush, V. (1945). As we may think. *Atlantic Monthly, http://www.csi.uottawa.ca/~dduchier/misc/vbush/as-we-may-think.html*

CERN, (1990). *http://www.cern.ch/*

Charlton, C., Little, J., & Woodward, M. R. (1994). A Retraining programme for unemployed IT professionals. *AXIS*, 1(3), pp 7-13.

Charlton, C. C., Leng, P. H., Little, J. M., & Woodward, M. R. (1994). *The Liverpool HP-UX Archive Software Library*, Paper presented at the Proceedings of the International Hewlett-Packard Users Conference, Maastricht.

Forrester Research. (1995). *http://www.forrester.com/*

Entickmap, N. (1992). Report on the Computer Weekly / SSP Survey: skills most in demand. Computer Weekly(26, November), p 25.

Genobject. (1996). *http://home.tig.com/cgi-bin/genobject/domainb/*

Gray, M. (1996). *http://www.mit.edu:8001/people/mkgray/mkgray.html*

Lotter, M. (1996). *Network Wizards: Internet Host Survey. http://www.nw.com/*

Mahajan, V., & Peterson, R. A. (1985). *Models for Innovation Diffusion.* London: Sage Publications Ltd.

Nelson, T. (1965). *http://www.aus.xanadu.com:70/Oh/xanadu/*

Nielson, J. (1993). *Usability Engineering:* Academic Press Inc.

Rogers, E. M. (1983). *Diffusion of Innovations.* (Third Edition ed.). New York: The Free Press, A division of Macmillan Publishing Company.

Rutkowski, A. M. (1996). *Internet Hosts 1989-1996. http://www.genmagic.com/*: General Magic Inc.

The Bangemann Report. (1994). *Europe and the global information society: recommendations of the European Council.* Brussels: European Commission.

The Economist. (1995). *http://www.economist.com/*

The European Commission. (1994). *Growth, competitiveness, employment: the challenges and ways forward into the 21st century* (White Paper ISBN 92-826-7423-1). Luxembourg: European Commission.

The European Community. (1994). *The EC Structural Funds and the development of Merseyside, 1994-1999:* EC Regional Policies Document, EC Publications, CX-87-95-765-EN-C.

The Internet Society. (1996). *http://www.isoc.org/*

Thomas, & Znaniecki. (1927). *The Polish Peasant in Europe and America.* New York: Knopf.

U.S. Small Business Administration: Office of Advocacy. (1995). *The Third Millennium: Small Business and Entrepreneurship in the 21st Century.* Paper presented at the White House Conference on Small Business, Washington, D.C.

Vail, S. (1993). Staff skills: new work for old hands. *Computing* (June 3), pp 23-25.

CHAPTER
13

Technology Diffusion in Software Development Processes: The Contribution of Organizational Learning to Software Process Improvement

Dirk Stelzer
Werner Mellis
Georg Herzwurm
University of Cologne, Germany

The improvement of software development processes has gained considerable attention both from academia and industry. Various concepts and standards have been published to facilitate software process improvement. Numerous software companies worldwide have established initiatives to improve their software development performance.

Empirical studies into the success of software process improvement initiatives report about remarkably positive results. However, most initiatives also had to handle several complex problems.

Despite the growing interest in the improvement of software development a profound knowledge of the enablers and inhibitors of software process improvement is still lacking. It is essential to achieve an understanding of the factors that cause success and failure of software process improvement initiatives. A thorough comprehension of the factors is a prerequisite for effective research that attempts to describe, to explain or to make recommendations for the improvement of software development processes. A complete understanding of the factors is also essential for practitioners that wish to implement software process improvement activities effectively and efficiently.

The chapter attempts to contribute to the comprehension of the factors that facilitate or impede the success of software process improvement initiatives. The chapter draws on findings of two empirical studies into success factors of improvement efforts in software organizations. We use technology diffusion to explain the success factors and we explore whether and how organizational learning might support the implementation of the factors.

The chapter has three objectives: to describe factors that facilitate or impede the success of software process improvement initiatives, to explain the factors from the perspective of technology diffusion, and to explore how organizational learning might contribute to the improvement of software development processes.

Background

The following paragraphs provide an overview of technology diffusion, software process improvement and organizational learning. We will define essential terms, describe key concepts, and provide an overview of the literature. Subsequently, we will explore the coupling of technology diffusion, software process improvement and organizational learning on a conceptual level.

Technology Diffusion

According to Rogers (1983, p. 12) "a technology is a design for instrumental action that reduces the uncertainty in the cause-effect relationships involved in achieving a desired outcome." New technologies are used to achieve desired results more effectively and/or more efficiently. "Diffusion is the process by which an innovation is communicated through certain channels over time among members of a social system" (Rogers, 1983, p. 5). "An innovation is an idea, practice, or object that is perceived as new by an individual or other unit of adoption" (Rogers, 1983, p. 11).

Various authors have discussed issues of technology diffusion in the context of information systems development. For example, Cooper and Zmud (1984) empirically examined research questions concerning the implementation of a production and inventory control information system. Basili and Musa (1991) discussed problems of technology diffusion in software engineering. Basili, Daskalantonakis, and Yacobellis (1994) described a model for technology transfer that was implemented in a software engineering environment. Fichman and Kemerer (1993) proposed a framework for the adoption of software engineering process innovations and applied it to structured analysis and design methodologies, fourth generation languages, relational database management systems, and object orientation. On the basis of an empirical study Iivari (1996) explained factors affecting the adoption of Computer-Aided Software/Systems Engineering tools. Zelkowitz (1996) investigated the adoption of several new software engineering technologies by an individual organization. He described the infusion process of object-oriented technology, of the cleanroom software development method, of formal inspections, and of an expert system within NASA. Fowler (1996) gives an overview of the literature on information technology transfer and diffusion.

However, there are no studies into technology diffusion in software process improvement initiatives that are based on the ISO 9000 standards or the CMM.

Software Process Improvement

A software process is "a set of activities, methods, practices, and transformations that people use to develop and maintain software and the associated products (e.g., project plans, design documents, code, test cases, and user manuals)" (Paulk, Curtis, Chrissis, & Weber, 1993, p. 3). The term 'software process improvement' denotes the "changes implemented to a software process that bring about improvements" (Olson, Humphrey, & Kitson, 1989, p. 53). The intent of software process improvement is improving software product quality, increasing productivity, and decreasing the cycle time for product development.

Worldwide, there are various approaches to implementing software process improvement. North American companies seem to prefer the Capability Maturity Model (CMM) (Paulk et al., 1993). European software organizations tend to use the ISO 9000 family to enhance their capabilities. (ISO 9000-1, 1994) Apart from the Capability Maturity Model and the ISO 9000 family there are various alternative approaches to software process improvement such as, BOOTSTRAP (Kuvaja & Bicego, 1994; Lebsanft, 1996), TRILLIUM (Coallier, 1995) and the emerging Software Process Improvement and Capability dEtermination standard (SPICE) (Dorling, 1993; Rout, 1995). The CMM and the ISO 9000 standards are the most prominent approaches to supporting software process improvement. Both concepts will be described briefly in the following paragraphs.

Capability Maturity Model

The objective of the Capability Maturity Model (CMM) is to guide software organizations in determining current process maturity and in identifying opportunities for process improvement. "Software process maturity is the extent to which a specific process is explicitly defined, managed, measured, controlled, and effective" (Paulk et al., 1993, p. 4). The CMM also provides organizations with guidance on how to steadily improve its software process to enable continuous and lasting gains in software process capability. "Process capability describes the range of expected results that can be achieved by following a software process" (Paulk et al., 1993, p. 4).

The CMM is based on the assumption that continuous process improvement should be based on evolutionary steps rather than on revolutionary innovations. The CMM provides a framework that structures these evolutionary steps into five maturity levels. The maturity levels define an ordinary scale for measuring the maturity of an organization's software process. The levels also help an organization prioritize its improvement efforts.

Each of the five levels describes a certain maturity level of a software process. Maturity levels 2 through 5 provide a roadmap for improving the software process. Each maturity level is decomposed into key process areas. A key process area "identifies a cluster of related activities that, when performed collectively, achieve a set of goals considered important for enhancing process capability" (Paulk et al., 1993, p. 30). Key process areas describe the issues that must be addressed to achieve a maturity level. To achieve maturity level 2, for example, an organization must establish the following key process areas: Requirements management, software project planning, software project tracking and oversight, software subcontract management, software quality assurance, and software configuration management.

ISO 9000 Quality Standards

In Europe, the ISO 9000 family is one of the most important standards in the field of quality management. Many software suppliers try to accomplish software process improvements by implementing a quality system based on the ISO 9000 standards.

The ISO 9000 family is a set of internationally recognized standards on quality management. The standards are intended "to provide a generic core of quality system standards applicable to a broad range of industry and economic sectors" (ISO 9000-1, 1994). A major purpose of quality management is "to improve the systems and processes so that continual improvement of quality can be achieved." The ISO 9000 family is based on the assumption that all work is accomplished by a process. Accordingly, quality management means managing and improving all processes in an organization.

The ISO 9000 family consists of various standards that are closely interrelated: The ISO 9004 series (ISO 9004-1, 1994) comprises standards for internal use by an organization. They provide guidance for the design and implementation of a quality system. A quality system is a set of "organizational structures, procedures, processes and resources needed to implement quality management" (ISO 9000-1, 1994).

ISO 9001 (1994), 9002 (1994), and 9003 (1994) are models for external quality assurance. They specify a set of requirements. The demonstration of conformance to these requirements is called certification or registration. ISO 9000-1 to 9000-4 (1994) are guidelines for the selection and application of ISO 9001, 9002, and 9003. ISO 9000-3 (1992), for example, is a guideline for the application of ISO 9001 to the development, supply and maintenance of software.

The ISO 9000 standards provide a framework for implementing quality systems. Quality elements describe the issues that must be addressed to implement a quality system. ISO 9000-3 (1992) suggests to apply a set of quality system elements. The elements in ISO 9000-3 are structured into three layers: a framework, life-cycle activities, and supporting activities that are not related to a specific phase of the software development life-cycle. The quality system framework describes management's responsibilities and activities when implementing software process improvement. ISO 9000-3 does not prescribe any specific life-cycle model, but it suggests nine life-cycle activities that should be implemented and maintained. The third layer comprises supporting activities, for example configuration management, product and process measurement, and procedures to ensure the quality of products supplied by subcontractors. The standards describe what elements quality systems should encompass but not how a specific organization should implement these elements. (ISO 9000-1, 1994)

Review of Empirical Research on Software Process Improvement

Thousands of software companies worldwide have established software process improvement initiatives. Experience reports and findings of empirical studies report about remarkably positive results. Herbsleb, Zubrow, Siegel, Rozum, and Carleton (1994) conducted a study on the benefits of CMM based software process improvement. They report about productivity gains of 35% per year, early defect detection gains of 22% per year, yearly reduction in time to market of 19%, and yearly reduction in postrelease defect reports of 39% per year. (Herbsleb, Carleton, Rozum, Siegel & Zubrow, 1994; Herbsleb, Zubrow, Siegel, Rozum, & Carleton, 1994) Loken and Skramstad (1995) performed a study of European software development companies. They found increased efficiency, improved product quality and higher revenues as consequences of ISO 9000 based software process improvement efforts.

Most software process improvement initiatives also had to handle a number of severe problems. For example, Herbsleb, Carleton, Rozum, Siegel, and Zubrow (1994) mention staff resistance and initial reluctance of software engineers to participate in the improvement initiative: Loken and Skramstad (1995, p. 4) mention increased overhead. Bellin, Stelzer, and Mellis (1996) identified several difficulties of ISO 9000 based software process improvement initiatives, for example: inadequate expectations of senior management, weak management commitment, lacking involvement of staff members, excessive bureaucracy, and insufficient organizational support for the improvement activities.

Previous empirical studies on software process improvement have mainly focused on problems of software engineering, the relationship between process and product quality, and the effects of improvement initiatives on business results. (Herbsleb, Carleton, Rozum, Siegel, & Zubrow, 1994; Herbsleb, Zubrow, Siegel, Rozum, & Carleton, 1994; Loken & Skramstad, 1995)

Most studies on software process improvement have mainly been descriptive or normative without attempting to explain factors affecting the adoption of innovations and thus the success of software process improvement initiatives.

Humphrey (1989, p. 17) describes some principles of software process change. He emphasizes the importance of change management and highlights some significant prerequisites for technology diffusion in software process improvement initiatives.

The studies and the empirical findings give us a better understanding of benefits, problems, and difficulties of software process improvement. However, Herbsleb, Carleton, Rozum, Siegel, and Zubrow (1994, p. 49) point out an important weakness of previous research work: "We are convinced, however, that it is time to move beyond the basic question of whether process improvement can pay off. We need to learn how, when, why, and for whom process improvement pays off. This includes an empirical examination of the assumptions underlying the models of process improvement such as the CMM, as well as an analysis of enablers and inhibitors not accounted for in [the] models. We need to achieve an understanding of the critical factors that cause success and failure."

Growing evidence shows that various critical success factors of software process improvement initiatives are not fully accounted for in the CMM, the ISO 9000 standards, and in other software process improvement models. The ISO 9000 standards, as well as the CMM, implicitly assume that new technologies will be readily adopted. Several factors that might impede the diffusion and adoption of innovations are not accounted for in software process improvement models and standards. Astonishingly, relatively little research has been done into the factors that contribute to the success of software process improvement initiatives.

Herbsleb, Carleton, Rozum, Siegel, and Zubrow (1994, p. 49) suggest to examine a broad sample of software process improvement projects in order to determine characteristics of successful and unsuccessful improvement efforts, and factors that distinguish successes and failures.

We have conducted two studies among 48 software organizations to identify success factors of software process improvement initiatives. The findings of the studies will be described in one of the following sections of this chapter. Subsequently, we will use Rogers' approach to technology diffusion to explain the factors that facilitate or impede the success of software process improvement initiatives.

Organizational Learning

The term "organizational learning" was popularized by Peter M. Senge in his book "The Fifth Discipline: The Art and Practice of the Learning Organization." Senge describes learning organizations as places "where people continually expand their capacity to create the results they truly desire, where new and expansive patterns of thinking are nurtured, where collective aspiration is set free, and where people are continually learning how to learn together" (Senge, 1990, p. 3). According to Senge, a learning organization is an "organization that is continually expanding its capacity to create its future" (Senge, 1990, p. 14). Garvin describes a learning organization as "an organization skilled at creating, acquiring, and transferring knowledge, and at modifying its behavior to reflect new knowledge and insights" (Garvin, 1993, p. 80).

Simon (1957), Cyert and March (1963), Argyris and Schoen (1978), Levitt and March (1988), Cohen and Levinthal (1990), and Schein (1992, 1995) to name only a few authors, have laid theoretical foundations for organizational learning. Rogers (1983) addresses the relationship between diffusion theory and social learning theory. Fiol and Lyles (1985), and Dodgson (1993) provide an overview of the literature on organizational learning.

Stata (1989), Senge (1990), Nonaka (1991), Garvin (1993), and others have made practical suggestions for the implementation of organizational learning in a managerial context.

Ohba (1993) described aspects of organizational learning in the context of software development. Thoresen (1996) analyzed processes of organizational learning in groups of workers adopting information technology. However, there are no studies into the potential contribution of organizational learning to software process improvement.

Although no theory or model of organizational learning has gained widespread acceptance (Fiol & Lyles, 1985), most authors agree on at least the following characteristics of organizational learning: It is a process of deepening knowledge and understanding of individuals in organizations that leads to improved behavior (Fiol & Lyles, 1985;

Dodgson, 1993; Garvin, 1993).

Most authors concede that organizational learning is a concept that has considerable analytical power. (Dodgson, 1993) We will explore whether and how organizational learning might contribute to the implementation of critical success factors in software process improvement activities.

Coupling of Technology Diffusion, Software Process Improvement and Organizational Learning

A comparison of the definitions of technology diffusion, software process improvement and organizational learning shows that the three concepts are closely interrelated. The definitions have at least three elements in common. Each of the concepts addresses

- a process of
- changing behavior
- to achieve improved results.

Technology diffusion is a process of adopting ideas, practices, or objects to achieve desired outcomes more efficiently and/or more effectively. Software process improvement is a process of changing software development activities to achieve improvements of product quality, productivity, and cycle time. Organizational learning is a process of modifying knowledge and understanding to achieve improved behavior.

Technology diffusion may help to explain software process improvement. Software process improvement is a special case of technology diffusion. Technology diffusion occurs when members of a social system adopt ideas, practices, or objects that are perceived as new. Software process improvement occurs when software organizations adopt new activities, methods, practices, etc. to improve software development.

The changes implemented to a software process when conducting an improvement initiative tend to be diverse. Most companies adopt several innovations in the course of a software process improvement initiative. For example, organizations that conduct CMM based improvement efforts usually implement six key process areas to reach maturity level 2. Each key process area consists of various practices and activities. Rogers call this a technology cluster. "A technology cluster consists of one or more distinguishable elements of technology that are perceived as being closely interrelated" (Rogers, 1983, p. 14).

On the initial level of the CMM the software process is characterized as ad hoc, and occasionally even chaotic. Few processes are defined, performance of projects is unpredictable and success depends on the capability of individuals. To achieve maturity Level 2, an organization must establish basic project management processes to track costs, schedule and product functionality. To achieve Level 3, an organization must establish activities to ensure that the software process is defined, documented and standardized. To achieve Level 4 the organization sets quantitative quality goals for both the software process and products. The organization establishes a measurement program and collects and analyzes data from the projects. To achieve level 5 an organization must establish activities to ensure that the software process is continuously improved by quantitative feedback from projects and from piloting innovative ideas and technologies.

When implementing an ISO 9000 based quality system a software organization must establish 20 elements, for example, internal quality audits, configuration management, procedures to control documents, product and process measurement, training, etc.

When a software organization implements an ISO 9000 based quality system or when an organization's software process moves from one maturity level to a higher maturity level technology diffusion occurs:the organization's top management, project managers, and software engineers adopt new activities, practices, tools, and techniques.

Software process improvements cannot be accomplished in a short time frame. Usually it takes several years to achieve substantial improvements. (Paulk et al., 1993, p. 14; Stelzer, Mellis, & Herzwurm, 1996, p. 710)

The improvements may be anticipated results or unanticipated effects that accompany the improvement initiative. The results may take the form of improved product quality, increased productivity, or decreased cycle time for product development. (Herbsleb, Zubrow, Siegel, Rozum, & Carleton, 1994) Process improvement efforts may also result in intangible benefits. For example, better morale, improved understanding of the corporate mission and vision, fewer crises, less stress, less staff turnover, and better communication within the organization have been reported to accompany software process improvement. (Herbsleb, Carleton, Rozum, Siegel, & Zubrow, 1994)

Organizational learning may help to design, implement and improve software process improvement initiatives. It may contribute to the elicitation, formalization, and distribution of knowledge. Organizational learning may facilitate thorough understanding and efficient communication of new technologies.

The three research fields, information technology diffusion, software process improvement, and organizational learning have attracted the interest of both the research and the practice communities. As we have shown in the previous paragraphs the three research fields are closely interrelated. Astonishingly, only little research has been done into the coupling of the three fields.

We will therefore use technology diffusion as a model to explain critical success factors of software process improvement. Furthermore, we will assess the potential contribution of organizational learning to the implementation of the factors.

Investigation Framework

Kautz and Pries-Heje (1996) identified five research directions in diffusion and adoption of information technology. The first direction attempts to identify or specify overall frameworks and models for the whole diffusion and adoption process. The second line of work is mainly normative. It gives recommendations on how to make diffusion and adoption happen in an organization. The third direction focuses on selected issues of adoption and diffusion. Representatives of this direction use qualitative methods, such as case studies, to explore and to identify important issues of diffusion and adoption of information technology. The fourth research direction is grounded in a quantitative research tradition. Representatives of this group use quantitative modeling and measuring to explain adoption and diffusion processes. The fifth research direction focuses on adoption and diffusion on a regional, national, or sectorial level.

Our research work uses a combination of the third and the fifth research direction identified by Kautz and Pries-Heje. We focus on issues of technology diffusion in the software industry. We use case studies to explore and to identify critical success factors of software process improvement activities. On a conceptual level we attempt to explain the success factors with the help of technology diffusion. Furthermore, we assess the potential contribution of organizational learning to the implementation of the success factors.

Research Objectives

The overall research objectives of this chapter are:

- to describe factors that may facilitate or impede the success of software process improvement initiatives
- to explain the factors from the perspective of technology diffusion, and
- to explore how organizational learning might contribute to the implementation of the factors and thus, to the improvement of software development processes.

We will identify critical success factors that are not fully accounted for in popular models for software process improvement, such as the CMM or the ISO 9000 standards. We will demonstrate that technology diffusion is an adequate model to explain the factors. We will show that organizational learning may contribute to implementing and maintaining the factors.

Research Questions

To achieve the research objectives we will consider the following questions:

- What are the key success factors of software process improvement initiatives?
- Does previous research into diffusion theory provide an adequate basis to explain the success factors?
- Which requirements must be met by an approach that attempts to support the implementation of the factors?
- Does organizational learning meet the requirements? Does it have the potential to provide helpful support for the implementation of the factors?
- Does the coupling of technology diffusion, software process improvement, and organizational learning provide promising research opportunities?

Research Methodology

We have conducted two studies to identify factors that contribute to failures and successes of software process improvement initiatives. The first study was carried out between October 1994 and May 1995 among 36 European software houses that had sought certification or were implementing an ISO 9000 quality system at this time. (Bellin et al., 1996; Stelzer et al., 1996) The results are based on interviews with quality managers and top executives and on progress reports and case studies published by these software companies. The study also draws on interviews with ISO 9000 auditors and consultants working in the software industry, and on the authors' experience gathered by consulting software houses that implemented a quality system and prepared for ISO 9001 certification.

Our major research hypothesis was that successful software process improvement initiatives implement and maintain various factors that are not fully accounted for in software process improvement models, such as the CMM or the ISO 9000 standards.

We asked the participants of the study to describe how exactly they have achieved improvements. We asked them to identify key success factors of the software process improvement initiative. The term "success factor" describes the aspects that the respondents of our study considered to be essential when implementing an ISO 9000 based improvement initiative. If a success factor is implemented, it facilitates the success of the improvement initiative. If it is not implemented - or not implemented correctly - it may cause failure of the improvement efforts. We described and categorized the answers and compiled a list of factors that the respondents of our study considered to be essential when implementing an ISO 9000 based improvement initiative.

Between March and May 1996 we conducted a second study to identify success factors of CMM based software process improvement. (Stelzer, 1996) We analyzed empirical studies conducted by the Software Engineering Institute and experience reports and case studies from a total of 12 companies that had undertaken CMM based improvement efforts. Once again, the main focus of our second study was to identify factors that cause failures and successes of improvement efforts. These factors are usually summarized as lessons learned in the published experience reports and case studies. The second study revealed factors that are very similar to the success factors of ISO 9000 based software process improvement activities identified in our first study.

The two studies are slightly different in various aspects: The first study investigates into ISO 9000 based improvements, the second study focuses on CMM based efforts. The first study comprises 36 companies, the second 12 organizations. The first study draws on interviews and on published reports; the second study only draws on published case studies. Nevertheless, both studies identified similar success factors.

We compared and categorized the success factors of the ISO 9000 and the CMM based improvement initiatives. Subsequently we elaborated a synopsis and compiled a list of 12 factors that contributed to success and failure of software process improvement initiatives in the organizations included in our studies.

The findings of our studies do not indicate that every software company that undertakes an improvement initiative does necessarily have to implement all of the 12 success factors. The findings also do not indicate that there are no other factors that cause success and failure of software process improvement. However, our studies give evidence to the assumption that several critical success factors are not fully accounted for in popular software process improvement models. These success factors are relevant in most software process improvement initiatives, irrespective of whether the initiative is based on the CMM or on the ISO 9000 standards. The findings also support an interesting hypothesis for further research work: Software process improvement initiatives have a greater likelihood of success to the degree that the 12 success factors are implemented.

As Herbsleb, Carleton, Rozum, Siegel, and Zubrow (1994) have pointed out it is necessary to achieve an understanding of the critical factors that cause success and failure of software process improvement. We will use Rogers' (1983) approach to diffusion of innovations as a model to explain the success factors. The explanation may help researchers to describe and to explain key issues in software process improvement. It may also help practitioners to gain a better understanding of the factors that facilitate or impede the success of software process improvement activities.

The explanation of the factors provides a basis to derive requirements for an approach that may support the implementation of the factors. We will derive 12 requirements to examine whether organizational learning may contribute to the success of software process improvement activities. The examination will be conducted in three stages: On the first stage, we will present Senge's and Garvin's suggestions to implement organizational learning. We have selected Senge's (1990) and Garvin's (1993) approaches to organizational learning because they have made practical suggestions of applying organizational learning in a managerial context. On the second stage, we will examine whether the recommendations have the potential to meet the requirements. On the third stage, we will present practical examples of organizational learning in software process improvement contexts. The examples will help to assess whether organizational learning is a feasible option to support software process improvement efforts.

At the end of the chapter, we will discuss future trends and suggest research opportunities within the domain of enhancing software process improvement by organizational learning.

Success Factors of Software Process Improvement

In the following section we will describe 12 factors that we have identified in our empirical studies. These factors have facilitated or impeded the success of software process improvement initiatives in the organizations included in our studies.

Setting clear, relevant and realistic goals

Successful improvement efforts have clear, relevant and realistic goals. (Herbsleb, Carleton, Rozum, Siegel & Zubrow, 1994, p. 39; Larner, 1995, p. 10; Haley, 1996, p. 40) Setting clear goals means that staff members are able to understand the objectives of the initiative. Reducing the amount of defects in software products delivered to customers by 50%, reaching CMM Level 2, or receiving an ISO 9001-certificate are examples for clear goals. However, the vision of total quality management, or of improving customer satisfaction seem to be too fuzzy for most software engineers. Setting relevant goals means that the improvement efforts should attempt to contribute to the success of the organization. Improvement initiatives usually contribute to the success of the organization if they help staff members to perform their tasks more efficiently or if there is a direct cause-effect relationship between the objectives of software process improvement and revenues, cash flow, or other business results. However, mere conformance to a standard usually is not a relevant objective. Setting realistic goals means that there should be a chance to achieve the goals in the foreseeable future and with a reasonable amount of resources. For most software companies, for example, it is not realistic to reach CMM Level 3 within one year. Managers that set excessive goals usually demotivate their subordinates. (Wohlwend & Rosenbaum, 1994, p. 838; Bellin et al., 1996)

Providing enhanced understanding

Software development interacts with many other parts of the business such as sales and marketing, hotline support, consulting, etc. Daily routine often prevents software engineers from reflecting on their work from a general business perspective. (Wohlwend & Rosenbaum, 1994, p. 838) Improvement initiatives that gave practitioners the opportunity to understand the business of their organizations better helped them to perceive problems that they had not noticed before. Subsequently, they understood why improvement efforts were important. This encouraged them to adopt innovations in software development. (Daskalantonakis, 1994, p. 18, Herbsleb, Carleton, Rozum, Siegel & Zubrow, 1994, p. 44) The production of process documents has helped software development teams to gain a better understanding of strengths and weaknesses of the current processes. (Wohlwend & Rosenbaum, 1994, p. 835) Internal audits and improvement workshops also contributed to an enhanced understanding. (Stelzer et al., 1996)

Unfreezing the organization

In software companies that have successfully conducted improvement initiatives at least some employees understood the need for improvement efforts. They had noticed process deficiencies and had realized that many practices were inadequate and inefficient, and they suffered from these weaknesses. These staff members were willing to change the traditional processes. Once they realized that managers were interested in achieving improvements, they made valuable proposals and committed themselves to the initiatives. (Wohlwend & Rosenbaum, 1994, p. 834; Stelzer et al., 1996) Organizations that have successfully conducted improvement projects observed a "cultural change." Staff members in these organizations had accepted the necessity to change the software development process or, to put it in Lewin's (1947, 1958) words, the organization had been unfreezed. Often, the entire organization committed to continuously improving software processes (Dion, 1993, p. 35; Seshagiri, 1996, p. 692) In other companies, employees had not noticed any problems. They did not accept the need for changes, and they were reluctant to participate in improvement activities. Of course this hindered the success of the CMM and ISO 9000 projects. (Herbsleb, Carleton, Rozum, Siegel & Zubrow, 1994, p. 39; Wohlwend & Rosenbaum, 1994, p. 835; Bellin, et al., 1996).

Acquiring knowledge

Successful improvement efforts in software companies have exploited various sources of knowledge. External sources are the CMM, the ISO 9000 standards, and case studies or experience reports published by other organizations. These documents provide helpful suggestions and useful insights. Both the CMM and the ISO 9000 standards require documenting the status quo of the software development process. Process description is helpful because it identifies knowledge that was not explicit before the improvement efforts were started. For example, managers usually had a general idea of the software process, but they did not have complete understanding of essential details. Employees usually did not know how colleagues fulfilled their tasks. During the documentation of the status quo it became clear that numerous individual software developers had invented their own procedures, methods, and tools for requirements elicitation, software testing, configuration management, code inspections, and other tasks. During the improvement initiatives the material was collected, analyzed, refined, and documented. This helped to standardize and to simplify routine procedures that subsequently resulted in savings of time, money, and energy. Thus, the acquisition of internal knowledge contributed to the success of software process improvement initiatives. (Humphrey, Snyder & Willis, 1991, p. 12; Dion, 1993, p. 29; Wohlwend, Rosenbaum, 1994, p. 835; Seshagiri, 1996, p. 690; Stelzer et al., 1996) .

Transferring knowledge

Knowledge that was acquired during the process of documenting the status quo in software development was subsequently transferred to all staff members. This was achieved by producing checklists and guidelines and by integrating best practices into the internal training scheme. (Dion, 1993, p. 29; Wohlwend & Rosenbaum, 1994, p. 835) Successful improvement initiatives used internal audits as consultancy service. The audits helped to distribute knowledge about new tools and techniques. Process workshops helped to present new ideas and to communicate process innovations to all relevant members of the organization. (Humphrey et al., 1991, p. 18; Dion, 1993, p. 29; Stelzer et al., 1996)

Tailoring improvement initiatives

Successful improvement initiatives focus on strengths and weaknesses of specific sites or small organizational entities. (Dion, 1993, p. 29; Daskalantonakis, 1994, p. 18; Wohlwend & Rosenbaum, 1994, p. 835; Larner, 1995, p. 10; Seshagiri, 1996, p. 690) They encourage local managers and engineers to conduct self-evaluations and to create their own action plans for improvements. This ensures grass-roots involvement of all staff members. It also helps to fight specific weaknesses and to expand on individual strengths that promise the greatest benefits. Other companies used a standardized and centralized approach. They did not tailor the improvement efforts to particular needs of different sites and teams. For example, some companies that have implemented ISO 9000 quality systems copied quality manuals and documents from other organizations. Other companies tried to enforce corporate-wide regulations for software engineering techniques. Often, staff members refused to work with these documents and to comply with the regulations. A considerable number of the companies that had used a standardized and centralized approach had to withdraw documents and regulations. They had to repeat their efforts and to revise the planning, design and implementation of the process improvement initiative. (Bellin et al., 1996)

Encouraging communication

Software process improvement initiatives are often accompanied by rumors, fears and resistance from staff members. Intensive communication helps to rectify rumors, to preclude misunderstandings, to overcome resistance, and to dispel software engineers' fears. Successful process improvement initiatives have encouraged communication among staff members. This was achieved, for example, by conducting internal audits, by introducing process workshops and by defining a common language. (Daskalantonakis, 1994, p. 18; Wohlwend, & Rosenbaum, 1994, p. 836; Stelzer et al., 1996) Internal audits provide an opportunity to discuss strengths and weaknesses of the process initiative, to submit proposals for improving the efforts, and to complain about disadvantages and drawbacks. Many software companies have introduced regular process workshops. The workshops are held to discuss problems and options for improving the software process. Quality managers and process specialists consider these workshops to be a very efficient way to ensure and to maintain staff motivation and to disseminate important information. In most software companies different teams speak different languages. Expressions and terms that are well understood in one team are confused by other teams. When two or more teams cooperate they usually have to resolve misunderstandings before they are able to work efficiently. The description of software processes forces the teams to define a common language. This common language facilitates communication in all projects. (Stelzer et al. 1996) Internal audits, workshops and the definition of a common language have facilitated communication among those involved in software process assessment and improvement initiatives.

Ensuring staff involvement

Staff participation is essential in improvement initiatives because employees have detailed knowledge and first hand experience of strengths and weaknesses of the current processes. (Haley, 1996, p.

41) To ensure grass-roots involvement successful process improvement initiatives have established local process groups, special interest groups, in-house training schemes, and forums for the exchange of ideas and for coordinating efforts among groups. (Humphrey et al., 1991, p. 21; Dion, 1993, p. 30; Larner, 1994, p. 10; Wohlwend & Rosenbaum, 1994, p. 835; Seshagiri, 1996, p. 684) Using the skills and experience of software engineers guarantees that the resulting process is a consensus that reflects the practical considerations of diverse projects. Staff members who participate in changing the process adopt innovations more readily. Employees who contribute to the process improvement initiative act as advisors, advocates, change agents, and communication liaisons. (Dion, 1993, p. 30; Herbsleb, Carleton, Rozum, Siegel & Zubrow, 1994, p. 39; Bellin et al., 1996)

Emphasizing teamwork and collaboration

Many teams and divisions in software companies suffer from inadequate cooperation with other groups. Software quality-assurance teams, for example, are often not adequately integrated in the process of software development. Staff members in most companies must learn to pull together as a team. Therefore, successful improvement activities emphasize collaboration. Collaborative projects include joint process descriptions, workshops, special interest groups, and consulting on process innovation. Joint activities help staff members discover unexpected similarities in products or processes. Successful improvement initiatives have established effective interfaces of various teams. This helps to exchange experience and results from teams doing similar work in different parts of the company. Close cooperation of business units provides natural feedback loops, enhances staff members' understanding and knowledge, encourages people to exploit synergies, and consequently improves productivity and quality. (Humphrey et al., 1991, p. 13; Wohlwend & Rosenbaum, 1994, p. 835; Herbsleb, Carleton, Rozum, Siegel & Zubrow, 1994, p. 35) The collective effort to design and implement a quality system or to conduct a process improvement initiative and the will to succeed in the ISO 9001-certification or the CMM-assessment process often creates a team spirit among all employees. This helps to improve cooperation, teamwork, and the working climate in future projects. (Humphrey et al., 1991, p. 21; Stelzer et al., 1996)

Management commitment

Software process improvement requires investment, time, risk taking and "the pain of cultural change" (Humphrey et al., 1991, p. 21). These obstacles cannot be overcome by delegation. The most important lesson learned in improvement initiatives is that management commitment is the key success factor of process improvement. (Humphrey et al., 1991, p. 21; Larner, 1991, p. 10; Wohlwend & Rosenbaum, 1994, p. 836; Paulish & Carleton, 1994, p. 57;

Daskalantonakis, 1994, p. 22; Haley, 1996, p. 40; Bellin et al., 1996) Progress is only possible if managers at all levels of the organization support the initiative. "Senior managers often dictate that an improvement change be made without completely realizing the time and effort ... required for the change. They assume that the change will occur without modification of other commitments" (Wohlwend & Rosenbaum, 1994, p. 838). As a consequence, middle management find themselves caught in a trap between the need to modify software processes without affecting deadlines and milestones, budget restrictions and functional requirements. Consequently, middle management often does not support software process improvement. (Wohlwend & Rosenbaum, 1994, p. 838) Senior managers must commit to developing, implementing and financing appropriate action plans in response to the assessment recommendations. They must also actively participate in assessment meetings and improvement workshops to demonstrate the importance of the initiative. Some software process specialists state that active and visible support from management is the biggest single key to the success of improvement efforts. (Humphrey et al., 1991, p. 12; Daskalantonakis, 1994, p. 18; Herbsleb, Carleton, Rozum, Siegel, & Zubrow, 1994, p. 39).

Change agents and facilitators

Successful improvement initiatives were supported by change agents and facilitators. In the context of CMM-based improvement projects members of software engineering process groups often served as change agents or as facilitators. (Humphrey et al., 1991, p. 21; Wohlwend & Rosenbaum, 1994, p. 838; Paulish & Carleton, 1994, p. 57; Larner, 1995, p. 10) In the context of ISO 9000 based quality systems, quality managers usually played this role. (Bellin et al., 1996) Often, they initiate the improvement projects, request resources, encourage local improvement efforts, and establish interfaces and communication channels between various groups. They also provide technical support and feedback, publish successes, help to overcome difficulties, and keep staff members aware of the improvement efforts. A significant lesson that the members of one software process improvement team learned was that they were "primarily acting as change agents and secondarily as software technologists" (Wohlwend & Rosenbaum, 1994, p. 839). Daskalantonakis (1994, p. 22) reports about various champions and facilitators that are needed to conduct process assessments and improvement activities effectively. The champions are competent individuals responsible for initiating, guiding and supporting the whole process, ensuring management's support, taking care of administrative items and supporting assessments; as well as production of the action-plan, and its implementation.

Stabilizing changed processes

Software process improvement efforts do not always have long lasting effects. On the contrary, changes are often short-lived. Within a short time, group conduct often slides back to the old habits (Wohlwend & Rosenbaum, 1994, p. 834; Bellin et al., 1996). Humphrey (1989, p. 63) calls this phenomenon in a similar context “software process entropy”. It is therefore essential to stabilize and to maintain changed processes to achieve long lasting improvements. Software engineers adopting a new process need guidance and support. Misunderstandings must be clarified. Initial problems and difficulties have to be overcome, weaknesses have to be modified to ensure smooth functioning of new processes. Software engineers need continuous feedback and motivation, recognition and reinforcement to stay involved in the improvement effort. (Wohlwend & Rosenbaum, 1994, p. 834; Larner, 1995, p. 10; Seshagiri, 1996, p. 692) To prevent software process entropy successful improvement initiatives provided comprehensive support and encouraged staff members to practice the new procedures.

Explanation of Success Factors From the Perspective of Technology Diffusion

In the previous section we have described 12 factors that contribute to the success or failure of software process improvement initiatives. The factors have been derived from our empirical studies into process improvement initiatives. The factors are essential for organizations that plan to base their improvement activities on the CMM, as well as for companies that use the ISO 9000 standards.

Organizations that wish to conduct successful improvement initiatives need a thorough understanding of the factors. However, most of the factors are neither explicitly accounted for in the CMM nor in the ISO 9000 standards.

In the following section, we will explain the success factors with the help of Rogers' approach to diffusion of innovations. The explanation will help researchers to better understand how technology diffusion occurs in software process improvement contexts. The improved understanding may give researchers helpful insights into future research opportunities. The explanations may also help practitioners to prepare, implement and maintain more successful software process improvement initiatives.

Setting clear, relevant and realistic goals

Diffusion theory emphasizes the significance of various characteristics of innovations for technology diffusion: relative advantage, compatibility, complexity, trialability, and observability (Rogers, 1983, p. 15). These characteristics help to explain the rate of adoption. The rate of adoption is the relative speed with which an innovation is adopted by members of a social system (Rogers, 1983, p. 23). The characteristics also help to explain the importance of setting clear, relevant, and realistic goals for software process improvement efforts.

"Relative advantage is the degree to which an innovation is perceived as better than the idea it supersedes" (Rogers, 1983, p. 15). A process improvement project that aims at relevant goals usually has a high degree of relative advantage. "Compatibility is the degree to which an innovation is perceived as being consistent with the existing values, past experiences, and needs of potential adopters" (Rogers, 1983, p. 15). Improvement efforts that have clear, relevant, and realistic goals are perceived as being compatible with existing values and needs. "Complexity is the degree to which an innovation is perceived as difficult to understand and use" (Rogers, 1983, p. 15). Improvement efforts aiming at clear and realistic goals are usually simple to understand. Potential adopters are motivated to develop new skills and understandings that are necessary to implement and maintain the innovations. "Trialability is the degree to which an innovation may be experimented with on a limited basis" (Rogers, 1983, p. 15). Setting realistic goals usually implies that the software process improvement efforts are not immediately introduced at the corporate level. Successful improvement efforts were tested at a local or site level with a reasonable budget. This provided the opportunity to improve the improvement activities before the initiatives were extended to the entire organization. "Observability is the degree to which the results of an innovation are visible to others" (Rogers, 1983, p. 16). Improvement efforts that attempt to achieve relevant objectives are highly visible and can easily be communicated. This stimulates peer discussions of the innovations, and improves the likelihood of adoption.

Software process improvement initiatives setting clear, relevant and realistic goals usually have a high degree of relative advantage, compatibility, trialability, and observability and a low degree of complexity. Thus, the innovations are adopted readily and rapidly. Rapid adoption of innovations contributes to the success of software process improvement initiatives.

Providing enhanced understanding

Rogers states that diffusion is one part of a larger process that begins with a perceived problem or need. (Rogers, 1983, p. xvi) If staff members do not perceive any problem or need, rapid diffusion of innovations is unlikely. Software engineers, however, who recognize problems or needs in the context of software development are more likely to adopt innovations. As already mentioned above, relative advantage is an important characteristic of innovations that explains the rate of adoption. The relative advantage of an innovation can only be recognized when the problems of the entire organization are understood. When certain issues are perceived and recognized as problems, staff members will more readily communicate and adopt the corresponding innovations. The more employees perceive problems, the higher the degree to which software process improvement initiatives are considered helpful. The higher the acceptance of improvement efforts the higher the probability that subsequent process innovations are widely adopted.

Unfreezing the organization

Lewin has introduced the importance of "unfreezing" organizations before substantial improvements can be achieved. (Lewin, 1947, 1958) He emphasizes that social processes usually have an "inner resistance" to change. To overcome this resistance an additional force is required, a force sufficient to break the habit, to unfreeze the custom. (Lewin, 1958, p. 209) In the organizations included in our studies two factors contributed to the additional force that was necessary to overcome the resistance to change: first, some employees realized deficiencies in current processes and they accepted the need for changes. Second, when management committed themselves to the software process initiative, they created a realistic opportunity to fight the deficiencies and to achieve substantial improvements. Perceived problems and management commitment contribute to unfreezing the organization. They stimulate the search for solutions, and the willingness for adopting innovations.

Acquiring & transferring knowledge

"The innovation-decision process is the process through which an individual (or other decision making unit) passes from first knowledge of an innovation to forming an attitude toward the innovation, to a decision to adopt or reject, to implementation of the new idea and to confirmation of this decision" (Rogers, 1983, p. 20). Rogers calls the first step in the innovation-decision process "knowledge." It "occurs when an individual (or other decision making unit) is exposed to the innovation's existence and gains some understanding of how it functions" (Rogers, 1983, p. 20). It is relatively easy to inform staff members about principles and contents of the CMM or of the ISO 9000 standards. Software process improvement standards and models are documented and easily accessible. However, exposing employees to internal knowledge, such as methods and tools designed by individual staff members, is not a simple task. Individual knowledge usually is not explicit or documented. It is not easily accessible for other staff members. Obviously, software engineers can only adopt an innovation if they know that it exists and gain some understanding of how it functions. This is why acquiring knowledge of individual methods and tools, and transferring it to all staff members is a critical success factor for software process improvement initiatives.

Tailoring improvement initiatives

Due to different strengths and weaknesses in various teams of a software company the relative advantage of software process innovations varies significantly throughout the organization. Tailoring means adapting improvement initiatives to the specific needs of various groups within an organization. Tailoring helps to improve the relative advantage of innovations in these groups. Tailoring also increases the compatibility of innovations. As already mentioned before, compatibility is the degree to which an innovation is perceived as being consistent with the existing values, past experiences, and needs of potential adopters. Tailoring provides the opportunity to adapt process innovations to the values, experiences and needs of various groups within the organization.

Encouraging communication

Diffusion theory teaches us that technology diffusion is a communication process. "Communication is a process in which participants create and share information with one another in order to reach a mutual understanding" (Rogers, 1983, p. 5). At its most elementary form, the diffusion process is the information exchange by which one individual communicates a new idea to one or several others. A communication channel is the means by which messages get from one individual to another. (Rogers, 1983, p. 17) Rogers explains the effectiveness of different communication channels: mass media channels are effective in creating awareness-knowledge. Interpersonal channels involve a face-to-face exchange between two or more individuals. Interpersonal channels are more effective in persuading an individual to adopt a new idea, especially if the interpersonal channel links two or more individuals who are near peers. Most people mainly rely on a subjective evaluation of an innovation that is conveyed to them from other individuals who have previously adopted the innovation. (Rogers, 1983, p. 17) These statements are also supported by Lewin's findings. (Lewin, 1947, 1958)

The ISO 9000 and the CMM mainly rely on written documents and training for knowledge transfer. Diffusion theory teaches us that interpersonal communication channels are more effective in persuading individuals to adopt new ideas. The findings of our studies support the importance of interpersonal communication channels. Internal audits, quality circles, and improvement workshops seem to be more efficient to communicate process innovations than written documents.

Ensuring staff involvement

Rogers distinguishes four types of innovation-decisions: optional, collective, authority, and contingent innovation-decisions. (Rogers, 1983, p. 29) Optional innovation-decisions are choices made by an individual to adopt or reject an innovation independent of the decisions of other members of the system. Collective innovation-decisions are choices made by a consensus among the members of a system.

Authority innovation decisions are choices made by relatively few individuals in a system who possess power, status or technical expertise. A contingent innovation-decision is a sequential combination of two or more of the three types of innovation-decisions. Generally, the fastest rate of adoption of innovations results from authority decisions. However, authority decisions are often circumvented during their innovation. (Rogers, 1983, p. 29)

Software process improvement initiatives are usually contingent innovation-decisions. Generally, top managers of a software company decide to implement an ISO 9000 quality system or to conduct a CMM based improvement initiative. However, development teams and individual software engineers have considerable influence in the decision whether and how to adopt process innovations. They may adopt innovations enthusiastically or they may comply to the suggestions reluctantly and without committing themselves to the proposed improvements. After the initial authority decision to implement software process improvement has been made, software development teams and software engineers must commit themselves to adopt the process innovations. The initial authority innovation-decision must be followed by various collective innovation-decisions. This is the reason why software process improvement efforts can only be successful if most software engineers are involved and agree to adopt the changes. (Dion, 1993, p. 35)

Emphasizing teamwork and collaboration

"Homophily is the degree to which pairs of individuals who interact are similar in certain attributes" (Rogers, 1983, p. 18). Individuals are homophilous when they share common meanings, work on the same subject or are drawn by the same interests. "More effective communication occurs when two individuals are homophilous" (Rogers, 1983, p. 19). The transfer of ideas occurs most frequently between two individuals who are alike. Communication between homophilous individuals usually has more significant effects in terms of knowledge gain, attitude formation and behavior change. When homophily is present, communication is likely to be more rewarding to all individuals involved.

Teamwork and closer collaboration have increased homophily of software engineers, quality managers, and project managers. The objective of succeeding in the ISO 9000 certification process or of reaching higher CMM levels creates a team spirit. It also supports teamwork, cooperation, and collaboration of individuals in different teams and departments of a software organization. Teamwork and improved collaboration increase the degree of homophily of staff members and thus contribute to the diffusion of technology innovations.

Management commitment

Diffusion theory emphasizes the importance of the social structure for technology diffusion. (Rogers, 1983, p. 24) The social structure of a software company gives managers in higher-ranked positions the right to issue orders to staff members. Although the orders of managers in software companies are not always carried out, they have a significant impact on technology diffusion. "Opinion leadership is the degree to which an individual is able to influence other individuals' attitude or overt behavior informally in a desired way with relative frequency. ... Opinion leadership is earned and maintained by the individual's technical competence, social accessibility, and conformity to the system's norms" (Rogers, 1983, p. 27). Managers that actively participate in software process improvement initiatives and provide active and visible support often serve as opinion leaders. The combination of power and technical competence is an ideal basis for promoting process improvements.

Change agents and facilitators

Opinion leaders are members of the social system in which they exert their influence. Change agents represent change agencies external to the system. They usually seek to obtain the adoption of new ideas and cooperate with opinion leaders (Rogers, 1983, p. 28) Quality managers and members of software engineering process groups often take the role of opinion leaders and/or change agents. Change agents and process facilitators are essential for ensuring that improvement efforts run smoothly. They may help to prevent process entropy.

12 Stabilizing changed processes

Lewin calls the process of stabilizing changed processes "freezing." (Lewin, 1958, p. 210) Freezing is necessary when permanency of the desired changes is included in the objectives of the initiative. In the context of software process improvement freezing prevents that an improved software process slides back to the old level. However, Lewin does not explain how freezing can be achieved.

Diffusion theory helps us to gain a better understanding of how processes can be stabilized. Rogers states that some innovations are changed as they are diffused. He calls this phenomenon "re-invention". Re-invention is the "degree to which an innovation is changed or modified by a user in the process of its adoption and implementation" (Rogers, 1983, p. 16). Implementing software process improvements usually implies re-invention. Tailoring improvement activities to the requirements of different organizational units is a kind of re-invention. Stabilizing changed processes implies to support adopters of innovations to manage re-invention. Stabilizing processes is necessary until the processes are adequately implemented and the need for re-invention finally disappears.

Potential Contribution of Organizational Learning to the Implementation of the Success Factors

In the previous section, we have shown that Rogers' diffusion theory provides an adequate model for explaining the factors that cause successes and failures of software process improvement.

The explanation of the factors may help researchers and practitioners to gain a better understanding of the factors that facilitate or impede the success of software process improvement efforts. The explanation of the factors also provides a basis to derive requirements for an approach that may support the implementation of the factors.

In the following paragraphs, we will derive 12 requirements. The requirements will help us to examine whether organizational learning may contribute to the success of software process improvement activities.

The examination will be conducted in three stages: on the first stage, we will present Senge's and Garvin's suggestions to implement organizational learning. On the second stage, we will examine whether the suggestions have the potential to meet the requirements. On the third stage, we will present practical examples of organizational learning in software process improvement contexts. The examples will help us to assess whether organizational learning is a feasible option to support software process improvement efforts.

Requirements for Implementing the Success Factors

In the following paragraphs, we will summarize requirements for an approach that aims at providing support for the implementation of the 12 success factors. Organizations that wish to conduct successful software process improvement initiatives need support in the following dimensions:

- setting clear, relevant and realistic goals,
- providing enhanced understanding,
- unfreezing the organization,
- acquiring and transferring knowledge,
- tailoring improvement initiatives,
- encouraging communication,
- ensuring staff involvement,
- emphasizing teamwork and collaboration,
- emphasizing management commitment,
- installing change agents and facilitators, and
- stabilizing changed processes.

Standards, concepts, models and recommendations that attempt to support software process improvement initiatives should meet these requirements. The requirements will help us to assess whether and how organizational learning might contribute to the success of software process improvement efforts.

We have selected Senge's (1990) and Garvin's (1993) approaches to organizational learning because they have made practical suggestions of applying organizational learning in a managerial context. Senge describes principles and practices of comprehension, communication, motivation and collaboration of individuals in organizations. Garvin focuses on knowledge acquisition and knowledge transfer in organizations.

Senge's Five Disciplines of Organizational Learning

Senge describes learning organizations as places "where people continually expand their capacity to create the results they truly desire, where new and expansive patterns of thinking are nurtured, where collective aspiration is set free, and where people are continually learning how to learn together" (Senge, 1990, p. 3). According to Senge, a learning organization is an "organization that is continually expanding its capacity to create its future" (Senge, 1990, p. 14).

The definition comprises several characteristics of organizational learning:

- Although organizational learning is a characteristic of organizations, individuals play a major role in the concept.
- Organizational learning addresses the cooperation of individuals in social systems.
- It is a continuous process.
- It aims at improving an organization's capacity to achieve its goals.

Senge suggests the use of five "disciplines" to implement organizational learning: personal mastery, mental models, shared vision, team learning and systems thinking.

Personal Mastery. Senge states that "organizations learn only through individuals who learn. Individual learning does not guarantee organizational learning. But without it, no organizational learning occurs" (Senge, 1990, p. 139). Personal mastery addresses individual knowledge, competence and skills that people need to produce superior results in business life. However, it goes far beyond these aspects. Personal mastery supports growth in all facets of human life, for example, emotional development, spiritual welfare, capacity to generate a personal vision, etc. The rationale for personal mastery is that an organization will only be able to continually expand its capacity to create its future if staff members strive for achieving their visions. People who practice personal mastery "are continually expanding their ability to create the results in life they truly seek" (Senge, 1990, p. 141).

Senge suggests three practices to implement personal mastery:

- Clarifying personal vision. Personal mastery requires creating a personal vision, or as Senge puts it, to continually clarify what is important to us. According to Senge, the vision generates energy to overcome obstacles and to create new ideas.
- Learning to see current reality more clearly. Personal mastery also requires a clear picture of current reality, because there usually is a discrepancy between current reality and personal vision. (Senge, 1990, p. 141)
- Holding creative tension. Focusing simultaneously on the personal vision and on current reality generates a creative tension. This tension is the challenging force that generates personal growth. "The essence of personal mastery is learning how to generate and sustain creative tension in our lives" (Senge, 1990, p. 142).

Mental Models. Mental models are assumptions that shape an individual's perception of reality and influence the way he or she feels, thinks, speaks and acts. Mental models are "internal pictures of how the world works" (Senge, 1990, p. 174). Senge states that most people *confuse* reality with *their* view of reality. This often leads to an inadequate perception of the world and to the "pursuit of simple answers to complex issues" (Senge, 1990, p. 185). Each individual has a mental model that is different from the mental models of his or her colleagues. The consequences are communication problems and misunderstandings. Often new ideas and promising suggestions are not implemented in organizations because they are in conflict with inadequate mental models of decision makers.

Mental models are implicit structures; they are not obvious. Specific efforts are needed to overcome the problems posed by mental models. Senge (1990, p. 186) suggests four practices to identify and to test mental models:

- Recognizing leaps of abstraction. Senge states that managers usually move too fast from observations to generalizations. They are unaware of their leaps from particulars to general concepts. To identify leaps of abstraction Senge suggests to question generalizations by explicitly separating it from the data that led to it.
- Exposing the left-hand column. Exposing the left-hand column is a technique for articulating thoughts, implicit assumptions and mental reservations. "It reveals ways that we manipulate situations to avoid dealing with how we actually think and feel" (Senge, 1990, p. 195). During the left-hand column exercise, managers recall a situation where they were interacting with one or several other people in a way that is not producing any progress or learning. By describing what was actually said (in the right-hand column) and what was left unspoken (in the left-hand column) managers learn to articulate

their hidden assumptions that prevent them from communicating effectively with other people.

- Balancing inquiry and advocacy. According to Senge, managers are often successful advocates. They have learned to "debate forcefully and to influence others" (Senge, 1990, p. 198). Top managers, however, often confront issues more complex than their personal experience. They depend upon insights from other people. Top managers need to blend "advocacy and inquiry to promote collaborative learning" (Senge, 1990, p. 198). The objective of combining inquiry and advocacy is to find the best argument. This includes to ask for data or experience that supports the views of communication partners.
- Distinguishing between espoused theories and theories-in-use. Learning does not only mean acquiring more information. Learning must eventually lead to improved behavior. It is vital to face up to distinctions between espoused theories (what is said) and theories-in-use (the implied theory in behavior). When managers recognize gaps between what they say and what they do they have identified discrepancies between their mental models and their current behavior. This gives them a chance to learn. Learning may lead to new mental models or to changed behavior.

Shared Vision. Shared vision describes a vision that all members of an organization have in common. According to Senge, a shared vision reflects more than the business goals of an organization. It provides the focus and the energy for learning because it reflects personal visions of all staff members. Shared vision creates a common identity.

Senge suggests four practices for building shared vision:

- Encouraging personal vision. Learning organizations continually encourage members to develop their personal vision because shared visions emerge from personal visions.
- Building shared visions from personal visions. The traditional top-down approach to writing vision-statements usually does not build on staff members' personal visions. Senge suggests to encourage visions that start in the middle of an organization. This requires ongoing conversation, where individuals feel free to express their ideas. It also requires listening to new insights and to other peoples' dreams. The task of managers is to create an environment that helps visions emerge.
- Spreading visions. The traditional command and control hierarchy requires compliance. Compliant members of an organization follow "the letter of the law" but they are not necessarily truly committed. However, building shared vision requires commitment. Staff members cannot be forced to commit because commitment requires freedom of choice. Senge recommends that managers should encour-

age motivating visions and give staff members the freedom to choose whether to follow the vision.

- Anchoring vision in a set of governing ideas. The governing ideas of an organization comprise its vision, purpose or mission, and core values. It is essential that visions are consistent with the organization's purpose or mission and with the core values that people live day by day. If the visions are not consistent with other governing ideas, the visions are useless or even counterproductive. They will not motivate or encourage staff members.

Team Learning. "Team learning is the process of aligning and developing the capacity of a team to create the results its members truly desire" (Senge, 1990, p. 238). Team learning is based on the assumption that teams have a greater potential to gain insights and to achieve superior results than individuals working on their own. However, teams must learn to function as a whole. Senge suggests three practices for team learning:

- Dialogue and discussion. Senge suggests to make use of two distinct types of discourse: dialogue and discussion. In a dialogue complex and subtle issues are explored. Dialogues do not necessarily seek agreement, but a better understanding of complex issues. In a discussion, different views are presented and defended. The objective of a discussion is to find the best views to support a decision. Teams must learn to distinguish between the two forms of conversation and to choose the appropriate form.
- Dealing with conflicts and defensive routines. Conflicts and defensive routines may oppose productive dialogue, discussion and problem solving. Teams must learn to deal with these forces and to cope with habits that prevent examining complex issues. "Defensive routines ... are entrenched habits we use to protect ourselves from the embarrassment and threat that come with exposing our thinking" (Senge, 1990, p. 250). Conflicts and defensive routines retain their power until they are made discussible. Therefore, team members must learn to recognize and to expose their assumptions. Team members must encourage each other to discuss conflicts and defensive routines.
- Practicing team learning. Team learning requires practice. However, unlike symphony orchestras or sports teams most management teams do not know how to learn together. Learning teams create opportunities to train working as a team. They reflect on how they might arrive at better decisions. They experiment with various types of discourses and they practice to resolve conflicts.

Systems Thinking. Systems thinking is the fifth discipline. It integrates the other four disciplines. The objective of systems thinking is to provide a systemic orientation that enhances understanding of the other disciplines. (Senge, 1990, p. 12) Senge states that most facets of behavior of organizations can be explained by certain patterns that control events. For example, systems thinking requires understanding reinforcing feedback (discovering how small changes can grow), balancing processes (discovering the sources of stability and resistance), and delays (discovering interruptions between actions and consequences). (Senge, 1990, p. 79-89) Learning requires perceiving these underlying structures to understand complex situations.

Senge suggests various practices for systems thinking, for example:

- Understanding interrelationships. Understanding interrelationships implies perceiving patterns of interrelated actions rather than linear cause-effect chains. (Senge, 1990, p. 73)
- Perceiving processes of change. Senge states that primary threats to survival come from gradual processes and not from sudden events. It is therefore essential to perceive processes of change rather than snapshots of current reality. (Senge, 1990, p. 73)

Garvin's Building Blocks of Organizational Learning

Garvin (1993, p. 79) remarks that most of Senge's recommendations "are far too abstract, and too many questions remain unanswered. Therefore, Garvin tries to present a framework that is more actionable and easier to apply.

"A learning organization is an organization skilled at creating, acquiring, and transferring knowledge, and at modifying its behavior to reflect new knowledge and insights" (Garvin, 1993, p. 80). Learning organizations are able to create new ideas and to translate them into new ways of behaving. "These companies actively manage the learning process to ensure that it occurs by design rather than by chance" (Garvin, 1993, p. 81).

Garvin presents five "building blocks" of organizational learning: systematic problem solving, experimentation with new approaches, learning from experiences and past history, learning from the experiences and best practices of others, and transferring knowledge throughout the organization (Garvin, 1993, p. 81).

Systematic Problem Solving. Garvin suggests to use systematic problem solving as it was described by the quality movement. (Deming, 1992) He gives several examples of how an organization can solve problems systematically:

- relying on scientific methods, rather than guesswork, for diagnosing problems,
- insisting on data, rather than assumptions for decision making, and
- using simple statistical tools to organize and analyze data.

Garvin emphasizes that systematic problem solving requires a shift of mind. Employees must learn that accuracy and precision are essential for learning. In a learning organization, employees try to understand underlying causes of problems instead of focusing on symptoms of the problem.

Experimentation. Experimentation aims at creating and testing new knowledge. Garvin describes two forms of experimentation: ongoing programs and demonstration projects:

- Ongoing programs. Ongoing programs "involve a continuing series of small experiments, designed to produce incremental gains in knowledge" (Garvin, 1993, p. 82). Garvin states that ongoing programs are especially common in continuous improvement programs on the shop floor.
- Demonstration projects. Demonstration projects "involve holistic, systemwide changes, introduced at a single site, and are often undertaken with the goal of developing new organizational capabilities" (Garvin, 1993, p. 83). Demonstration projects are often used to establish, to test, and to refine new approaches that the company plans to adopt on a larger scale.

Experiments are closely related to systematic problem solving. Learning organizations use experiments as a way to proceed from superficial knowledge of symptoms to a deep understanding of underlying cause-effect relationships.

Learning from past experience. Learning organizations analyze their successes and failures to learn from past experience. Garvin describes three phases of gaining knowledge from past experiences:

- description and review of successes and failures,
- systematic assessment of underlying causes of successes and failures, and
- documentation of lessons learned in a form that employees find open and accessible.

Although a review of failures is usually more productive than celebrating successes, decision makers often fail to reflect on failures. Learning organizations establish processes that encourage learning from mistakes.

Learning from Others. A significant part of learning comes from reflection and self-analysis. However, gaining a new perspective often requires looking from outside one's immediate environment. Garvin suggests two ways of gaining an outside perspective:

- Benchmarking. Benchmarking is "a disciplined process that begins with a thorough search to identify best-practice organizations, continues with careful study of one's own practices and performance, progresses through systematic site visits and interviews, and concludes with an analysis of results, development of recommendations, and implementation" (Garvin, 1993, p. 86).
- Learning from customers. Learning from customers may take the form of conversation or observation. Conversations with customers may provide feedback about products or services and helpful insights into changing preferences. If customers cannot articulate their attitude an organization must observe them in action.

Learning from others requires replacing the "not invented here" syndrome by "enthusiastic borrowing." It also requires a receptive environment and managers that cultivate the art of open, attentive listening.

Transferring Knowledge. To become effective, knowledge must be spread quickly and efficiently throughout the company. Garvin presents various mechanisms for transferring knowledge:

- Written, oral, and visual reports summarize findings, describe lessons learned, provide checklists, and describe important success factors.
- Site visits and tours may be tailored to the needs of different audiences such as upper managers or lower ranks.
- Personnel rotation programs provide first-hand experiences for individuals.
- Education and training programs are powerful mechanisms for transferring knowledge if they are linked directly to implementation.
- Standardization programs help to proliferate best practices throughout the organization.

Applying Organizational Learning to Software Process Improvement

In the following paragraphs, we will explore whether and how organizational learning might help to implement and maintain the success factors of software process improvement. We will examine each of the success factors and discuss which issues of organizational learning described by Senge or Garvin might support the implementation of the factors.

Setting clear, relevant and realistic goals

Setting clear, relevant and realistic goals for a software process improvement initiative requires understanding strengths and weaknesses of current software processes, and developing objectives for process changes. In some software organizations, managers did not sufficiently understand essential details of software development. Consequently, they were not able to set relevant and realistic goals for CMM or ISO 9000 based software process improvement efforts.

Senge's suggestions to identify and to test mental models might help managers to identify inadequate perceptions of software development in their organization. Recognizing and discussing mental models might help managers to understand problems and opportunities of current software processes more clearly. This will provide a helpful basis to propose and to communicate challenging objectives for software process improvement efforts.

Setting relevant and realistic goals requires focusing on current reality of software development as well as on a challenging vision for the improvement efforts. Senge's discipline of personal mastery addresses the issue of holding the creative tension between vision and current reality.

Successful improvement efforts require that all staff members pull together as a team and jointly try to accomplish the objectives. Senge's discipline of building shared vision aims at encouraging, building and spreading personal visions of staff members. Building shared vision might support the evolution of objectives into a shared vision for the software process improvement initiative.

Providing enhanced understanding

Providing enhanced understanding is a key issue of organizational learning. Senge addresses it in two of his five disciplines: mental models and systems thinking. Identifying and testing mental models helps to improve managers and staff members' perception of reality. It helps employees to reflect on their daily work from a general business perspective. It also helps managers to gain a better understanding of strengths and weaknesses of current software processes. Systems thinking assists managers and software engineers to understand interrelationships of software process improvement activities and business success. Systems thinking may provide systemic orientation on the complex issues of software process improvement initiatives.

Garvin relies on systematic problem solving, experimentation, and learning from past experience to provide enhanced understanding. Systematic problem solving and experimentation support staff members to understand underlying causes of problems in the software process instead of focusing on "quick fixes." In software process improvement initiatives, systematic problem solving may take the form of carefully diagnosing problems with the help of scientific methods. Experimentation may assist software process specialists in implementing, testing and refining new software engineering methods or tools at a single site before they are adopted on a larger scale. Analyzing successes and failures of software projects provides an opportunity to learn from past experience. It provides knowledge on what may work, and what may lead to failures of software development. Learning from past experience enhances understanding of future risks and opportunities of software process improvement initiatives.

Unfreezing the organization

Neither Senge nor Garvin explicitly mention the need to unfreeze organizations before substantial improvements can be achieved. However, providing enhanced understanding may support the process of unfreezing the organization. Staff members that have understood the current weaknesses of software development will more readily accept the necessity to change software development processes. Senge recommends to spread inspiring visions throughout the organization. Inspiring visions for software process improvement initiatives will help to overcome organizational resistance to change, generate a creative tension and provide the energy for learning and achieving improvements.

Acquiring knowledge

Senge recommends to make use of the discipline's mental models, team learning and systems thinking to acquire knowledge that is relevant for improvement efforts. In the course of software process improvement initiatives managers often confront complex issues. Usually they are not able to solve these problems alone. The success of software process improvement efforts depends upon insights of various staff members. Therefore, Senge suggests that managers should balance inquiry and advocacy to find the best views and arguments on complex issues. The goal of combining inquiry and advocacy is to promote collaborative learning. Senge also recommends to make use of dialogue and discussion. Using dialogue and discussion assists teams in exploring subtle issues of software process improvement initiatives. Choosing the appropriate form of discourse may help teams to acquire knowledge that is useful for improvement activities.

Garvin recommends that organizations should acquire knowledge by learning from past experience and by learning from others. Acquiring knowledge by describing, reviewing, analyzing and documenting past successes and failures provides software organizations with highly significant information for future improvement activities. Garvin describes various techniques of learning from others; for example, benchmarking and conversation with or observation of customers. Software process improvement networks provide an effective and efficient environment for software specialists to learn from experience with software process improvement initiatives in various organizations.

Garvin's recommendations to systematic problem solving and experimentation may help to test and refine knowledge that was identified by knowledge acquisition techniques, such as Senge's inquiry or dialogue.

Transferring knowledge

In Senge's approach to organizational learning, transferring knowledge is a function of team learning. Senge suggests to practice dialogue and discussion to spread knowledge throughout the organization. Garvin presents a specific building block of organizational learning that is dedicated to transferring and distributing knowledge to staff members. Reports, site visits, personnel rotation programs, training programs and standardization efforts may help to distribute knowledge necessary to conduct successful software process improvement initiatives.

Tailoring improvement initiatives

Tailoring involves understanding the specific strengths and weaknesses of an organizational unit that applies a software process improvement program. It requires understanding the factors that contribute to success or failure in the particular environment. Tailoring improvement initiatives also requires that the members of the organizational unit agree upon common objectives and that they pull together as a team to achieve the objectives.

Senge's suggestion to identify and to test mental models may help software managers to achieve a more appropriate picture of software development. Refining mental models, recognizing leaps of abstraction, and distinguishing between espoused theories and theories-in-use may provide a helpful basis for tailoring improvement efforts. Systems thinking might help to perceive and to analyze interrelationships between planned changes, staff members' attitudes, the social and organizational structure of the software process, and potential effects on business results. Garvin's building block of learning from past experience may help to gain better insights into what may and what may not work in different teams or departments of the organization. Senge's team learning may support cooperation and collaboration in teams. Cooperation and collaboration are essential when teams tailor innovations to their specific environment.

Encouraging communication

Communication is essential for acquiring and for transferring knowledge. It plays an important role in clarifying and spreading the objectives of a software process improvement initiative. It may help to overcome staff members' resistance and to strengthen software engineers' commitment to continuously improving the software process.

Senge and Garvin make various suggestions of how to implement effective communication. Garvin suggests to use mass media channels to spread knowledge throughout the company. Mass media channels, such as reports, checklists or standardization programs may improve staff members' awareness of the improvement efforts. Senge emphasizes the importance of interpersonal communication channels. He suggests to use inquiry and advocacy, and dialogue and discussion to achieve advances in internal audits, meetings, and workshops.

Ensuring staff involvement

Visions that start in the middle of an organization are more likely to be successful than visions that start from the top. Senge relies on encouraging personal visions of staff members and on building a shared vision to ensure grass-roots involvement. Encouraging personal visions includes encouraging staff members to clarify what is important to them. In the context of a software process improvement initiative, this requires that employees recognize weaknesses of current development processes and generate a personal vision of improved processes.

Senge recommends intensive communication to create an environment that helps visions emerge. When staff members choose to follow an emerging vision gradually, a shared vision of the improved software process is formed. Software engineers that actively participate in assessing, discussing and refining the vision will more readily adopt the innovations brought about by software process improvement efforts. Building shared visions contributes to collective innovation-decisions.

Emphasizing teamwork and collaboration

Senge has dedicated one of his five disciplines to teamwork and collaboration. He suggests to practice team learning. Team learning implies experimenting with dialogue and discussion, learning how to resolve conflicts, and creating opportunities to train working as a team. Teams usually have a greater potential to achieve superior results than individuals working on their own. Consequently, team learning contributes to the success of software process improvement efforts. Team learning might help to increase homophily of staff members. Increasing homophily contributes to efficient communication and eventually enhances teamwork and collaboration. Improved cooperation usually facilitates the success of software process improvement efforts.

Management commitment &
Change agents and facilitators

We have identified management commitment as one of the key success factors for software process improvement initiatives. Management commitment in a learning organization requires more than active participation and visible support of improvement efforts. It requires continually creating an environment that encourages learning. Senge describes the role of leaders as mentors, coaches, and teachers that help others to understand complex situations. (Senge, 1990, p. 339) Neither Senge nor Garvin explicitly suggest to use change agents.

Organizational learning depends on competent managers. In a learning organization managers serve as opinion leaders and facilitators. They provide support and feedback, publish successes, and keep staff members aware of the improvement efforts.

Evidently managers that act as mentors, coaches, teachers, and facilitators will enhance the success of software process improvement efforts. They will help staff members to understand weaknesses and strengths of current processes. They will motivate employees to suggest options for improving software processes. They will animate software engineers to adopt process innovations, and encourage software engineers to commit themselves to continuously improving software development.

Stabilizing changed processes

Stabilizing changed processes requires understanding the forces that lead to process entropy. Senge's disciplines of using mental models and systems thinking and Garvin's suggestion to experiment with ongoing programs and demonstration projects may lead to a better understanding of factors that cause software process entropy. Organizational learning is based on the assumption that re-invention is a continuous process. Permanent efforts are needed to prevent process entropy and to stabilize changed processes. Consequently, proponents of organizational learning do not suggest to freeze processes. However, they recommend to permanently adapt processes to the needs of the specific situation.

In the context of software process improvement this seems to be a helpful advise because user requirements, software and hardware technology, the competitive position of software suppliers, and other significant issues are changing rapidly. Freezing software processes might be an impediment to success. However, stabilizing and adapting processes to the requirements of the specific situation will improve the competitive advantage of software suppliers.

Examples of Organizational Learning in Software Organizations

Senge's and Garvin's approaches to organizational learning provide several helpful suggestions of how to establish a learning organization. On a conceptual level, we have shown that Senge's disciplines and Garvin's building blocks of organizational learning meet the requirements for approaches to support the implementation of software process improvement success factors. Senge's and Garvin's suggestions address most of the success factors that we have described in a previous section of this chapter.

Senge's and Garvin's approaches to organizational learning complement one another very well. Garvin addresses organizational learning from the perspective of an organization. He describes techniques for knowledge acquisition and knowledge transfer. Senge focuses on individual aspects of organizational learning. He outlines principles and practices of comprehension, communication, motivation and collaboration of individuals in organizations.

In the section that follows, we will present practical examples of companies that have applied elements of organizational learning in a software development context. The examples will help to assess whether organizational learning is a feasible option to support software process improvement efforts.

Postmortem Reviews. A postmortem review, or post implementation audit, is an activity where people come together to review a previous project and to learn from collective experience (Kerth, 1996). The objective of postmortem reviews is to analyze strengths and weaknesses of software development projects and to learn from mistakes and successes. Brady and DeMarco (1994) included postmortem reviews in a list of what they call "best-practice candidates" for software managers. Glass (1995) calls postmortem reviews "best-of-practice concepts" that gradually win favor in software companies.

Kumar (1990) conducted a survey of 92 MIS organizations. He found that 30% of the organizations surveyed evaluated 75% or more of their computer-based information systems. 35% of the organizations evaluated between 25 and 74% of the management information systems. Another 14% of the organizations evaluated between 1 and 24% of the installed systems. 21% did not conduct any post implementation audits. Those companies that conducted post implementation reviews were asked their opinion about the most important benefits of the audits. To "provide feedback to system development personnel" was regarded to be a very important benefit of the reviews. To "provide feedback for modification to development methods" and to "provide feedback for modification to project-management method" were other important benefits of conducting post implementation reviews. However, Kumar also found that several companies do not use postmortem reviews for improving development processes.

Johnson (1990) describes post implementation audits performed at Halmark. The audits are held to determine user satisfaction and the performance of the development department. Another objective of the post implementation audits at Halmark is to reveal options for process improvements.

Collier, DeMarco, and Fearey (1996) describe a defined process for postmortem reviews. Their suggestions are based on experience made during 22 software projects at Apple Computer and various other software companies.

The examples show that postmortem reviews have helped software organizations to identify strengths and weaknesses of past projects. The reviews provide an opportunity to learn from successes and failures, and they reveal options for process improvement. Postmortem reviews are a helpful tool for acquiring and transferring knowledge and they provide enhanced understanding of current software processes.

Spreading Best Current Practices. Everett and Musa (1993) describe an approach to identifying, documenting and publishing best current software development practices. The efforts are part of an initiative launched by AT&T to improve its product development processes. Best current practices describe the "best currently known way to accomplish a set of tasks" (Everett & Musa, 1993, p. 77). The identification of best current practices is based on case studies submitted by several projects.

Best current practices are documented in a handbook that outlines costs and benefits of applying the practices. The handbook recommends a sequence of implementation steps and provides a list of conformance questions to help project managers assess how well they are following the practice. The handbook also comprises a list of key factors that contribute to successful implementation of best current practices.

AT&T offers company-wide support to projects to implement best current practices. Three-day courses provide technology training needed to implement a best current practice. A one-day management course provides information for decision makers who evaluate the practice. Another course provides an overview for managers and staff members who do not directly perform the best current practice but who may be affected by it. Consultants help project managers to integrate best current practices into their development environment. User groups publish newsletters and sponsor workshops that give practitioners an opportunity to exchange information. (Everett & Musa, 1993)

The initiative of spreading best current practices facilitates the transfer of knowledge that is essential for improving software projects. The handbook that documents best current practices provides enhanced understanding. Courses, user group meetings, workshops, and newsletters encourage communication and help to involve staff members. Management courses contribute to management commitment. Consultants that help software engineers to integrate best current practices into their development environment act as change agents. Support activities help to stabilize changed processes and to fight software process entropy.

Measurement Programs. Several software companies have implemented measurement programs as a means to quantify product quality and process productivity. The objective of the measurement programs is to gain a better understanding of the factors that influence product quality, development costs and cycle time.

Grady and Caswell (1987) describe a company-wide measurement program that was implemented at Hewlett-Packard. They present metrics and measurement tools, suggest a strategy to plan and implement measurement programs, and discuss elements of the change process that is necessary to implement a measurement program. Grady (1992) makes several suggestions how to learn from measurement results and how to use the lessons learned to achieve software process improvements.

Walrad and Moss (1993) discuss the role of measurement in pursuit of software development quality and productivity. They present efficiency and effectiveness measures and group them according to the maturity levels of the CMM. They also discuss organizational and cultural issues associated with implementing a measurement process. Walrad and Moss gathered experience with measurement processes at IBM.

Pfleeger (1993) describes lessons learned in building a corporate software metrics program at Contel, a telecommunications firm. The objective of the metrics program was to contribute to software process improvement. Pfleeger describes activities and guidelines that help to learn from the results of the measurement program.

Basili (1992) proposes a framework called an experience factory. The experience factory is an organizational unit that has the task to collect experiences obtained through software projects and to package them in a reusable form for future projects. The experience factory is embedded in a conceptual framework that gives guidance to software companies that plan to implement organizational learning through conducting measurement programs.

Measurement programs help managers to set clear, relevant, and realistic goals for improvement efforts. Measurement programs help to collect, analyze and document knowledge that is essential for software process improvement efforts. Measurement results may provide an enhanced understanding of strengths and weaknesses of software processes.

The examples show that several software companies have implemented elements of organizational learning. Postmortem reviews, spreading best current practices, and conducting measurement programs contribute to implementing and maintaining several of the success factors of software process improvement. They support implementing organizational learning in software organizations. The examples show organizational learning is not only a theoretical concept, but also a feasible option to support software process improvement efforts.

Future Trends

Growing evidence shows that managerial, organizational and social factors, such as comprehension, communication, motivation and collaboration, are essential for the success of software process improvement efforts. Various software organizations have realized that it is fundamental to implement and to maintain these factors. However, the factors are not sufficiently accounted for in current software process improvement models and standards. Future suggestions to improve software processes will have to be complemented by recommendations to enhance technology diffusion and adoption, to manage change, and to facilitate learning.

We have shown that the coupling of technology diffusion, software process improvement, and organizational learning provides promising research opportunities. In the following paragraphs we will outline some recommendations for further research into the coupling of the three fields.

Daly, Miller, Brooks, Roper, and Wood (1995) propose an evolutionary multi-method approach to performing empirical research in a software process improvement context. The approach helps to formulate and to refine hypotheses from a series of studies using different empirical techniques. The studies may include collection of industrial data, structured interviews with academics and practitioners, questionnaire based surveys, laboratory experiments, etc. If two or more different empirical techniques produce similar results it is more likely that the findings are valid and reliable. The empirical techniques are used in an evolutionary program: Important issues are discovered by an initial explorative study. These issues are used to formulate hypotheses for further investigation. In the next study these hypotheses are tested, refined, and investigated further. The process is repeated until the hypotheses have been thoroughly investigated.

Our studies represent initial, explorative research into the factors that contribute to the success of software process improvement initiatives. In our first study we identified 12 factors that contribute to success and failure of ISO 9000-based software process improvement initiatives. In our second study on CMM-based improvement efforts we tested and refined the findings of our first study. The second study revealed that success factors of CMM-based improvement activities are very similar to the factors of ISO 9000 based software process improvement initiatives. The findings support the hypothesis that software process improvement initiatives have a greater likelihood of success to the degree that the 12 success factors identified in our studies are implemented.

In the paragraphs that follow, we have listed some interesting issues for further research work:

- More research is needed to find out whether our studies failed to identify any significant success factor.
- Some of the success factors identified in our studies are closely interrelated and overlap each other. Further investigation is necessary to gain a better understanding of the relations between the factors.
- Further empirical studies are needed to examine whether the success factors are also valid for software process improvement efforts that

are based on other models, for example BOOTSTRAP or SPICE, or for initiatives that are conducted in other cultures, for example in Japan.
- We need to understand the effects of improved software engineering techniques on the performance of software organizations and compare them to the effects of improved social skills such as communication, motivation, and collaboration.
- It would be useful to identify and to describe more practical examples of software organizations that apply elements of organizational learning. A greater number of case studies would help to describe successes and failures, strengths and weaknesses, and to assess costs and benefits of these efforts.
- Various elements of knowledge and understanding are necessary to conduct a successful software process improvement initiative. More research is needed to identify the most effective mechanisms for acquiring, analyzing, and transferring knowledge in a software organization. We need to conduct experiments to help evaluate, understand, control, and improve different elements of organizational learning in software development environments.
- Organizational learning consists of various interrelated elements. Most organizations will not be able to implement all elements simultaneously. We need to assess different strategies for implementing organizational learning in software companies.
- The CMM distinguishes five maturity levels. Further investigation is necessary to examine whether software companies should apply different approaches to organizational learning depending on the maturity level of their software processes. If it is advisable to use different approaches, we need to know which elements of organizational learning are applicable on which maturity level.
- Organizational learning should improve the business success of a company. We need to design methods and tools to identify and to quantify the effects of organizational learning on improved performance.

To address these issues it would be useful to develop a more detailed theoretical model of software process improvement that identifies and explains important factors, their interrelationships and effects on software development performance. This model could serve as a basis for further research work.

We have shown that diffusion theory helps to explain the success factors of software process improvement. We have explored the contribution of organizational learning to improvement efforts in the software industry. An interdisciplinary approach that integrates insights from various research fields might help to gain an even more comprehensive understanding. Examples of research fields that might contribute to improving software development are: sociology, psychology, economics, management sciences, organizational theory, software engineering, and software management.

Conclusion

Initiatives to improve software development must implement various critical success factors. However, several key factors are not fully accounted for in popular software process improvement models and standards, such as the CMM or the ISO 9000 standards. The models do not sufficiently support, for example, principles and practices of comprehension, communication, motivation and collaboration of individuals in organizations.

The list that follows summarizes recommendations that representatives of software organizations considered to be the most helpful when implementing a CMM- or ISO 9000-based improvement initiative.

- Set clear, relevant and realistic goals for improvement initiatives.
- Provide enhanced understanding of strengths and weaknesses of current software processes.
- Communicate deficiencies and create a willingness to change current processes.
- Identify and describe knowledge that is relevant for software development and software process improvement.
- Transfer knowledge to all staff members that might benefit from the knowledge.
- Tailor improvement efforts to particular needs and specific conditions of organizational units.
- Encourage communication.
- Encourage grass-roots participation in improvement projects.
- Train staff members to work effectively and efficiently in teams.
- Ensure management commitment and provide active and visible management support for all improvement activities.
- Facilitate the change process by installing change agents.
- Stabilize changed processes, provide comprehensive support and encourage staff members to practice new processes.

Implementation of these factors requires a solid understanding of how the factors contribute to the success of software process improvement activities. The implementation also requires practical guidelines that help practitioners to implement and to maintain the factors.

We have shown that diffusion theory helps us to explain and to understand how the factors facilitate or impede the success of software process improvement initiatives. The explanation helps researchers and practitioners to better understand how technology diffusion occurs in software process improvement contexts.

We have also shown that organizational learning provides suggestions of how to implement and to maintain the success factors. Senge's and Garvin's approaches to organizational learning address issues of knowledge acquisition and transfer, comprehension, communication, motivation and collaboration. Senge's and Garvin's suggestions address most of the factors that are missing in contemporary standards and models for software process improvement. Improvement initiatives in the software field will probably be more successful when they are complemented by principles and practices of organizational learning. Practical examples of software companies that have implemented elements of organizational learning show that organizational learning is not only a theoretical concept, but also a feasible option to support software process improvement efforts.

Further research into the interrelationship of technology diffusion, software process improvement, and organizational learning may lead to considerable advances in improving software development. Further advances could probably be achieved by integrating insights from various research fields, such as sociology, psychology, economics, management sciences, organizational theory, software engineering, and software management.

References

Argyris, C., & Schoen, D. A. (1978). *Organizational learning: A theory of action perspective.* Reading, MA: Addison Wesley.

Basili, V. R. (1992). The experience factory: Packaging software experiences. In H. Woda & W. Schynoll (Eds.), *Proceedings of IPSS-Europe International Conference on Lean Software Development* (pp. 3-1 - 3-22). Stuttgart, Germany.

Basili, V. R., Daskalantonakis, M. K., & Yacobellis, R. H. (1994). Technology transfer at Motorola. *IEEE Software,* 11(3), 70-76.

Basili, V. R., & Musa, J. D. (1991). The future engineering of software: A management perspective. *IEEE Computer,* 24(9), 90-96.

Bellin, G., Stelzer, D., & Mellis, W. (1996). ISO 9000: Ein Qualitaetsstandard ohne Auswirkung auf die Softwareentwicklung? In H. C. Mayr (Ed.), *Beherrschung von Informationssystemen. Tagungsband der Informatik '96* (pp. 347 - 366). Vienna: Oesterreichische Computer Gesellschaft, Oldenbourg.

Brady, S., & DeMarco, T. (1994). Management-aided software engineering. *IEEE Software,* 11(11), 25-32.

Coallier, F. (1995). Trillium: A model for the assessment of telecom product development & support capability. *Software Process Newsletter,* 2(2), 3-8.

Cohen, W. M., & Levinthal, D. A. (1990). Absorptive capacity: A new perspective on learning and innovation. *Administrative Science Quarterly,* 35, 128-152.

Collier, B., DeMarco, T., & Fearey, P. (1996). A defined process for project postmortem review. *IEEE Software,* 13(7), 65-72.

Cooper, R. B., & Zmud, R. W. (1984). Information technology implementation research: A technological diffusion approach. *Management Science,* 36(6), 727-738.

Cyert, R. M., & March, J. G. (1963). *A behavioral theory of the firm.*

Englewood Cliffs: Prentice-Hall.

Daly, J., Miller, J. Brooks, A., Roper, M., Wood, M. (1995). A multi-method approach to performing empirical research. *Software Process Newsletter*, 2(4), 10-12.

Daskalantonakis, M. (1994). Achieving higher SEI levels. *IEEE Software*, 11(7), 17-24.

Deming, W. E. (1992). *Out of the crisis: Quality, productivity and competitive position*. Cambridge: University Press.

Dion, R. (1993). Process improvement and the corporate balance sheet. *IEEE Software*, 10(7), 28-35.

Dodgson, M. (1993). Organizational learning: A review of some literatures. *Organization Studies*, 14, 375-394.

Dorling, A. (1993). SPICE: Software process improvement and capability determination. *Software Quality Journal*, 2(4), 209-224.

Everett, W. W., & Musa, J. D. (1993). A software reliability engineering practice. *IEEE Computer*, 26(3), 77-79.

Fichman, R. G., & Kemerer, C. F. (1993). Adoption of software engineering process innovations: The case of object orientation. *Sloan Management Review*, 34(1), 7-22.

Fiol, C. M., & Lyles, M. A. (1985). Organizational learning. Academy of *Management Review*, 10(4), 803-813.

Fowler, P. (1996). The challenge of information technology transfer and diffusion. In K. Kautz & J. Pries-Heje (Eds.), Diffusion and Adoption of Information Technology. *Proceedings of the first IFIP WG 8.6 working conference on the diffusion and adoption of information technology* (pp. 8-11). London: Chapman & Hall.

Garvin, D. A. (1993). Building a learning organization. *Harvard Business Review*, 71(4), 78-91.

Glass, R. L. (1995). In search of self-belief: The 'BOP' phenomenon. *IEEE Computer*, 28(1), 55-57.

Grady, R. B. (1992). *Practical software metrics for project management and process improvement*. Englewood Cliffs: Prentice-Hall.

Grady, R. B., & Caswell, D. L. (1987). *Software metrics: Establishing a company-wide program*. Englewood Cliffs: Prentice-Hall.

Haley, T. J. (1996). Software process improvement at Raytheon. *IEEE Software*, 13(11), 33-41.

Herbsleb, J., Carleton, A., Rozum, J., Siegel, J., & Zubrow, D. (1994). *Benefits of CMM-based software process improvement: Initial results*. (Technical Report CMU/SEI-94-TR-13. ESC-TR-94-013). Pittsburgh: Software Engineering Institute.

Herbsleb, J., Zubrow, D., Siegel, J., Rozum, J., & Carleton, A. (1994). Software process improvement: State of the payoff. *American Programmer*, 7(9) 2-12.

Humphrey, W. S. (1989). *Managing the software process*. Reading, MA: Addison-Wesley.

Humphrey, W. S., Snyder, T. R., & Willis, R. R. (1991). Software process improvement at Hughes Aircraft. *IEEE Software*, 8(7), 11-23.

ISO 9000-1 (1994). *Quality management and quality assurance standards. Part 1: Guidelines for selection and use*. Geneva: International Organization for Standardization.

ISO 9000-3 (1992). *Quality management and quality assurance stan-*

dards. Part 3: Guidelines for the application of ISO 9001 to the development, supply and maintenance of software. Geneva: International Organization for Standardization.

ISO 9001 (1994). *Quality systems. Model for quality assurance in design, development, production, installation and servicing.* Geneva: International Organization for Standardization.

ISO 9002 (1994). *Quality systems. Model for quality assurance in production, installation and servicing.* Geneva: International Organization for Standardization.

ISO 9003 (1994). *Quality systems. Model for quality assurance in final inspection and test.* Geneva: International Organization for Standardization.

ISO 9004-1 (1994). *Quality management and quality system elements. Part 1: Guidelines.* Geneva: International Organization for Standardization.

ISO 9004-4 (1993). *Quality management and quality system elements. Part 4: Guidelines for quality improvement.* Geneva: International Organization for Standardization.

Iivari, J. (1996). Why are CASE tools not used? *Communications of the ACM,* 39(10), 94-103.

Johnson, J. R. (1990). Hallmark's formula for quality. *Datamation,* 36(Feb. 15), 119-122.

Kautz, K., & Pries-Heje, J. (1996). Research directions on diffusion and adoption of information technology. In K. Kautz & J. Pries-Heje (Eds.), Diffusion and Adoption of Information Technology. *Proceedings of the first IFIP WG 8.6 working conference on the diffusion and adoption of information technology* (pp. 3-7). London: Chapman & Hall

Kerth, N. L. (1996). An approach to postmorta, postparta & post project reviews. Unpublished manuscript.

Kumar, K. (1990). Post implementation evaluation of computer-based information systems: Current practices. *Communications of the ACM,* 33(2), 203-212.

Kuvaja, P., & Bicego, A. (1994). BOOTSTRAP - A European assessment methodology. *Software Quality Journal,* 3(3), 117-128.

Larner, C. (1995). Practical experiences and lessons gained by a major European bank from a software development process improvement program based on the Capability Maturity Model. In *American Society for Quality Control (ASQC), European Organization for Quality (EOQ),* Society for Software Quality-USA (SSQ), & Union of Japanese Scientists and Engineers (JUSE) (Eds.), Proceedings of the First World Congress for Software Quality (pp. 1-11). San Francisco, CA.

Lebsanft, E. (1996). BOOTSTRAP: Experiences with Europe's software process assessment & improvement method. *Software Process Newsletter,* 3(5), 6-10.

Levitt, B., & March, J. G. (1988). Organizational learning. *Annual Review of Sociology,* 14, 319-340.

Lewin, K. (1947). Frontiers in group dynamics: Concept, method and reality in social science, social equilibra and social change. *Human Relations,* 1, 5-41.

Lewin, K. (1958). Group decision and social change. In E. E. Maccoby, T. M. Newcomb, & E. L. Hartley (Eds.), *Readings in social psychology* (3rd ed., pp. 197-211). New York: Holt, Rinehart and Winston.

Loken, C. B., & Skramstad, T. (1995). ISO 9000 certification - Experiences from Europe. In *American Society for Quality Control (ASQC), European Organization for Quality (EOQ)*, Society for Software Quality-USA (SSQ), & Union of Japanese Scientists and Engineers (JUSE) (Eds.), Proceedings of the First World Congress for Software Quality (pp. 1-11). San Francisco, CA.

Nonaka, I. (1991). The knowledge-creating company. *Harvard Business Review*, 69(6), 96-104.

Ohba, M. (1993). The impact of organizational knowledge on quality. *American Programmer*, 6(6), 28-39.

Olson, T. G., Humphrey, W. S., & Kitson, D. (1989). *Conducting SEI-assisted software process assessments* (Technical Report CMU/SEI-89-TR-7). Pittsburgh: Software Engineering Institute.

Paulish, D. J., & Carleton, A. D. (1994). Case studies of software-process-improvement measurement. *IEEE Computer*, 27(9), 50-59.

Paulk, M. C., Curtis, B., Chrissis, M. B., & Weber, C. V. (1993). *Capability Maturity Model for software*, version 1.1 (Technical Report CMU/SEI-93-TR-024). Pittsburgh: Software Engineering Institute.

Pfleeger, S. L. (1993). Lessons learned in building a corporate metrics program. *IEEE Software*, 8(5), 67-74.

Rogers, E. M. (1983). *Diffusion of innovations* (3rd ed.). New York: The Free Press.

Rout, T. P. (1995). SPICE: A framework for software process assessment. *Software Process - Improvement and Practice*, 1(8), 57-66.

Schein, E. H. (1992). *Organizational culture and leadership* (2nd ed.). San Francisco: Jossey-Bass.

Schein, E. H. (1995). On dialogue, culture, and organizational learning. *IEEE Engineering Management Review*, 23(1), 23-29.

Senge, P. M. (1990). *The fifth discipline: The art and practice of the learning organization.* New York: Doubleday/Currency.

Seshagiri, G. (1996). Continuous process improvement - Why wait till level 5? In H. El-Rewini & B. D. Shriver (Eds.), *Proceedings of the Twenty-Ninth Annual Hawaii International Conference on System Sciences:* Vol. 1 Software Technology and Architecture, 29, 681- 692.

Simon, H. A. (1957). *Administrative behaviour* (2nd ed.). New York: Macmillan.

Stata, R. (1989). Organizational learning - The key to management innovation. *Sloan Management Review*, 30(Spring), 63-64.

Stelzer, D. (1996). Enablers and inhibitors of CMM-based software process improvement efforts. A synopsis of empirical findings. Unpublished manuscript.

Stelzer, D., Mellis, W., & Herzwurm, G. (1996). Software process improvement via ISO 9000? Results of two surveys among European software houses. In H. El-Rewini & B. D. Shriver (Eds.): *Proceedings of the Twenty-Ninth Annual Hawaii International Conference on System Sciences:* Vol. 1. Software Technology and Architecture, 29, 703-712.

Thoresen, K. (1996). Learning at work. In K. Kautz & J. Pries-Heje (Eds.), Diffusion and Adoption of Information Technology. *Proceedings of the first IFIP WG 8.6 working conference on the diffusion and adoption of information technology* (pp. 176-201). London: Chapman & Hall

Walrad, C., & Moss, E. (1993). Measurement: The key to application development quality. *IBM Systems Journal*, 32(3), 445-460.

Wohlwend, H., & Rosenbaum, S. (1994). Schlumberger's software improvement program. *IEEE Transactions on Software Engineering*, 20(11)

833-839.

Zelkowitz, M. V. (1996). Software engineering technology infusion within NASA. *IEEE Engineering Management Review*, 43(3), 250-261.

CHAPTER
14

Technology Diffusion and Adoption in Small, Rural Firms

Arild Jansen
Finnmark College, Norway

Studies of regional technology policies have shown that efforts aimed at the effective stimulation of the adoption of information technology (IT) in rural areas are plagued with difficulties. However, our experience after visiting such areas is that IT is being implemented and used successfully in most branches and sectors of the local economy. We were thus interested in examining how small rural enterprises manage with new technology, considering their usually limited experience, their tendency to suffer from a weak local infrastructure and the lack of adequate support services.

This chapter presents the results of a study of how the three contingency factors technology, organization and infrastructure affect the diffusion and adoption of IT in rural areas. Our findings indicate that all three factors play important roles, but that the organizational characteristics are most influential for how acquisition and implementation processes are accomplished. We have identified a number of quite distinct patterns of technology diffusion among rural enterprises, with correspondingly different requirements for assistance and support. These needs should be addressed seriously by IT suppliers, educational institutions and R&D institutions.

Rural economies are traditionally characterized by small— often very small—enterprises, a general lack of infrastructure and technical support, few technical staff, and limited experience, or none at all, in the use of computers. Organizations trying to implement new technology have experienced barriers other than those offered by large, centralized organizations. Evaluations of such technology diffusion initiatives tell a number of stories about less successful implementations of new technology (Remø,1989; Levin and Nilsen,1990; Finne,1991; Torvatn and Rønningsbakk,1993). This paper presents the results of a study[1] on IT diffusion in a rural area in northern Norway which less than ten years ago was a technological laggard (Jansen,1995). Today, we find that IT is being implemented successfully and is used increasingly. Why is it that some technology diffusion efforts succeed while others fail, and what are the most important factors for managing new technology?

A framework is needed for understanding the rural diffusion of IT

We know that the successful introduction of new technology in organizations is dependent on both external and internal factors. However, precisely which factors are most important seems to vary from case to case. A review of the literature shows that most research efforts are directed towards what is stimulating the introduction or growth of IT in large organizations, while less attention seems to be paid to the implementation and use of IT in smaller businesses, in particular in a rural context (Cragg & King,1993; Lind et al., 1989; Thong and Yap, 1995). Heikkilä et al (1991) point out that many results of empirical studies are contradictory because different situations need different methodologies. It is questionable whether general results from information systems (IS) research can be directly applied in the small business context. We then need more adequate models for analysing the adoption of IT in small, and in particular rural, enterprises.

Small organizations do not in general have the competence to develop their own software, or even to make major modifications to existing ones. They have to acquire and implement standard software. Experience indicates that there is no single best way to carry out these activities, and that different types of organizations follow different patterns. A useful approach in the analysis of these activities may be 'risk analysis', that is, to identify uncertainty factors which require proper management in order to avoid failure in IT implementations. The specific objective of this study is to contribute to the development of a more appropriate framework for analyzing the diffusion and adoption of IT, by focusing on how:

- to describe and analyze the different types of system implementation processes that are currently in effect in small, rural organizations

- to identify the salient dimensions that can be used to characterise successes and failures of IT diffusion in a rural context
- to identify a set of 'risk factors', internal and external to the organizations, that are important, or even crucial, for the implementation of IT.

The empirical base is a study of the diffusion and use of IT in northern Norway. The data has been collected through an evaluation of an IT program during 1993-94 and through student projects in the period 1992-96. This was followed up by interviews in 42 enterprises. These were carried out in 1996.

This chapter is organized in the following way. Section 2 presents the research background, followed by a description of the overall framework and research methodology in section 3. Section 4 and 5 give the empirical findings, while the last section discusses the relevance of the research model.

Adoption and Adaptation of Information Technology

In his classical model, Rogers (1983) describes technology diffusion as a simple, linear model of communication by which messages are transferred from a source to a receiver. In his recent book, however, he emphasizes diffusion as information exchange among participants in a communication process (Rogers, 1995). He describes the diffusion of innovations as a social process by which subjectively perceived information about a new idea is communicated. It includes the handling of technical, financial and social uncertainty.

Herbig and Kraemer (1993) claim that diffusion is the process of adoption of an innovation by potential users. This may be conceived as having two stages: initiation (information gathering, conceptualizing, planning, etc.) and implementation (all the events, actions and decisions involved in putting an innovation into use). Veryard (1995) points to two different perspectives of technology diffusion. To the supplier, it is the delivery of a technological product and associated services, while to the user organization, this process is typically regarded as implementation of the technology. He points out that this implementation process implies adapting the technical solution to the organization. This constitutes a technical innovation. However, it also requires an adaptation of the organization to that specific solution, which is an organizational innovation.

Technology diffusion as a system development process

The traditional model for system development depicts the process as a 'waterfall' process covering all steps from problem identification up to the point when the system is up and running. The model has been heavily criticised for many reasons (e.g., Boehm,1976). One major objection is related to the stage-by-stage methodology and its emphasis on fully elaborated documents as completion criteria for early requirement and design phases, (e.g., Boehm,1988; Dalhbom and Mathiassen,1993). Andersen et al. (1986) emphasize the close relation between product development activities (such as analysis, design and implementation) and the management and organizational development actions (as planning, evaluation and regulations). They emphasize the importance of involving the user in all steps of the development cycle, and of ensuring that the necessary resources, such as time, knowledge, tools etc., are available, in order to support effective user participation.

For small enterprises, this time-consuming life cycle methodology is neither feasible nor desirable. Typically, small enterprises, in particular in a rural context, have a low degree of formalisation and specialization and often little experience with IS development. The manager may be the most experienced or motivated user of IT. The planning horizon is rather short and decisions are often made in an informal way. Small businesses seldom have the need for formal communication through IT, but rather use computers as tools for performing specific tasks. Their possibilities for developing applications on their own are often minimal and the small business has to rely more on software packages than do larger organizations (Heikkilä, 1991).

The challenge of IS implementation in a rural context

Although a number of descriptive surveys have been made on the use of computers by small firms, very little is known about the evolution of computing in the small firm. Some studies suggest that Nolan's growth stages model is worth testing in small firms, but so far no such studies have been reported (Cragg and King, 1993). In their exploratory study of IT evolution in six small firms, Cragg and King found that factors that encourage IT growth include relative advantage, competitive pressure, consultant support and managerial enthusiasm. Major factors that discourage IT growth were found to be inadequate education, lack of managerial time, economic factors and lack of technical support.

Thong & Yap (1995) claim that the following factors are correlated with the adoption of IT in small businesses: level of vendor support and consultant effectiveness, length of experience with IT, sufficiency of resources, level of management support and level of user participation. Friedman and Mønsted (1989), in a study of the diffusion of computer services and equipment into small Danish businesses, concluded that network relations, both formal and informal, are of great, if not decisive, importance for successful organizational change: 'The problem is not to be small, but to be isolated'. Further, they draw attention to internal factors such as the standardization of tasks, the introduction of better data discipline, the definition of user requirement and the development of competence in general.. However, none of these studies addresses issues of IT implementation in particular.

An evaluation of an IT program which aimed to stimulate the diffusion of IT into small and medium-sized businesses (SMEs) in the north of Norway (Jansen, 1995b) concluded that some implementations of IT had been successful, but that a number of implementations had more or less ended in failure. The problems reported included lack of functionality of the technical solutions, standards that were too rigid and the failure to integrate the new software with business operations. Also, various internal inhibitory factors were highlighted, such as low IT competence, weak management, shortage of time and resources in the implementation processes and the inability to accomplish organizational changes. Businesses often experienced inadequate local technical support and services which were not always compensated by support from the vendor.

IS success or failure - managing uncertainty

Most small organizations implementing IT systems are mainly concerned with costs, time, resources and risks (Veryard, 1995). Veryard defines risk as "the likelihood of delayed or non-existent benefits, or the risks of unwanted side-effects." The focus here is on how to reduce the uncertainty in this respect.

Jensen and Junge (1995), in an analysis of different risk management approaches, found that several system development methods do include risk factors in their models. By comparing the various models, they identified which of these factors are the most relevant ones for small enterprises. These were: i) personnel shortage and lack of IS user experience; ii) insufficient or inadequate functionality and usability of the software; iii) small computer resources; iv) lack of management involvement and competence, and v) unrealistic budgets and time schedules.

Gross and Ginzberg (1984) divided the risks associated with the acquisition of a software package into three uncertainty categories: i) whether the package can be modified when needed; ii) whether the vendor is reliable and the package has a good reputation, and iii) whether the package satisfies the user requirement. In a survey of 56 small businesses, Lees (1987) found that user satisfaction and the usage of IS increased when formal system analysis, design and implementation procedures, internal technical personnel and/or hired consultants, and extensive vendor services were employed. Delone (1988) summarized the results of earlier studies of the use of IS in small businesses by affirming that the following factors are important: experience with computers, the ability to plan and control computer use, management knowledge and involvement, personal acceptance of computers, the use of external support, the level of training and the type of computer use.

Heikkilä et al. (1991) studied the use of software packages in small organizations. They summarize the main differences between using standard software and tailored software as follows:

- the user may have to change work procedures to be able to work with the package;
- the user is likely to need changes to some programs in the package to meet his unique requirement;
- the user becomes dependent on the software vendor - this is due to the need for assistance and for updating his package;
- the cost of acquisition, implementation and use of the standard package can be predicted quite reliably, and are lower than that for tailored software;
- the package is available without development delay. The user may check with existing user sites beforehand, and can have the package implemented instantly.

Learning at different levels

It is being increasingly accepted that learning within the organization (or *organizational learning*) is critical for changing the organization. Argyris (1992) claims that organizational learning occurs under two conditions. First, learning occurs when an organization achieves what is intended through adjustments of individual activities (*single-loop learning).* Second, learning occurs when a mismatch between intentions and outcome is identified and it is corrected. The mismatch is turned into a match through changes in general goals and strategies (*double-loop learning).* Argyris emphasizes that both types of learning are essential, but "that many organizations, often without realizing it, create systems of learning that suppress double-loop inquiry and make it very difficult for even a well-designed information system to be effective".

The role of the infrastructure

The infrastructural perspective of IT is essential, as it captures the characteristics of IT that are related not to one specific system, but to the whole range of IT applications across the organization or even the society. George et al. (1991) identify an organization's computer infrastructure as 'all the resources and practices required to help people adequately use computer systems to do their primary work, which is a major area in which organizations make key social choices about computerization.' Gunton (1989) takes a similar approach in defining 'an enabling mechanism, aiming at supporting all of the information systems activities taking place across the organization'.

This chapter will address the infrastructural perspective at the *enterprise* level and the *local (community)* level. Small businesses in general have a weak internal technical organization, and must rely on outside facilities and support services. In a rural environment the local infrastructure is likely to be inadequate. While branch offices of larger organizations have access to a strong, corporate (business-wide) infrastructure that can compensate for the shortage of local facilities, it is likely that independent firms are more vulnerable to deficiencies in the local infrastructure.

Research Framework and Model

From the previous discussions it is evident that small, rural organizations face problems related to IT diffusion other than those normally focused upon in IS research. Even if there exist interesting and relevant findings from studies of small businesses, these results cannot be applied directly to rural organizations. There is, therefore, a need for a framework that addresses both the implementation and the use of IT and that aims to understand the relationships between the various factors that influence the various phases of the IT diffusion process.

Research approach: a contingency model of IT innovations in organizations

Kraemer et al (1989) developed a process model for IT innovations on the basis of an extensive study of several organizations. Their work concluded that both internal and external factors affect the IT system structure and its use, but they do so indirectly (p. 54). Their main conclusion was that managerial actions are the most influential factor for determining the result of change in IT use, and that managerial actions produce different results in different organizations. Even if these conclusions derive from studies of complex public organizations, it might be interesting to study if their general framework is relevant also in the analysis of IT innovations in a rural context.

This study is influenced by both frameworks discussed above. A contingency-theoretical model has been developed on the assumption that the outcome of IS implementations in small, rural organizations is dependent upon both internal and external factors. Our approach is based on the idea is that there is 'no best way' to achieve a successful IT implementation, but that there are some minimum requirements which must be satisfied to avoid failure. We furthermore assume that the functionality and usefulness of the technical solution in small organizations are, contrary to larger organizations, largely a matter of external factors. Departing from these assumptions, our framework will distinguish between four dimensions or classes of variables:

1) **Technology,** which in general is what is available on the market. More specifically, it is defined by the technical solution that is implemented in each individual firm. It is characterized by its functionality, technical qualities, usability and other capabilities related to the application area and specific user requirements.
2) **Organization,** described by internal characteristics of the enterprises and their inter-organizational relations. The main focus is on factors that we assume are relevant for the management of new technology, such as previous IT experience and competence.
3) **The local infrastructure**, comprising the technical facilities and support services related to IT available in the region, including local IT vendors, consultants and other types of relevant competence which is available.
4) The **IT implementation and adoption processes,** which includes the acquisition and installation of hardware and software, including necessary tailoring, and how the organization adapts to the new tools.

The fifth dimension is the **environment,** which covers a broad range of factors. We will limit ourselves to studying the environment through its impact on available technology, on the enterprises and their external relations and on the local infrastructure. Other factors, for example, the regional economic structure, the market and eco-

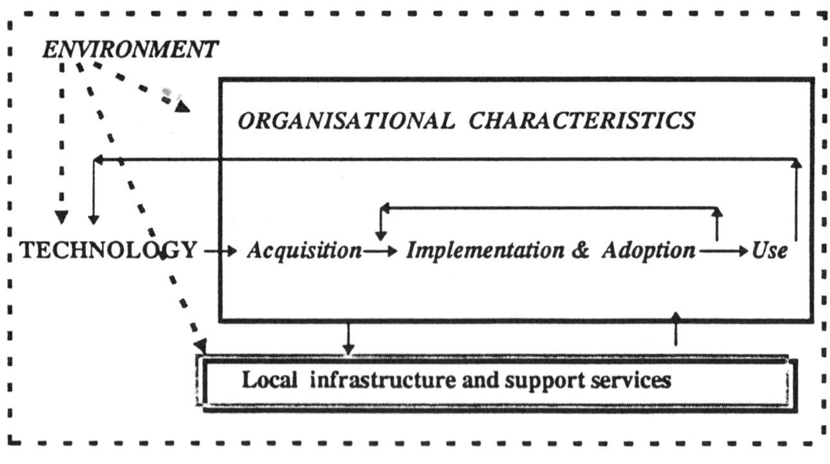

Figure I: A general framework for IS implementation in a rural context

nomic relations, the institutional framework, education and R&D institutions, which are important for the diffusion of technology in general, are not considered in this study. These distinctions permit us to organise our work around a framework with this overall structure in Figure I.

The model indicates that two types of organizational learning may take place. One type occurs when the the organization learns to operate and use the technical solutions, and to adjust the routine work to match the requirement of the system. The other type occurs when the organization learns to accomplish more fundamental changes in its way of working, to integrate the new solutions with the existing systems in order to utilize the potentials of the new system, and to move the focus towards a business-wide strategy for information management. These two types of learning correspond roughly to what Argyris (1992) defines as single- and double-loop learning.

Research issues and methodology

The general framework depicted above is rather open. The specific aim is to examine how factors such as technology, organizational characteristics and local infrastructure affect the adoption and use of IT in different types of small rural organizations, and in particular how these organizations cope with the uncertainties related to such challenges. In particular, we will address these questions:

- How was the acquisition process carried out, and what were the most important sources of information upon which the decision was based?
- How was the implementation accomplished? Did the users take part in the implementation process and was there any kind of modification of software (e.g. in prototyping and tailoring)?
- What type of support was required most from the local community infrastructure?
- What were the most important requirements for the technical solution and in particular to the application software.

In the selection of the methodology for this study, we had to consider the current state of knowledge in the field and the specific objectives of the study. As the available literature in this specific field is limited, an explorative research approach was chosen. All data was collected from enterprises in the northernmost region in Norway[2] in the following way:

- Through an evaluation of an IT program for Finnmark County (see Jansen, 1995, 1995b)
- Through student projects, in which current IT usage was studied in a number of enterprises in the region (1993-1996)

- Through interviews in 42 organizations, by visits (33), or telephone interviews (9).

Our research approach included both qualitative and quantitative methods of data collection and analysis. The model described in Figure 1 was developed and refined during the two first phases of this work. In the last phase, data was collected through semi-structured interviews with one or two employees from the selected organizations. The coding was done during the interview by the author.

The aim was not to select a representative sample of enterprises in the region, but to collect data from organizations covering a broad range of branches and sectors. The research objective was to find different patterns of IT implementation and use and to see how their importance varies across different types of organizations. Our findings need to be validated in later studies.

The reason for the choice of this particular research approach were:

- Current knowledge about the specific challenges and problems related to the diffusion and adoption and use of IT in a rural context is limited. We found that none of the existing research models were fully suited to the rural context.
- The level of formalization was low in several of the small enterprises in this study. Their ability to plan IT investment and their ability to assess the achieved benefits may be limited. It is our experience that strict survey-oriented approaches are likely to give low response rates, particularly among those most relevant to our study.
- The available measurement instruments do not work very well in this particular environment, as the understanding of the specific problems we were studying was low, and the quality of some of the responses turned out to be rather poor.

The contingency variables[3]

The independent factors in Figure 1 were operationalised as follows: *technology*[4] was characterized by: type of hardware and software in use and their application areas; where the application was developed; whether it was used for communication, etc. The *organization* was characterized by: business type and size, interorganizational relations and previous experience with IT. The *local infrastructure* was characterized primarily according to: how it was perceived by the firms, the availability of hardware/software and support services, the extent to which local vendors were used, whether the support was adequate, and the quality of telecommunications.

The implementation and adoption processes

These processes were described in terms of the acquisition, implementation and use of IT and IS. We asked a representative from each enterprise, usually the manager, to state the grounds for acquiring new technology. The representative was asked what the most relevant sources of information were, whether any requirement analysis had been carried out, and so on. Further, the person was asked to describe the implementation process in terms of success of installation, how the software conformed to the tasks, the perceived lack of competence and how organizational adaptations were managed.

The contribution of IT usage to an organization's performance is difficult to measure. We considered various measurements (e.g. Miller, 1989, Pitt et al., 1995, Heikkilä, 1991), but we found that none would be appropriate for our study. Some of these variables have thus been used in this study.

Empirical Findings and Analysis

The next two sections present the empirical findings. In the first part we discuss the data from the survey, followed by descriptions of some selected cases. In Tables I-V, statistical tables are given for the independent and dependent variables.

Number of employees (n=42)	*Mean: 10.5*	*Median: 9*	
IT experience (in years) (n=39)	*Mean: 6.3*	*Median: 6.5*	
Type of enterprise			
Manufacturing	14 %		
Sales of goods	62 %		
Sales of services	24 %		
Degree of org. dependence			
Independent and no co-operation	43 %		
Voluntary co-operation	17 %		
Linked to supplier/ importer/ etc.	21 %		
Branch office of corporation	19 %		

	Degree of organizational dependence		
Level of comp. of IT person	**Independent**	**Dependent**	**Total**
No specific education	73 %	25 %	64 %
Vocational training	23 %	31 %	24 %
College or university education	4 %	44 %	12 %
No of women as IT-person'	43 %	30 %	38 %
Number of employee (median = 9.4)			
0 - 9 employees	56 %	41 %	50 %
10 + employees	44 %	59 %	50 %
IT experience (Median 6.5 years)			
1 -3 years	27 %	13 %	21 %
4 - 6 years	35 %	30 %	33 %
7 +	38 %	57 %	46 %

Table I: Organisational Characteristics

Organization

Most of the 42 organizations surveyed were small firms. Their average size was about nine employees. Eleven firms had less than six employees, while only five firms employed more that 30 people. Fifty-seven percent were *independent* firms, while 43% either were branch offices of nation-wide corporations or were connected to one major supplier. (These are subsequently referred to as *dependent* firms.) Sixty percent sold to a local or regional market, while 20% also sold internationally.

The average size of the dependent firms was greater than the average size of the independent firms. However, if the one dependent manufacturer employing 150 employees is ignored, the difference in size is not substantial. More than half of the firms had been using IT for more than 6 years, and the larger, dependent firms tended to have longer experience with IT. On the other hand, in two thirds of the companies, the employee responsible for the IT system (the "IT manager") did not have any formal education, while only 12% had a college or university level of education. We find, as expected, that the larger organizations tended to have a higher level of IT competence that the smaller ones. Perhaps more surprisingly, about 40% the 'IT managers' were women.

Technology

Of the 42 firms surveyed, we found that 39 had some type of computer facilities. In total, 72% of the firms used IT for daily operational routines. Not surprisingly, we found that the type of technical solution and how it is used depended on the organizational characteristics. Most of the organizations had PC based systems in-house, but there was a significant difference between the proportions of independent and dependent firms (83% and 59% respectively) with in-house systems. A large majority (81%) of the dependent firms used central computer systems in their daily work, either in production or in sales, while the independent firms with access to such facilities were fewer (17%). Independent firms most frequently used generally available application packages (81% of the independent firms), while 53% of the dependent firms obtained their software from suppliers or as a part of a corporate IT infrastructure. Few firms had developed software based on own specifications.

Software for administrative work was mostly standard software across all categories of firms. However, a number of firms used a local accounting or auditing bureau to take care of their accounting. In some firms, such functions were taken care of at their main office. We also observed that the fax machine was still frequently used for placing orders, making acquisitions etc.

Number of PCs/terminals	***(n=39)***	*Median* **4**		
Number of applications	***(n=39)***	*Median* **4**		

	Degree of organisational dependence		
Type of main computer solutions (n=39)	*Independent*	*Dependent*	*SUM*
PCs	44%	13%	28%
PC network	43%	6%	33%
Link to central IS	17%	81%	39%
Application area of IT (n=39)			
Production of goods/provision of services	58%	46%	54%
Acquisition/ logistics	58%	93%	72%
Sales/offers	58%	93%	72%
Accounting, payroll etc.	71%	87%	77%
Financial planning/MIS	21%	60%	36%
Communication	25%	6%	15%
Type of software			
Standard/trade spec.	81%	39%	60%
Provided by supplier	9%	16%	12%
Corporate solution	1%	37%	19%
Own development	10%	8%	9%

	Type of software			
	Standard/trade spec software	*Provided by supplier*	*Corporate solution*	*Spec development*
Application Area				
Prod. goods/prov. serv.	35%	25%	20%	20%
Acquisition/ logistics	52%	25%	15%	7%
Sales/offers	61%	16%	18%	7%
Accounting, payroll etc.	86%	0%	10%	6%
Financial planning/MIS	57%	0%	36%	7%

Table II: Technology

The use of local infrastructure and support services

Table III shows that about 38% of the firms purchased all hardware and software from vendors outside the region, while 23% acquired the whole package from a local dealer. However, organizational dependence seems to influence their choice. Seventy-five per cent of the independent firms purchased software locally in contrast with only 40% of the dependent firms. This is probably because software was provided to the dependent firms by the main supplier, or because the acquisition of the software was corporation strategy. About 50% of the independent firms that had requested local support did not have their needs met. As expected, the dependent firms had

	Degree of organisational dependence		
Acquired hardware locally (n=39)	Independence	Link supplier	TOTAL
Yes, everything	25%	20%	23%
Some	50%	20%	38%
No	25%	60%	38%
Why not purchased locally(n=32)			
Hardware/software not available	83%	21%	56%
Lack of competence	5%	7%	6%
Provided through supplier	11%	29%	19%
Support from corporation or chain	0%	37%	19%
Need for local support			
No	12%	33%	18%
Yes, and satisfied	38%	40%	39%
Yes, but inadequate	50%	27%	41%
Modification/Tailoring of software (n=32)			
Yes, minor	36%	50%	43%
No	59%	31%	49%
Yes, substantial	5%	14%	8%
Telecommunication services adequate			
Yes: **81%**			
No: **19%**			

Table III: Use of local infrastructure

less need for local support. About half of the firms did not modify their software.

The acquisition and implementation process. The major reasons for investing in new technology were expected savings in costs (80%) and increased efficiency, reported by about 74% of the firms. Dependent firms were more likely to mention an improvement in logistic functions as a reason (80%), and compliance with corporate or supplier strategy (64%). The need for better communications facilities was mentioned by some firms (31%).

Source of information. Here we see great differences. Dependent firms reported suppliers/importers or corporations as the most important sources of information (93%), while information from trade organizations and from other enterprises was considered most influential by the independent enterprises (61%). Offers from local IT vendors had minor importance.

Requirement. Not surprisingly, functionality (100%) and user-friendliness (68%) were reported as the most important criteria for IT equipment. Local support was regarded as useful by half of the independent firms. Technical quality, costs and tailorability seemed to be of less importance. Less than half of the enterprises had specified their requirement prior to acquisition (46%).

Implementation and adoption. Our findings indicate that, in

	Organizational dependence		
Most important grounds (n=39)	Independent	Dependent	Total
Cost savings/management	67%	100%	80%
Efficiency	92%	53%	74%
Improve quality of goods/services	58%	53%	56%
Logistics	29%	80%	49%
Adapt to corporate or supplier strategy	17%	54%	31%
General communication	21%	40%	31%
Main source of information			
Supplier or corporation	30%	93%	56%
Internal requirement specifications	43%	50%	46%
Trade organization	61%	25%	46%
Offers from local IT vendor	39%	31%	38%
Cooperating/competing firms	43%	13%	31%
Most important requirement			
Functionality	100%	100%	100%
User-friendliness	65%	69%	68%
Local vendor/support	52%	19%	38%
Costs	30%	13%	23%
Technical quality	17%	25%	22%
Tailorability	14%	20%	15%
Reference to other	14%	7%	10%

Table IV: Acquisition

general, technical installation did not cause major problems. Only 25% reported experiencing a substantial problem. However, when it comes to software, close to 50% of the independent firms experienced inadequate software, while somewhat fewer of the dependent firms felt so.

Lack of IT competence was felt to be a problem by two thirds of the independent firms, but only about 25% considered it a serious problem. Thirty-one percent reported they had experienced a substantial lack of time or resources, and did not receive the requested assistance or support from local suppliers or consultants.

Organizational learning. A number of firms reported that they lacked the time and personnel with which to accomplish organizational changes and to integrate the IT functions into daily operations. We found that the various software packages in use at a firm were often isolated, so that manual routines were required for the conversion of data. Tailored systems were not usual, and most firms had no general strategy for information management.

Our data seemed to show that organizational dependence is important, if not determining, for how small rural firms manage with IT. In our study, the size of the firm and the type of firm seemed to be of similar importance. However, the analysis also revealed some methodological weaknesses. Some problems in collecting relevant and complete data have already been pointed out. In addition, we experi-

	Degree of organizational dependence		
	Independent	Link supplier	TOTAL
Success of installation (N=37)			
No/minor problems	61%	75%	67%
Substantial technical problems	39%	25%	30%
Inadequate software (N=35)	47%	31%	41%
Suffered from lack of IT competence (N=35)			
No/minor	40%	57%	47%
Substantial, but manageable	43%	35%	40%
Severe, did not manage	16%	8%	13%
Perceived problem in organisational adaptation (N=35)			
No/minor	28%	35%	31%
Some, but managed	40%	43%	41%
Lack of time &resources	32%	22%	28%
The effects of implementations (n=32)			
Improved quality of goods/services	59%		
Logistics	43%		
Resource saving	8%		
Efficiency	75%		
Control/management	47%		
Improved communication	25%		
Less satisfactory results (n=20)			
Not improved efficiency	18%		
High costs	12%		
High resources			
Perceived/expressed critical factors (N=28)			
Adequate functionality	81%		
Technical quality	34%		
Internal IT competence	52%		
Local support	43%		
External support	25%		

Table V: Implementation

enced that most of the firms had not recently installed new hardware or software. This implied that the process of implementing the new technical solution, including the adaptation of the organization to new work routines, had reached different phases in different firms. The data thus only indicates the total perceived problems and benefits of the IT packages used, and it is difficult to associate specific positive or negative experiences to the use of *each individual application program.* Further, our respondents may have used their 'selective' memory to filter their experiences of the implementation process, such that 'troubles' were forgotten.

Another interesting question is how existing technical solutions influence the acquisition and implementation process. We suspect that such a relationship exists, but our data did not tell us whether the acquisition of hardware was done prior to the choice of application package.

Some Cases of Success and Failure

The data collected from the interviews, often through more than one visit in each firm, provides us with much more information than the statistical figures can provide. We have presented below some case stories from different categories of enterprises, as more can be learned from such descriptions.

A local painting store

The E. B. painting store is an independent, family-owned company with 11 employees. Of these, 6 work in sales and administration, and the rest are craftsmen. The company has been using IT for some years, but is still a moderate IT user.

The major *IT system* is a specialized paint-mixing system (Multicolour), delivered by the major painting manufacturer in Norway. This system caused some initial problems, but is now functioning satisfactorily. The company is dependent on the vendor for maintenance. The vendor usually responds within one day. In addition, the company has a standard, PC-based office package for accounting, invoicing and managing the payroll. This system is not integrated with the Multicolour system, nor is it linked to external networks. It was purchased from a local IT supplier, in compliance with a policy of buying from local firms when possible.

The *organization* did not change dramatically since the implementation of these systems. The company had no IT plan, nor do any of the employees have special responsibilities for the IT systems. Some of the staff have been offered courses on how to run the systems and they take care of their basic operation. The company receives some support from a local IT supplier for maintenance of their office information system.

The *effect* of these implementations has been improved customer service, more effective inventory management, more effective economic control and generally better administrative routines. However, the system is not used for marketing, sales operation or more advanced stock administration. The owner explains it this way : "Our intention was not to maximize the efficiency and the profit. If this was the case, we would have fired two or three employees immediately".

Another painting store in the district has the same Multicolour system and uses in addition a stand-alone PC for some applications. The manager clearly expressed that limited IT experience and scarce personnel resources did not allow for competence build-up among the staff. So far, they could not afford the support for the customizing of standard applications to meet their specific needs.

On the other hand, an elderly owner of another small painting shop in a small coastal village reported that her business was doing fairly well. She used a standard PC/Windows system for administrative routines and a cash register. Her shop had acquired hardware and software from an IT supplier in the region, and the support, often by phone, was sufficient for their needs.

A local snowmobile dealer

Alta Snowmobile and MC dealer is an independent company with five employees. It sells and repairs snowmobiles from one particular supplier. The company was established four years ago, and has been using IT for about two years. Its main IT system is used for order entry and inventory control, and can be considered a branch specific system. However, it was initially incomplete and had several bugs. The small, non-local software house failed to finish the development work, and a new IT consultant took over. The system is still not in use; instead, the staff use the fax while waiting for more reliable versions to be installed.

In addition, they bought a standard PC network from a local IT company, to be used for administrative routines. This system is functioning well, but the application areas it is currently used for are limited. Accounting, for example, is not managed by the system. They have received useful help for running the standard systems from the local IT company, but the snowmobile supplier provides virtually no assistance.

Quite another story was told by a car dealer. The dealer applies a centrally maintained system for order entry, inventory control and other logistic functions. This system is provided by the main supplier (the car importer) and is running smoothly. In addition, the dealer has a standard PC Windows network for administrative work and sales support, also provided by the importer.

The case of North Cape Fish

North Cape Fish was a fishing plant in a small, coastal community close to the Barents Sea. The situation in the fishing industries in the 1980s was difficult, characterized by great uncertainty in the delivery of raw fish. The level of technology in the production processes was in general inadequate, resulting in low utilization of the fish resources. Moreover, the routines for financial control functioned poorly. The IT program provided support for investment in IT. Its aims were to improve the effectiveness and quality in the production processes and in particular to improve management routines. An economic analysis, carried out by an external consultant, formed the basis for the IT implementations, which were initiated in 1991.

Organization. The staff had initially very limited IT competence. One IS officer was hired for the project. Due to problems in the implementation work, the time schedule became tight. Small resources had been allocated for training and preparing changes in work routines. Management did not involve itself to a satisfactory degree, so that the required organizational changes were not taken seriously. Specifically, data collecting routines were reported to be inadequate, resulting in data of poor quality.

The technical solution was a PC network along with a standard office package, and in addition a system for material flow and production management. This included a payroll system which, in the case of the fishing industry, is rather complicated. The system, being a trade-specific solution, was developed by a software house. The system had satisfactory functionality and user interface. However, as the first installed versions turned out to be unstable with several bugs its implementation was delayed considerably.

The *local infrastructure* was of no real help, in that there was no local IT supplier or relevant competence available for support or exchange of experience. Vendor support was basically adequate, but limited by the long distances and the difficulty of travel.

The *acquisition and implementation* process included drawing up the requirement specifications and development work. To some degree, this was carried out in dialogue with the users. The project contributed some positive achievements, such as improved administrative routines and increased effectiveness in daily operations. The delays in the implementation, however, caused several problems. The tight timetables, that did not allow for proper training and planning for organization change, were particularly difficult to work. Because the data gathered was of poor quality, the production control system did not work as well as expected, and management control proved to be inadequate.

While it would be unfair to say that the project was a complete failure, it was certainly no real success either, at least when viewed from a short term perspective. The problems were attributed to deficiencies in the technical solutions. This was compounded by the absence of locally available persons with the relevant technical competence to deal with the problems. The major weaknesses were to be found in the organization itself, and in that there was no adequate support available locally.

The Tourist Hotel

The tourist hotel is a small independent enterprise located in a small inland community. It had no IT system in 1990. A general study of the enterprise was carried out, which concluded that an investment in IT was feasible. The requirements were defined before the technical solution was acquired and installed.

The organization had no IT system or any IT experience prior to the implementation. However, the manager, who was newly appointed, had received a higher education that included some IT courses, and he had a special interest in IT. The vendor, in close dialogue with the manager, adapted the system to the local environment. The manager took the responsibility for planning and organizational changes. Adaptations to the various work routines were made in a stepwise manner, allowing for learning to take place.

The *technical solution* was a PC network with a standard front-end software package for hotels, including a booking system and a system for the restaurant. The functionality and flexibility of the system was satisfactory, and the operations caused no major problems. The internal network has since been extended to include other co-operating firms in the community. The system has not, so far, been linked to any nationwide booking or marketing network. Although local support was virtually absent, sufficient assistance was provided by the vendor, located in Oslo.

In sum, the implementation was quite successful. A standard but adequate, solution was implemented without any serious problems and the availability of IT competence in the organization could compensate for the weak local infrastructure. The ambitions, in terms of project objectives, were at a realistic level. The *outcome* was satisfactory, with significant benefits both in terms of increased quality of the service and enhanced effectiveness. Management operations became more efficient, indicating that data collecting routines were satisfactory. The way organizational changes were managed allowed for learning to take place.

The manager of another small hotel also manages fairly well with a standard PC network, including a specific hotel booking and management system. This system was acquired from a centrally located software house in co-operation with a local IT supplier. The local supplier provides assistance both in technical maintenance and 'first aid' in the use of the booking system. Problems that are not handled by the local IT supplier are usually solved by the software house through 'telesupport.' IT usage is on a modest level, but the hotel manager felt fairly happy with the situation, although a number of possible improvements were pointed out.

The local consultant engineer

P. RI is another independent local company. It has 13 employees. Three are in administration, while the remainder are engineers. Typical work is planning and project management, public-work and quality assurance. IT is used extensively in all types of work.

The *technical solutions* are modern PC-networks and CAD work stations, configured with rather advanced peripheral equipment (plotters, printers, digitizing board, etc.), and software packages for various purposes. The software packages are advanced, though standard, and designed for the specific applications. In the past, they had experienced a number of problems with some software packages, causing delays in accomplishing tasks. For this reason, internal quality assurance now includes monitoring IT equipment and has become an important part of all routines.

Organization. The staff consist of rather experienced users who have used IT for many years. Daily operation and minor maintenance work are taken care of by the staff. More substantial updates and maintenance support are mostly provided by non-local IT vendors. Staff members do not, however, undertake any development work themselves. The IT usage is an integrated part of their daily work, including co-operation with customers and subcontractors. Local support is not critical for normal routines.

Outcome. Today, the company benefits greatly from the use of IT, as the quality of the systems is of decisive importance for the success of the company. The company, which is totally dependent on a high quality IT solution, has learned to survive in an environment with a relatively weak infrastructure. Company strategy is to use fairly advanced, well-proven hardware and software, and make sure that the staff can deal with daily problems.

A similar pattern was evident from visits to local electrical installation firms. These have developed a high level of IT competence and use their IT skills to give them a competitive advantage, at least in a rural environment.

Organizational implementation of IS in public administration.

FLID[5] is a major system development and organizational transformation project in the Norwegian Revenue Authority, an organization with offices in each of the 435 municipalities in Norway (Sørgaard, 1994). The history of the project spans more than ten years, and has involved several distinct phases of planning, pilot projects and system implementations at central, regional and local levels.

The *technical solution* is based on UNIX systems, installed in each office, which are integrated in a nation-wide network providing access to central databases. The implementations in the local offices had been planned in detail. The operational procedures are standardised, such that all the 270 smaller tax offices have almost identical installations. This makes it easier to have centralized user support, which reduces the need for computer operations at local tax offices. The local tax offices do not use the local community infrastructure except for standard telecommunications services.

Organizational implementation. In its initial stages, the project was technologically oriented, aiming at the modernisation of the technical infrastructure and the provision of modern tools to all employees. Today, the FLID is a strategic element in the overall business strategy, and is strongly integrated in the management of the whole organization. The primary goal is to increase the quality of tax assessment work, and to reallocate the use of resources through new management and reporting principles. Much emphasis has been placed on developing competence at every level in the organization. Some local offices have complained that staff training had been somewhat limited, as were the resources allocated for local organizational adaptation. The lack of local infrastructure has not, in general, caused significant problems for this project, mainly because the central organization had government backing.

The *outcome* of this large project was largely as predicted, in that improvements were made in work quality and effectiveness[6]. The general conclusion is that the project proved to be a success and that its history, including a shift from technological orientation to organizational orientation with strong management involvement shows that learning has taken place at higher levels in the organization. However, seen from the perspective of the local offices, this project did not allow for local adaptation. The corresponding learning processes had been impeded.

A somewhat similar case was a local office of a nationwide bank and assurance company. It used centrally developed and maintained corporation systems in service production and administrative routines. It made use of few, if any, services from local IT suppliers.

Discussion

The primary intention of this study was to describe both successes and failures of IT implementations as seen from the perspective of small local organizations. A full analysis of all the factors that may affect the diffusion and adoption of IT in rural areas was beyond the scope of the study. We deliberately selected a wide range of enterprises, of various sizes, ownership, trade membership, interorganizational structures and so forth, aiming at variety and differences rather than similarities.

Our overall goal was to provide insight to the following questions:

1. Which organizational factors are critical for accomplishing an IT implementation i.e. what factors decide whether we will see an acceptable implementation or not?
2. What minimum requirements should the local infrastructure of a rural area satisfy to be capable of supporting IT implementation and use?
3. Is it possible to define the 'appropriate technology' for a given organization i.e. a technical solution with functionality, usability and other characteristics appropriate to the specific needs?

We found that about 93% of the enterprises in the survey had some type of computer facilities, and that all used IT in primary business operations, but not necessarily in administrative routines. These findings show that Norwegian rural areas are quite heavily computerized, and that carrying out daily routines is heavily dependent upon IT. In our sample, the average length of IT experience was about six years, but rather few enterprises had a person with formal IT competence. This varied, depending on organization size and type.

Contingencies in IT implementations

However, despite the many differences, the results of our study indicate that there are interesting patterns of how small, rural firms manage with new technology, and different trajectories in the diffusion and adoption of IT. From our analysis, it seems fruitful to distinguish between four different classes or categories of enterprises according to the way they acquire and implement IT.

Independent, very small enterprises. Their fields of cooperation are usually voluntary, and are with suppliers and/or loosely coupled value chains of other firms. The enterprises are characterised by a low degree of formalization and have limited IT experience. The acquire standard PC-based systems, standard application packages and generally available trade-specific systems. Often, their involvement in adapting or tailoring the systems is low. They have to rely on external support, either directly from the local supplier or from a distant source using telecommunications. Typical examples of such firms can be found in the retail trade, among small scale production companies and among service providers. IT is not a competitive weapon, at least so far.

Independent, small enterprises which have gained substantial IT experience. Typically, they have implemented their second or third generation system, and their various software packages are becoming more integrated. Some of their applications are being tailored to met specific business needs, even as a competitive advantage. They acquire new systems based on well-defined requirements. They still benefit from exchange of information and experience within trade organizations. Local IT suppliers are of limited help in providing software and development work. Typical examples are electrical installers, engineering firms and more advanced manufacturers.

Enterprises having close links to suppliers or importers, etc. and members of chains. Their IT systems are linked to central systems and software is often provided by the central organization. Still, their application systems used in shop floor operations and administration may not be integrated. Internal IT competence is low. The acquisition and implementation of at least part of their IT systems is influenced, or even controlled, by central decisions. They are less dependent on local support, but will usually purchase items of hardware and software locally, and ask for local support within selected areas.

The branch offices of centrally located corporations. Their IT systems are to a large extent integrated into a nationwide network, and local offices run identical application systems. Virtually all hardware and software is acquired and implemented as part of a corporate strategy. The firms are not well integrated into the local economy.

These approximate categories should not be understood as static pictures. All four groups will increase their IT experiences and capabilities to acquire and integrate more advanced systems. However, their strategies for doing so are different and they need different types of support.

The small, independent and low-competence enterprise needs help in the acquisition and standard implementation work, including basic customisation of standard software. They also need assistance for establishing operational routines and basic 'troubleshooting'. They rely on external support, which may be based on telecommunications.

The more experienced and sophisticated IT user need access to more advanced IT support, primarily for the modification and tailoring of software systems to specific needs, and to integrate separate systems into more homogenous systems.

Minimum requirement to the technical solution

Our findings indicated that the functionality and user-friendliness of the technical solution as delivered by the vendor is the most important factor. Standard software or 'trade specific systems' are acquired most commonly. This can be easily explained in that the capabilities for adapting the application in small businesses are very limited. Support for this type of work is usually not available from local IT suppliers.

Tailorability seems to be of less importance. However, several of the enterprises experienced that when purchasing standard systems, the software package was 'too large' or 'was not easily adapted to the organization'. One business manager claimed, "The system gave more information that we needed" or, "..we were not able to utilize the full potential of the system" (Christensen, 1993). Others claimed that the real needs related to the IT system were not analysed sufficiently before the acquisition of the IT system.

We will claim that whereas functionality and user-friendliness are most important in the initial phases, tailorability increases in importance as the level of user competence increases. We may then define the *appropriateness* [7] of the technical solution as the degree to which the solution fulfills the requirements related to functionality, flexibility and usability, such that it fits the level of competence and maturity related to IT in the organization.

Risk factors - critical internal

The important and sometimes main goal is, in most cases, to improve quality in production or service provision and to reduce costs by improving efficiency in administrative routines. In most of the enterprises studied, the acquisition of a particular IT solution had not been determined by technical requirement. The ability of the internal organization to manage the integration of the IT solutions into the organization was evidently an important, and often decisive, factor. Many reported that implementation required more time and resources than predicted, and that internal training and learning were not given sufficient attention. The availability of a high level of IT competence seemed to be less critical where the organization had selected an appropriate technical solution.

It was interesting to note that many of the organizations seemed to manage without significant support from the local infrastructure, given that the technical solutions were appropriate. Furthermore, given that information and help from remote IT consultants or from co-operating business partners was adequate. Similarly, there were many cases where local assistance, if it were available, would have compensated for low internal IT competence, and lack of available support clearly hampered the efficient use of IT.

Conclusions

First of all, this study shows that IT has been extensively implemented and adopted across all sectors and branches in Norway, such that IT now has become an indispensable tool in all parts of the economy, even in very small communities. This is primarily not the effect of public IT policies. It is rather resulting from the growth in the rural economies in Norway in general.

The diffusion of IT in rural enterprises has predominately been demand-driven: the goals were reasonably well defined, but did not always correspond to clearly outlined requirements for the IT solutions. The large majority of the enterprises purchase and implement standard or well-proven technical solutions. The most reliable sources of information were trade organizations, competitors or co-operating companies. If adequate software and support is not available locally, they have to rely heavily on trade specific systems or software provided by the main supplier. There is a general need for more appropriate software, tailored for small businesses.

Not surprisingly, we found that a high level of IT competence is *not* the most critical factor in managing with IT. The ability of the manager to understand and accept changes in the organization that are necessary to reap a benefit from the new technology, and the motivation and resources for accomplishing such changes, seem to be of greater importance. Strong internal technical skill and competence alone do not guarantee the success of IS implementations (e.g. Prekumar & King, 1994).

Defining user-specific requirements for the technical solution, and identifying a solution that is realistic and manageable in the rural context, are critically important points. Time is very often a critical factor, along with sufficient resources that allow for organizational learning, including user involvement.

Our findings provide evidence that the contingency relationship model presented in this study is worth further examination; that the degree of organizational dependence is a decisive factor for the type of IT diffusion that take place. We identified quite different patterns of technology diffusion among rural enterprises, and correspondingly different needs for assistance and support. These needs should be addressed seriously by IT suppliers and by educational and R&D institutions.

Endnotes

[1] This study was part of the "Rural Development Program" funded by the Norwegian Research Council.

[2] The data presented here is only a part of all data collected in the complete study, as the primary focus is on small, private enterprises.

[3] These operationalisations conform to a large extent with Heikkilä (1995) , but also borrow from Cragg & King (1993).

[4] It should be noted that technology can be viewed as an independent or dependent variable. Technology available on the market is independent, while the selected technical solution, including the application package, is dependent upon decisions made during the acquisition process and upon the application area. The choice of the technical solution will clearly depend on previous experiences with IT implementations.

[5] FLID is the Norwegian acronym for "National Register and Tax Office, Introduction of Data Processing".

[6] More detailed studies show significant differences in achievement between offices, depending on various factors (Sørgaard,1994).

[7] This term is derived from the term "appropriate technology," which usually is found in literature on rural development.

References

Andersen et al (1986) *Professional systemutvikling.* Teknisk Forlag a/s, København.

Argyris, C. (1992) *On Organizational Learning.* Blackwell, Cambridge Massachusetts.

Boehm, B.W. (1976) *Software Engineering.* IEEE Transactions on Computers, Vol. C-25 (12), Dec. 1976.

Boehm, B.W. (1988) *A Spiral Model for Software Development and Enhancement.* In IEEE Computer, May 1988.

Bunker, D.J. and MacGregor, R.C. (1995) *User participation and the successful computerisation of the business environment.* In Kautz et al (eds.) Proc. from First IFIP 8.6 WC, *Diffusion and Adoption of IT,* NR, Oslo. ISBN 82-539-0398-7.

Christensen, R. (1993) *Informasjonsteknologi i Finnmark.* Sluttrapport (In Norwegian) . FINNUT, Vadsø.

Cragg, P. and King, M. (1993) *Small Firm Computing. Motivators and Inhibitors.* MIS Quarterly, March 1993.

Dalbom, B. and Mathiassen, L. (1993) *Computers in Context.* Basil Blackwell

Finne, H. (1991) *Dette viste vi fra før sa disponenten.* (IN Norwegian) IFIM/ ORAL, SUM Report 91-03, Trondheim

Friedman, L.and Mønsted, M. (1989) Small Enterprises Coping with the Challenge of IT. In Borum and Hull Kristensen (eds.). *Technological Innovations and Organizational Change.* Copenhagen Business School. ISBN 87-7034-252-0.

George, J., Iacone, S. and Kling, R. (1994) *How do office workers learn about computing?* Information Technology and People 6 (4), 249-269.

Gross, P. H. and Ginzberg, M.J. (1984) *Barriers to the adoption of application software packages.* Objectives, Solutions, 4, 211-226.

Gunton, T. (1989) *Infrastructure. Building A Framework for Corporate Information Handling.* Prentice Hall Int.

Heikkilä, J., Saarinen, T. and Sååksjårvi, M. (1991) *Success of software packages in small businesses: an exploratory study.* European J. of Information Systems. 1 (3), 159-169.

Herbig, P.A. and Kraemer, H. (1993) *Low Tech. Innovations Resurveying The Basic Meaning of Innovations.* Management Decisions, 31/3, 4-7.

Jansen, A. (1995a) *Rural development through diffusion of Information Technology.* Scandinavian Journal of Information Systems, 7 (1), Aalborg.

Jansen, A. (1995b) *Informasjonsteknologi som vekstfaktor i utkantstrøk?* (In Norwegian) Report to Finnmark County, Vadsø.

Jensen, J. and Junge, P. (1995) *Risk Management Approach. A Socio-Technical Analysis.* Aalborg University.

Kraemer, King, Dunkle and Lane (1989) *Managing Information Systems.* Jossey-Bass Inc., San Francisco.

Kwon, T.H. and Zmud, R.W. (1987) *Unifying the Fragmented Models of Information Systems Implementations.* In Boland and Hirscheim (eds.) Critical Issues In Information Systems Research. Wiley and Son, New York.

Lees, J.D. (1987) *Successful development of small businesses information systems.* Journal of Systems Management, 38:3, 32-39.

Levin, M. and Nilssen, T. (1990) *Etter et år med BUNT: fem kritiske suksessfaktorer for den videre utvikling* (In Norwegian) IFIM/ORAL SUM report 90-02, Trondheim

Lind, M.R., Zmud, R.W. and Fischer, W.A. (1989) *Microcomputer adoption - the impact of organizational size and structure.* Information and Management, 16, 157-162.

Lucas, H.C.,Walton, E.J. and Ginzberg, M.J. (1988) Implementing packaged software. *MIS Quarterly,* 537-549.

Lucas, H.C. (1981) *The key to successful Information Systems.* Columbia University Press, New York.

Miller J.(1989) *Information system effectiveness in fit between business needs and system capabilities.* In proceedings of the 10th. Int. Conference on Information Systems. Boston, Mass. Dec. 4-6, p 273-288

OECD (1989) *IT and Growth Opportunities.* ICCP, 19, Paris.

Pitt, L., Watson, R. and Kavan, C. (1995) Service Quality: A Measure of

Information Systems Effectiveness. *MIS Quarterly*, June.

Prekumar, G. and King, W.R. (1994) *Organizational Characteristics and Information Systems Planning: An Empirical Study.* Information System Research, 5 (2).

Remøe, S.E. (1989) *Industriell Fornyelse og Teknologi-overføring: Et sammendrag av evaluering av NTNF's program for spredning av DAK/DAP.(In Norwegian)* ØF-notat, 02.0282.00. Hamar.

Rogers, E. (1983) *Diffusion of innovations*, The Free Press, New York, 3rd edition.

Rogers, E. (1995) *Diffusion of innovations*, The Free Press, New York. 4th.edition

Sørgaard (1994) *Organizational Implementation of IS in The Norwegian Revenue Authority: A Case Study.* In Kerola (ed.) Proceedings of the 17th. Iris, University of Oulo.

Thong, J.Y.L. and Yap, C.S. (1995) An IT adaption model for small businesses. In Kautz et al (eds.) Proceedings from 1. IFIP 8.6 WG *Diffusion and Adoption of IT.* Oct. 1995, NR, Oslo. ISBN 82-539-0398-7.

Torvatn,H. and Rønningsbakk, B.(1993) Teknologioverføring på norsk. (In Norwegian) In Lorentzen et al (1993) *Teknologiutvikling og regional forandring* (In Danish) AKF Forlaget, Copenhagen.

Veryard, R. (1995) *IT Implementation or Delivery? Thought on Assimilation, Accommondation and Maturity.* In Kauts et al (eds.) *Proceedings from First IFIP 8.6 WG Diffusion and Adoption of IT.* Oct. 1995, NR, Oslo. ISBN 82-539-0398-7.

Section IV: The Conceptualization of Innovation and Diffusion Processes

In the fast changing business and technological world we experience today, what is the business IT innovation potential and what processes should organizations use? Clearly, the chapters in the first three sections offer views and solutions to these questions. The four chapters in this sections are more general in nature.

Joseph and Swanson take a look at the couplings among packaged software, inhouse development, organizational efficiency, and core business IS/IT support. They observe from their data that application software packages to a large extent replace older existing, mostly home-grown systems. However, the utilization of software packages may result in standardization across firms within an industry sector. It is commonly held that competitive advantage implies doing something that makes the organization and its products favorably special in the eyes of the customers. Standardized IT solutions may not have the potential to foster uniqueness. The intuitive implication is that organizations may have to develop separate IS/IT strategies for efficiency and core business activities. However, the present and future increased software package flexibility may very well blur this distinction.

O'Callaghan states that diffusion theory does not have the power to explain why and how organizations adapt IS/IT or the impacts on organizational performance. He suggests a framework that integrates diffusion theory with business process reengineering (BPR). The framework includes the observation that a decision to adopt a particular IS/IT solution is followed by a diffusion process. The diffusion process must be based on BPR considerations. Also, BPR must be founded on clearly articulated desired outcomes.

Larsen offers the observation that innovation theory, as applied to IS/IT, at present does not explain the obvious richness in IS/IT development and use. He claims that there is a basic need for defining core innovation issues and how actors interpret those issues. This chapter presents an issue and interpretation framework that is used to develop a model for a single IS/IT innovation process from initiation to system shutdown. The author observes that many IS/IT innovations occur simultaneously in most organizations. Additionally, the usage phase of any information system contains many smaller innovation processes. A model for multiple and simultaneous innovation processes are offered. This model is used to discuss couplings among the strategic IS/IT level and IS/IT applications.

McGuire reflects on the growing body of information systems literature that suggests viewing modern organizations and information technology development through the lens of complexity and chaos theory. This approach, many authors now claim, provides much insight into the non-linear, multi-layered innovation process that occurs in organizations dependent on information technology. Issues of non-linearity and chaos are especially relevant when discussing software development organizations and technology transfusion within these organizations. Organizations successfully operating at the "edge of chaos" may well be providing a preview of new operational paradigms.

CHAPTER
15

The Package Alternative in System Replacement: Evidence for Innovation Convergence

Thomas Joseph
E. Burton Swanson
University of California, Los Angeles

For many firms, packaged software bought in the marketplace appears to be an increasingly appealing alternative to custom-built code, when extant application systems come to be retired and replaced. Here we explore this phenomenon in terms of innovation "convergence" in the broader industry and in the light of selected findings from a recent survey. We also identify certain limits to innovation convergence and find them to be encountered most frequently with the firm's core value activities, with implications for the adoption and diffusion of packages.

While information systems in organizations are often long-lived, they are by no means forever. Eventually, systems and their application software come to be retired, usually when maintenance becomes increasingly problematic and new, more effective solutions are envisioned (Swanson, 1995b). In most cases, systems that are retired are replaced with new systems, as the need for applications support of the organization's business processes does not simply go away.

In replacing business application software, organizations face the buy-versus-build dilemma (Wreden, 1995). While bespoke, custom-built software can be made to fit a business nearly perfectly, it typically takes considerable amounts of time and resources, and frequently over-runs schedules and budgets. It also requires effortful building and sustaining of consensus among different stakeholders. Increasingly, notwithstanding the presumed advantages of custom-tailoring the software to the organization, many enterprises have been turning instead to (pre)packaged software (Mason, 1996), developed and marketed to many prospective adopters and users.

Because packaged software is favored by economies of scale, its costs and projected maintenance expenses for the individual prospective adopter may look to be comparatively acceptable and certain. The apparent ability to acquire and implement the package in a shorter time frame may also be attractive. Nonetheless, business application packages pose risks to the prospective adopter. They usually require changes to local business processes or else costly adaptation of the software. They must further be integrated into the existing portfolio of applications, often with the assistance of costly consultants. Their vendors must usually be relied upon for future updates and services (Gross and Ginzberg, 1984), yet local maintenance and continued readaptation cannot be entirely foregone. Hence, the choice between replacing a system with custom code or with a package is rarely an easy one.

Still, for many firms, packages appear to be an increasingly appealing replacement alternative. In the present chapter, we suggest that innovation and its diffusion among firms in the marketplace play an important role in explaining this phenomenon. While the desire to innovate might at first be thought to motivate and favor custom development by firms seeking unique competitive advantage, we observe less of this in present practice than might be expected. Rather, many firms facing system replacement appear to be looking to the market for their innovations.

Here we explore this phenomenon in the light of findings from a recent survey of information systems in North American organizations (Swanson 1995a). Broadly we investigate trends in systems replacement decisions. We focus in particular on the extent to which organizations innovate locally through the purchase of packages for replacing extant systems. Such packages may reflect innovation "convergence" in the broader industry (Brady, Tierney, & Williams, 1992). We explore this convergence and the associated factors that may explain in part the adoption of the package alternative. We also identify certain limits to innovation convergence. We find these limits to be encountered most frequently with the firm's core value activities.

Innovation Convergence and the Emergence of Packaged Software

We begin by observing that certain innovations in the application of information technology (IT) to organizations and their business processes may develop, evolve and eventually become entrenched in the economy through packaged products and services offered in the marketplace. While system utilities and personal productivity applications have traditionally been the domain in which software packages have dominated (Mason, 1996), packages for business applications have more recently been rapidly growing in adoption and are taking hold in almost all industry sectors. In addition, vendors such as SAP AG, Oracle, and PeopleSoft have created a vibrant market for client/server based integrated enterprise applications featuring packages that support a wide range of business processes in organizations in various industry sectors.

Brady, *et al.* (1992), following a model by Ward (1987), describe a characteristic product life cycle for a business application package. New applications typically emerge in individual organizations, in response to an opportunity for operational or strategic innovation. As the nature and benefit of the innovation becomes evident to other organizations, the innovation is imitated. Where sectoral significance of the innovation is recognized, convergence in software occurs around the business function. (The concept of convergence is borrowed from Rosenberg, 1976.) In our terms, organizations employ an increasingly common business logic. This frequently results in productization of the innovation in the form of packaged software. Finally, as the product(s) becomes widely adopted and implemented, it potentially becomes entrenched as infrastructure for the industry sector, as with ATM technology in banking (Ward, 1987).

Nevertheless, Brady, et al. (1992) caution against assuming that all types of software, including complex industry applications, will become commodities once their importance to a sector is demonstrated. Markets may not develop or endure for a variety of reasons, including lack of sufficient numbers of potential customers. One important consideration is that software products are not stable artifacts—their technical characteristics, patterns of usage and organizational significance are likely to change significantly over time (Brady, et al. 1992). Their periods of "entrenchment" may turn out to be relatively short lived, should they occur at all. Another consideration is that not all types of business activities may lend themselves equally to convergence; some may tend to be specific to the firm and therefore resistant to convergence.

Scope for Innovation Convergence in Business Applications

The emergence of packaged application software depends on the extent to which there is convergence around one or more business functions across organizations. We conjecture that there are greater opportunities for such convergence in firm infrastructure activities such as accounting, human resources, and other administration, as opposed to core value activities such as operations, distribution, marketing, and customer service. Indeed the earliest packaged business applications were in payroll, accounting, and other firm infrastructure activities. Many routines in firm infrastructure activities are relatively common across organizations, in part due to legal and regulatory requirements and/or conventional professional practices, as with accounting. Firms apparently also rarely seek to achieve strategic advantage by distinguishing themselves in firm infrastructure activities (see, however, Palvia and Perkins' (1992) description of Federal Express Corporation's PRISM system in support of its human resources function). This provides opportunities for the emergence of packages to support such functions. Widespread adoption of packages which often embody best practices drive further convergence in these activities across organizations.

In core value activities, opportunities for convergence are likely to be fewer due to the strategic advantage that firms derive from differentiation and specificity in these activities, with their direct implications for the firm's products and services. In recent years, firms have in particular sought to derive such strategic advantage through IT-based innovation in core value activities, which for first-movers likely requires custom-built software. Again, where the strategic advantage derived through an innovative IS application is recognized, competitors will attempt to replicate this success by deploying similar applications. However, sustainable competitive advantage is not often derived solely from IT innovations, but through IT innovations leveraged with resources or capabilities which are not easily replicated by other firms (Clemens & Row, 1991; Mata, Fuerst, & Barney, 1995).

Nevertheless, where an initial competitive advantage derived is not dependent on key resources or capabilities which cannot be replicated by other firms, innovation diffusion and convergence is likely around the application within the industry sector. Sectoral significance is then likely to lead to packaged applications over time. These are then adopted by the "laggards" seeking to achieve competitive parity. Manufacturing applications offer a good example of such convergence in core value chain activities. Yet opportunities for convergence may be fewer in core value chain activities than in firm infrastructure activities.

Limits to Innovation Convergence

In summary thus far, where there is innovation convergence around a business function, packages increasingly become an alternative to custom code to support that function. Nevertheless, there are inherent limits to innovation convergence in the applications domain, which follow from the dynamism and organizational complexity of industry/firm-specific activities (Brady, *et al.* 1992). In particular, packaged applications are often unable to cater to the diversity of firm-specific activities. Moreover, packaged applications developed to support organizationally complex activities will necessarily be based on important design assumptions about a target organization's size, structures, operations, business processes and environment. These assumptions can often be understood in the light of the developmental history of these packages, but may not be evident to the organizations that evaluate them for adoption. Additionally, individual systems within a firm's applications portfolio often have interdependencies reflecting the coordinative linkages between the value chain activities that they support. The individual package must thus coexist and be integrated into the existing portfolio of applications in the organization adopting it. Hence, where contingent local conditions are significant, packages often have to undergo considerable customization before they can be successfully deployed within organizations. Because such customization often demands significant organizational resources, the choice between custom code and packages for organizationally complex activities is rarely straightforward.

As already suggested, we conjecture that limits to innovation convergence and corresponding barriers to the adoption of packages will be encountered most frequently with firms' core value activities, where firm specificities are most likely to prevail.

System Replacement: Observations from an IS Census

We have discussed the emergence of business application packages as a viable alternate to custom code in support of activities where innovation convergence occurs. We have also addressed the limits to such convergence, where there are barriers to the adoption and implementation of packages. Here we analyze the role that packages play in the systems replacement plans reported by organizations as part of a recent survey. We consider the planned use of packages in the light of the above discussion.

In the winter of 1993-94, the Information Systems Research Program at UCLA's Anderson School undertook a survey which it termed an IS Census. Its purpose was to learn more about the life-cycles of the present installed application software base among North American organizations. A diverse sample of 54 North American organizations participated in the census, completing and returning the survey questionnaire. The participants were mostly medium to large enterprises averaging 8000 employees and 260 IS staff. A substantial 46% are located in California, with the rest distributed across North America. Some 33% percent come from the manufacturing sector, with 17% from the service sector, 15% from the finance sector, and the rest from other industry sectors. The respondents described a total of 758 systems among their application systems portfolios. The census and its findings are further described in Swanson (1995a).

With respect to our present interest, the census included data on organizations' systems then planned for near-term replacement, that is, within the next three years, as well as data on innovations planned for the replacement systems, including whether a package would be adopted. Each existing system was further described by respondents in terms of the business function(s) it supported, thereby associating it with the firm's value activities.

Porter and Millar (1985) categorize a firm's value activities into nine generic categories: firm infrastructure (including general management, legal work and accounting), human resource management, technology development, procurement, inbound logistics, operations, outbound logistics, marketing and sales, and service. Furthermore, they classify the last six categories as primary value activities of a firm; the others are classified as support activities that provide the inputs and infrastructure that allow the primary activities to take place. In the context of the present article, we are primarily interested in distinguishing activities that are likely to be specific to a firm, those that directly provide value, from those that are likely to be generic across firms, those that broadly constitute infrastructure. We label the former as core value activities (CVA) and the latter as firm infrastructure activities (FIA). Thus we include general management, legal work, accounting, and human resource management under the FIA category and the other activities, including procurement and technology development, under the CVA category.

Accordingly, we classified the application systems reported in the census as either FIA or CVA systems based on the business functions supported. From our data, we observed that a system supports between two and three business functions on average. In many cases, a system supporting one or more core value activities was also reported to support a firm infrastructure activity, in particular, accounting. In such cases, we assume that support for the firm infrastructure activity is likely a secondary function of the system and we therefore classify it as a system supporting a core value activity. Overall, among 679 systems, we therefore classified 471 as CVA systems and 208 as FIA systems. (The respondents provided no information on the business function supported for 35 systems. Additionally, the information provided on the replacement of 44 systems was not unambiguously associated with replacement in the short-term. Hence, in the present analysis, we include only 679 of the 758 systems.)

Analysis

About one third (230) of the 679 systems are reported as planned for near-term replacement (within the next three years). Table 1 summarizes replacement plans with package versus custom code. We observe that almost half (112) of the planned replacements will be with a package, suggesting a significant trend. More than two-thirds (163) of the replacements will involve a migration to client/server architectures, and more than half of these (85) will also involve a package. We observe further that nearly two thirds (150) of the planned replacements are of existing mainframe-based systems. More than half of these (79) will involve a package. Overall, we see a notable association between the shift to client/server and the adoption of packages to replace existing host-centric applications. We note that the paradigm shift to client/server computing has been accompanied by a spurt in the availability and adoption of commercial client/server based packages for business applications. As one indicator of this phenomenon, total revenues for the top ten vendors of client/server packages grew 110% from 1993 to 1994 (Xenakis, 1996).

Table 2 draws from the census data to compare existing CVA and FIA systems in terms of several basic characteristics. We observe that average age, expected remaining life, and the percentage of purchased code in existing systems in both categories are about the same. However, FIA systems are on average substantially smaller in size and have fewer staff assigned to their maintenance. FIA systems also

Mode of Replacement	Number of Systems	Systems Currently Mainframe Based	Systems Currently Other Based	Replacement with Client/Server Architecture	Replacement with Other Architecture
Package	112	79	33	85	27
Custom Code	118	71	47	78	40
Aggregate	230	150	80	163	67

Table 1: Systems Scheduled for Near-Term Replacement with Package Versus Custom Code, N=230

System Type	Existing Systems	Age (yrs)	Est. Life Remaining (yrs)	Size (KLOC)	Code Purchased (% of total)	Maintenance Staff (FTE)	Businesss Coordination (BCI)
CVA	471	6.97	4.90	612	31.96	2.59	3.76
FIA	208	7.70	4.42	377	32.11	1.79	0.31

Notes: BCI (Business Coordination Index) measures the cross-functional business links supported. The business functions here are: purchasing and supply, operations, distribution, marketing and sales, customer service, accounting and finance, personnel, administration, and engineering. Where only one function is supported, BCI=0; where two business functions are supported, BCI=1; where three functions are supported, BCI=3; where four functions are supported, BCI=6; and so on.

Table 2: Average Characteristics of Systems Based on Core Value Activities (CVA) Versus Firm Infrastructure Activities (FIA), N=679

support fewer cross-functional links across business functions, although this, in part, may be attributed to the definition of the two categories.

As seen in Table 3, in both the CVA and FIA categories, about the same proportion of systems (one third) is planned for near-term replacement. CVA and FIA systems appear to be following equivalent life-cycles in their historical development and replacement. However, when we look more closely, we see an interesting difference.

We observe from Table 4 that 70% of FIA systems planned for near-term replacement are to be replaced with packages. In sharp contrast, only 39% of CVA systems are to be replaced with packages. Following our discussion above, we suspect that this provides evidence for the limits to innovation convergence in CVA systems. Recall that core value activities are likely to be more specific to a firm than are infrastructure activities. Packages to meet a firm's CVA requirements may simply not be available, where there is substantial specificity and diversity of core technology and processes within an industry. Even when available, packages may require significant customization, unless the organization commits to redesigning its business processes around the assumptions inherent in the package. As we have seen, CVA systems also often coordinate laterally across several business functions; packages that replace existing CVA systems must either provide comprehensive support for these business functions or else interface with other supporting systems. This may further explain why fewer CVA systems are to be replaced with packages than are FIA systems.

System Type	Number of Systems	Systems to be Replaced in Near Term	Systems Not to be Replaced in Near Term
CVA	471	160 (33.97%)	311 (66.03%)
FIA	208	70 (33.65%)	138 (66.55%)
Aggregate	679	230	449

Table 3: Scheduled Replacement of Existing CVA Versus FIA Systems, N=679

System Type	Systems to be Replaced in Near Term	Systems to be Replaced with Custom Code	Systems to be Replaced with Package
CVA	160	97 (60.62%)	63 (39.38%)
FIA	70	21 (30.00%)	49 (70.00%)
Aggregate	230	118	112

Table 4 : Package and Custom Code Replacement for CVA Versus FIA Systems, N=230

As we have observed earlier from Table 2, CVA systems are on the average much larger than FIA systems. From prior research, it is known that complexity of systems tends to increase with their size (Banker, Datar, Kemerer, & Zweig, 1993). The complexity of CVA systems might therefore also contribute to the difficulty in replacing them with packaged code. Accordingly, we might expect larger systems to be candidates for replacement not with packages, but rather with custom code. However, we observe from Table 5 that the CVA systems to be replaced with packages are substantially larger on average than systems to be replaced with custom code. This, on the surface, appears to be counter-intuitive. However, we speculate that this may be explained by the recent emergence in the market of comprehensive enterprise packages that support the entire value chain of the firm.

Leading business software vendors now provide suites of packages that support a variety of business functions. Packages from a single source are often tightly integrated. In many industry sectors, firms can create entire application portfolios from such packages, potentially benefiting substantially from economies of scale achieved by the vendor. Thus, where replacements with packages are feasible, one or more large systems are likely to be candidates for replacement with suites of packages. Larger systems may be more likely to represent major core applications targeted by package vendors, while smaller systems may be more likely to represent peripheral applica-

System Type	Systems to be Replaced with Package (KLOC)	Systems to be Replaced with Custom Code (KLOC)	Systems not Scheduled for Replacement (KLOC)
CVA	819	440	641
FIA	391	467	352

Table 5: Average Size (in thousand lines of code, KLOC) of CVA Versus FIA Systems

System Type	Systems to be Replaced with Package (purchased code as % of total)	Systems to be Replaced with Custom Code (purchased code as % of total)	Systems not Scheduled for Replacement (purchased code as % of total)
CVA	36.64	10.99	37.52
FIA	35.14	3.00	35.60

Table 6a.: Average Amount of Purchased Code (as % of total) in CVA Versus FIA Systems

tions specific or idiosyncratic to the firm. Moreover, the paradigm shift in computing to distributed architectures offers opportunities for firms to replace several legacy applications with suites of packages, often from a single vendor. Thus, several larger systems may together be replaced as part of a strategic move to client/server. In most (637) cases, respondents also provided data on the proportion of purchased code in their existing systems. In Table 6a, we see the average proportion of purchased code in systems to be replaced with custom code is much less than in those systems to be replaced with packages or in those systems not planned for short-term replacement. But the averages do not adequately tell the story.

In Table 6b, we show the distribution of systems in each of the same categories, broken down by three groupings of the percent of extent purchased code. We can infer the likely source of the existing systems from this data. An existing system which is reported to have 0-5% purchased code consists basically of locally built custom code. An existing system with 95-100% purchased code is most likely a package (most packages require at least some customization to a firm). (Outsourced systems, custom built by a service provider, may also be indicated as having 100% purchased code but this is very unlikely. Outsourced custom development is rarely undertaken entirely independent of the firm in which a system is to be implemented.) An existing system with 6-94% purchased code may or may not involve a package (it may well also incorporate custom code purchased from a service provider).

We note from Table 6b that of the 112 systems planned for replacement with a package, 22, about 20% (adjusting for missing data) are apparently already implemented with packages. Similarly, 93, about 22% of 449 systems not due for near-term replacement are apparently packages. In contrast, only 5, less than 5% of 118 systems to be replaced with custom code are apparently packages. This provides further evidence for the argument that innovation convergence leads to the emergence of packages. A higher proportion of purchased code signals a greater degree of innovation convergence in

Extant % Purchased Code	Systems to be Replaced with Package			Systems to be Replaced with Custom Code			Systems Not Due for Replacement			Total
	CVA	FIA	Both	CVA	FIA	Both	CVA	FIA	Both	
0%-5%	31	29	60	77	17	94	156	72	228	382
6%-94%	16	12	28	9	3	12	68	27	95	135
95%-100%	14	8	22	5	0	5	68	25	93	120
Not Reported	2	0	2	6	1	7	19	24	33	42
Total	63	49	112	97	21	118	311	128	449	679

Table 6b: Distribution of Purchased Code in CVA Versus FIA Systems, N=679

applications that has already occurred as indicated by the availability of software from external sources. If a system due for replacement already uses a considerable amount of purchased code, it is more likely that it will be replaced with a package. Replacement with custom code is more frequently undertaken where such convergence has not yet occurred.

Concluding Discussion

As indicated by the overall trend towards replacement of older existing, mostly home-grown, systems with packages, our analysis of the census data suggests substantial innovation convergence in business application software. However, our analysis also provides evidence for limits to innovation convergence in business functions that exhibit high degrees of firm-specificity. We have seen that FIA systems which support activities that are often generic across firms are more likely to be replaced with packages than are CVA systems that support activities that are often specific to a firm.

Nevertheless, even in core value activities, a potential for innovation convergence in application software exists, centered around a corresponding convergence in an industry sector's core technology and processes. Moreover, where packages then become widely available and adopted in an industry sector, they may well promote further convergence in business processes and practices. Because of their affordability and the likelihood that they can be more quickly implemented, prospective buyers may choose to modify their business processes to fit the package, rather than attempt to custom build their own software. Thus, the business processes and practices of an industry sector may follow the example of inventory management in manufacturing, which is commonly based on material requirements planning (MRP) and its associated software (Anderson, Schroeder, Tupy, & White, 1981). Here, the convergence in business practice around MRP set the stage for the emergence of packaged applications to support this function. The implementation of MRP systems based on dominant packages resulted in further convergence in business practices. Indeed, MRP then became feasible for many manufacturing firms largely due to the availability of supporting packages.

The data from the census are also consistent with the present trend towards replacement of large systems with suites of packages, often from a single vendor. Leading vendors now provide comprehensive packages that support a variety of business functions in different industry sectors. Packages from a single source may be tightly integrated to provide more or less an entire application portfolio for a firm. The data from the census also provide evidence that firms are replacing several mainframe applications with client/server based packages. The obsolescence of host-centric applications brought about by the advent of systems based on new industry standards motivate organizations to replace several systems in their current applications portfolio with suites of packages from a dominant applications vendor. Thus, for example, Eastman Kodak Co. has replaced 70 systems and databases, most of them homegrown, with an integrated package suite from SAP America (Caldwell, 1994).

While innovation convergence in business applications is inherently more difficult in core value activities, it is obviously happening in several industries as evidenced by the growing numbers of packages to support such activities. The dominant packaged applications vendors compete heavily on the breadth of their product line (Andren, 1994). Today, these vendors offer packages for almost every major industry segment. The adoption of packages for core value activities involve decisions that have significant long-term business implications for a firm. Does the firm adopt the business model reflected in the architecture of a dominant vendor product and lose its flexibility to adapt its business practices to respond to competitive pressures (Scofield, 1995)? Or does the firm attempt to modify the underlying architecture of the package to fit its business model at an expense often

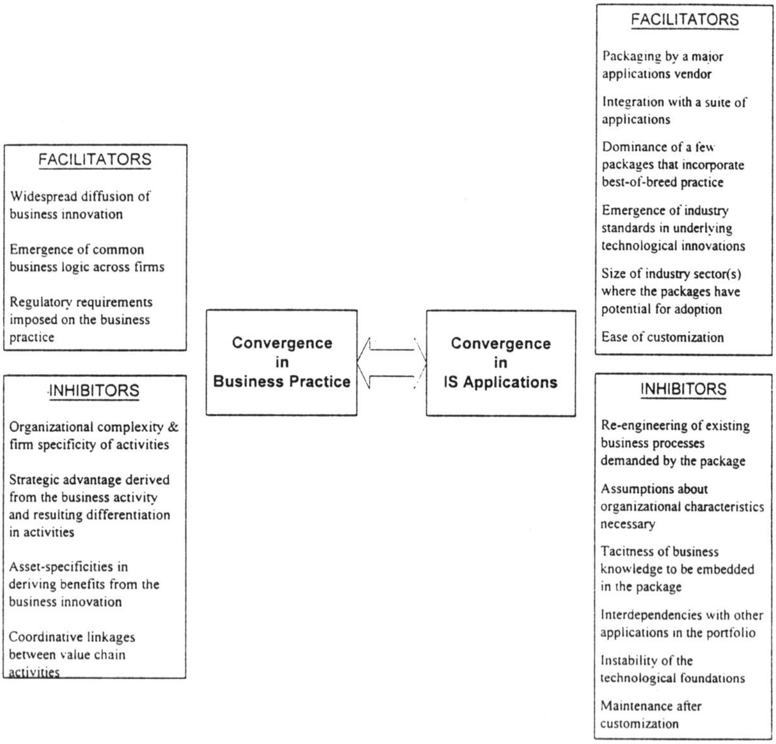

Figure 1: Convergence Facilitators and Inhibitors

exceeding that of custom development and involving substantially greater maintenance effort? In many cases, where packages become industry standards, organizations may have little choice but to adopt and adhere to a package and its associated practices. In such instances, package vendors become licensers of business processes and compete among each other to incorporate best of breed business practices in their software. This suggests that most firms in the future will derive little if any strategic advantage from differentiation on these value activities. Understandably, therefore, firms are increasingly becoming more selective in resorting to custom development; often this is considered only when the package alternative does not exist or is not viable.

In conclusion, we suggest that there are subtle relationships between innovation convergence in business practices, and innovation convergence in systems and their application software. Drawing from our overall discussion, we summarize these relationships for the reader's convenience in Figure 1. We identify facilitators and inhibitors of both forms of convergence. We thereby provide the broader perspective for the principal assertion examined here, namely, that the limits to innovation convergence and corresponding barriers to adoption of packages will be encountered most frequently with firms' core value activities.

Finally, we add a cautionary note on the likely future of packages, observing that object-oriented software design has in recent years become very popular. Indeed, from our census data, of the 118 planned replacements of existing systems with custom built code, we are told that 42 will be object-oriented. We note that industry standards for distributed object architectures are emerging currently that promise to facilitate seamless integration of objects from disparate sources (such as Distributed OLE, CORBA, and OpenDoc). With the emergence of one or two widely accepted industry standards, it may be speculated that firms will have the ability to build custom applications based on objects that may be either developed in-house or obtained from external sources. A whole new industry segment may be emerging around off-the-shelf application software components (Gartner Group, 1995). Here, we might expect to observe convergence in software to support highly granular business activities common across several industries. This phenomenon would be different from convergence in applications at the level of business functions. These newer software components embody fewer assumptions about the business and are highly customizable. By integrating such software components, firms would potentially have the ability to rapidly build and deploy applications tailored to fit their own business models and processes. Still, packaged software may not be easily replaced with granular objects. Vendors also see opportunities to pre-package and market object components vertically for specific application domains (Foley, 1996). Overall, therefore, the future for the integrated package market as it now exists appears highly uncertain.

We close by reminding the reader that our findings and discussion are more suggestive than conclusive. We have used data from a rather small sample of firms and their systems. We have merely attempted to highlight current trends in systems replacement and to venture an explanation for our own survey findings using the concept of innovation convergence. As very little is presently known about the full life-cycle economics of packages from the viewpoint not only of vendors, but of adopter organizations, in particular, much remains to be accomplished by further research.

References

Anderson, J. C., Schroeder, R. G., Tupy, S. G., & White, E. M. (1981). *MRP: A Study of Implementation and Practice*, American Production and Inventory Control Society (APICS).

Andren, E. (1995). It starts with people, *InformationWeek*, November 14, 72-76.

Banker, R.D., Datar, S.M., Kemerer, C., & Zweig, D. (1993). Software complexity and maintenance costs, *Communications of the ACM*, 36, 11, 81-94.

Brady, T., Tierney, M., & Williams, R. (1992). The commodification of industry applications software, *Industrial and Corporate Change*, 1, 489-513.

Caldwell, B. (1994). Who needs programmers?, *InformationWeek*, April 25, 23-30.

Clemens, E.K. & Row, M.C. (1991). Sustaining IT advantage: the role of structural differences, *MIS Quarterly*, 15, 275-292.

Foley, M. J. (1996). Objects for business apps are coming, but slowly, *Datamation*, 42, May 15, 52 ff.

Gartner Group (1995). Conference presentation.

Gross, P.H.B., & Ginzberg, M. (1984). Barriers to the adoption of application software packages, *Systems, Objectives, Solutions*, 4, 211-226.

Mason, T. C. (1996). SIC7372: Software, Prepackaged, *Encyclopedia of Global Industries*, Computers and Data Processing, 190-200. Detroit: Gale Research.

Mata, F.J., Fuerst, W.L. & Barney, J.B. (1995). Information technology and sustained competitive advantage: A resource-based analysis, *MIS Quarterly*, 19, 487-505.

Palvia, P. C., & Perkins, J. A. (1992). The PRISM system: a key to organizational effectiveness at Federal Express Corporation, *MIS Quarterly*, 16, 277-292.

Porter, M.E. & Millar, V.E. (1985). How information gives you competitive advantage, *Harvard Business Review*, 63, 4, 149-160.

Rosenberg, N. (1976). *Perspectives on Technology*. Cambridge: University Press.

Scofield, M. (1995). Packages: one size fits all?, *InformationWeek*, October 2, 136.

Swanson, E.B. (1995a). Findings from an information system census, Information Systems Working Paper #8-95, The Anderson School at UCLA.

Swanson, E.B. (1995b). System maintenance and expected life, Information Systems Working Paper #9-95, The Anderson School at UCLA.

Ward, J. (1987). Integrating information systems into business strategies, *Long Range Planning*, 20, 19-29.

Wreden, N. (1995). Build, buy, or modify?, *Beyond Computing*, January/February, 47-49.

Xenakis, J.J. (1996). Taming SAP, *CFO*, March, 23-30.

CHAPTER
16

Technology Diffusion and Organizational Transformation: An Integrative Framework

Ramon O'Callaghan
Tilburg University

The Information Systems literature has been concerned with diffusion and implementation of technologies within organizations for a long time. As the use of computer-based systems grew, researchers became concerned with the gap between systems potential and actual use. This gap began to be investigated under the heading of "Implementation Research." Implementation research, however, suffered from an identity problem: When is a system said to be implemented?

Traditionally, research has focused on technology adoption mostly as a single decision point. While most researchers would agree that implementation goes beyond the single decision point, how far beyond is a matter of contention. This problem has become more apparent in recent years as technology applications have become both more pervasive and more important for the core business. Today, the organizational impacts of IT are far more significant than ever before, and the success of technology implementation is increasingly associated with organizational learning and with effective change management. Furthermore, IT is often used as a tool for organizational design and/or as enabler of organizational transformation. In such cases, both technology adoption and technology implementation are far more complex than traditionally described in the literature.

Despite the existence of a large data base in the form of both case studies and survey data, implementation research lacks a cohesive framework for binding empirical research. This paper is an attempt to develop such a framework by bringing together the classic paradigms on adoption and diffusion of technology with the more recent views on organizational transformation and reengineering. It starts with a review of the traditional views of innovation diffusion (e.g., Rogers, 1983) along with some more recent perspectives that place organizational learning and know-how at the core of the diffusion process (Attewell, 1992). Then, some relevant literature on implementation and organizational change research is discussed (e.g., Radnor et al., 1970; Keen, 1977; Markus, 1984; Leonard-Barton & Kraus, 1988; Walton, 1989; McKersie & Walton, 1991). Based on these views, an integrating framework is then developed that incorporates elements of technology adoption, technology diffusion, organizational learning, technology implementation, organizational change and reengineering. The utility of the framework is illustrated through its application to EDI as a case in innovation diffusion with major implications for business reengineering.

Background

Empirical research on adoption and diffusion of innovations has been a fertile ground for many researchers. Two basic paradigms have inspired such research. The first one views diffusion as a process of communication and influence whereby potential users become informed about the availability of new technology and are persuaded to adopt through communication with prior users (Rogers, 1983). The second one regards the process as an economic one. It views diffusion in terms of cost and benefit: i.e., the higher the cost of the innovation, the slower the diffusion will occur; and the higher the perceived benefit, the faster adoption will occur (Mansfield, 1968). Both paradigms advise a style of diffusion research that focuses on adoption by individuals or by single firms.

Adoption of innovations

Rogers and Shoemaker (1971) and Rogers (1983) provide a comprehensive review of adoption and diffusion research. One of their key findings is that the rate at which innovations are adopted is related to the attributes of the innovation as perceived by potential adopters. Rogers and Shoemaker (1971) concluded that most of the variance in the rate of adoption can be explained by five "universal" attributes: Relative Advantage, Compatibility, Complexity, Trialability and Observability. In a meta-analysis of 75 research studies, however,

Tornatzky and Klein (1982) found the strongest support for the relationships of Relative Advantage, Compatibility, and Complexity with the rate of adoption, and less support for Trialability and Observability.

It is worth pointing out that these factors do not depend exclusively on the innovation itself. Advantage, Compatibility and Complexity are all "relative", i.e., they depend also on the characteristics of the individual or organization adopting the innovation (Leonard-Barton, 1987; More & Benbasat, 1991). For example, the way an organization is managed (including its strategy, structure, processes, and systems) will determine the extent to which it can take *advantage* from a given new technology. The literature on alignment of IT and business strategy also supports this view (e.g., Henderson and Venkatraman, 1993). Likewise, the specific organizational context (e.g., processes, systems, procedures) in which the new technology is introduced will determine the degree of *compatibility* with existing practices. Finally, the level of technological sophistication, the skills and experience of the staff (i.e., the existing know-how) will determine the perceived *complexity* of the innovation.

Implementation has been described as the mutual adaptation between technology and organization (Leonard-Barton & Kraus, 1988). This mutual adjustment helps in reducing the complexities and incompatibilities encountered in the introduction of a new technology by an organization. In summary, the factors that influence the initial adoption decision (perceived attributes) are not solely "attributes of the innovation." They also depend on the organizational context and, therefore, can be managed subsequently during the implementation phase.

Finally, an innovation's rate of adoption is affected by the extent of promotional efforts of the sponsor. The relationship between rate of adoption and promotional efforts, however, is not usually direct and linear. There is a greater pay-off from a given amount of promotion at certain stages in an innovation's diffusion. Empirical research has shown that the greatest response occurs when opinion leaders are adopting, which usually occurs somewhere between 3 and 16 percent adoption in most systems (Petrini, 1966).

Not only do promotional efforts have an effect at different stages in the sequence of an innovation's adoption, but the system's self-generated pressures toward adoption become important as an increasing proportion of members of the system adopt. Rogers (1983) refers to this increasing pressure from interpersonal networks as the "diffusion effect."

The diffusion effect: social pressure

The diffusion effect is the cumulatively increasing degree of influence upon an individual or an organization to adopt or reject an innovation, resulting from the activation of peer networks about an innovation in a social system. For example, when only 5% of all the potential adopters are aware of a new idea, the degree of influence upon an individual to adopt or reject the innovation is quite different from when 95% have adopted. In other words, the norms of potential users toward adoption change over time as the diffusion process proceeds, and the new idea or new technology is gradually incorporated into the life-stream of users.

This diffusion effect is well known in the marketing literature as well. Bass (1969) developed a growth model for the timing of initial purchases of new products. His theory stems mathematically from the contagion models in epidemiology (Bartlett, 1960). Behaviorally, the assumptions are similar in certain respects to the theoretical concepts in the literature on new product adoption and diffusion (Mansfield, 1961; Rogers and Shoemaker, 1962; King, 1966). The basic assumption of the Bass model is that the probability of an adoption of a new product at a certain point in time given that an adoption has not yet taken place is an increasing linear function of the number of previous adopters. Bass differentiates two classes of adopters: innovators and imitators.

Innovators decide to adopt an innovation independently of the decisions of other members of the social system (i.e., the pressure for adoption does not increase with the growth of the adoption process). *Imitators* are influenced in the timing of adoption by the decisions of other members of the social system (i.e., the pressure for adoption increases with the number of previous adopters). The Bass diffusion model has been successfully demonstrated in retail service, industrial technology, agriculture, and consumer-durables sectors.

Traditionally, however, diffusion research and adoption research have followed separate paths. Diffusion modeling efforts have generally resulted in deterministic interpretations of the time-dependent aspects of the diffusion process; e.g., the S-shaped curve (Mahajan and Peterson, 1985). Most diffusion models tend to regard all innovations as equivalent units, i.e., they ignore the impact of potential adopters' perception of innovation attributes (Rogers, 1983). Attempts have been made to develop innovation diffusion models that incorporate potential adopters' perceptions of the relevant innovation attributes. For instance, Srivastava et al. (1985) developed and tested multi-attribute diffusion model for forecasting the adoption of investment alternatives for consumers. The multi-attribute model is reported to provide better results than the naive diffusion model.

Research on the diffusion of technology innovation among business organizations has focused on the competitive factors influencing diffusion. Robertson and Gatignon (1986) extend the behavioral paradigm (Rogers, 1983) by incorporating competitive factors as explanatory variables. These include the supply side competitive environment (i.e., competitiveness, reputation, standardization, vertical coordination, resource commitments in R&D and marketing) and the adopter industry competitive environment (i.e., competitive intensity, demand uncertainty, heterogeneity, and communication factors). An empirical test of this conceptualization (Gatignon and Robertson, 1989) suggests that firms most receptive to innovation are in concentrated industries with limited price intensity and that supplier incentives and vertical links to buyers are important in achieving adoption. The results also suggest that adopters can be separated from non-adopters by their information-processing characteristics.

The technology life cycle

One conclusion that emerged from the early implementation research was the hypothesis of a life cycle pattern associated with the long-term implementation of systems in organizations (Radnor et al., 1970; Nolan, 1973; Nolan and Gibson, 1974; Keen, 1977; Nolan, 1979). That stream of research viewed technology implementation as an intra-organizational diffusion process. The concept of life cycle, however, can be generalized by looking at technology diffusion beyond the boundaries of a given organization.

The diffusion of a technology within a given industry and, ultimately, within the society at large, may affect the perceptions of potential adopters in a way which is similar to the diffusion effect described above. The reason, however, is different. It is the increasing knowledge about the technology (instead of peer pressure) that influences the adoption decision. This better understanding of the technology comes from the collective experience with the technology as the number of users grows (Bass, 1980). This is particularly true of complex technologies where adopters face what Rosenberg (1982) calls "learning by using." Users spend several years developing an understanding of the strengths and weaknesses of the technology. In turn, the knowledge gained by the users becomes very important to the innovator in order to enhance the technology and generate new designs (Rice and Roger, 1980; Eveland and Tornatzky, 1990).

The life cycle of a technology (*technology life cycle*) can thus be associated with the cumulative distribution of adopters over time. The hypothesis is that learning increases over time with collective experience and then it affects the likelihood of adoption. This is in line with Attewell's conceptualization of technology diffusion in terms of organizational learning, skill development, and knowledge barriers (Attewell, 1992). For example, as the technology and its effects become better known, the perceived risks appear to be lower (e.g., proven benefits and lower complexity), and then adoption becomes more likely.

Know-how and organizational learning are thus potential barriers

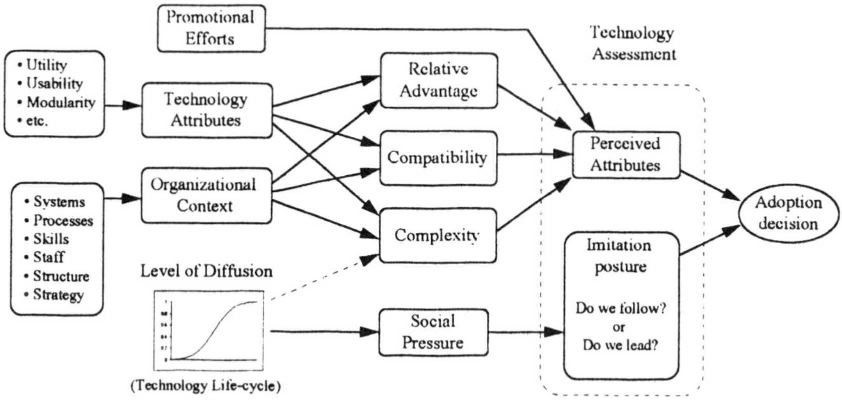

Figure 1: Technology Adoption

to adoption. Firms may delay adoption of complex technologies until they obtain sufficient technical know-how to implement and operate successfully. When knowledge barriers are overcome, diffusion speeds up. Figure 1 integrates the preceding views on adoption, diffusion, learning and the technology life cycle.

Beyond Adoption: Implementation and Organizational Change

The Organizational Change literature provides a perspective for viewing implementation as a dynamic process involving the introduction of a technical change into an existing social system. Kotter et al. (1979) and Beer (1980) state that organizations need to look at the introduction of technology as a change strategy. They suggest that successful management of change requires management commitment, political support, and resources. Lewin (1952) proposes a three-stage prescription for bringing about organizational change: 1) Unfreezing (i.e., getting people to examine and question their present way of doing things); 2) Change; and, 3) Refreezing (i.e., making policies and procedures consistent with the behavior needed under the new system).

Radnor et al. (1970), Nolan and Gibson (1974), and Nolan (1973, 1979) identified several stages in the utilization of computer-based information systems. The intra-organizational diffusion begins with a "missionary" phase, in which the IS department seeks to sell its services, leading to a "contagion" stage in which demand for such service grows dramatically, through to "institutionalization" of the service.

Curley (1981) looked at implementation of word-processing technology from a user's viewpoint. She examined the process by which a selected sample of organizations decide to purchase and implement the technology. Curley sought explanations for why an apparently worthwhile technology may fail to become widely diffused or live up to its expected potential. Her framework on "the dynamic relationships which exist in bringing an innovation to the market, diffusing it, and subsequently implementing it within an organizational setting" borrows some concepts from the literature on innovation diffusion.

Markus (1984) developed a framework for identifying, explaining, and predicting the effects of information systems on people, organizations, and performance. The framework is based on a theory, drawn from the social and behavioral sciences, that the impacts of systems are not caused by systems technology or by the people and organizations using them but by the interaction between specific design features and the related features of the system's organizational setting. This view is incorporated in the adoption model depicted in Figure 1.

Leonard-Barton and Kraus (1988) provide a similar, but more general, approach in describing implementation as mutual adaptation between technology and organization. Implementation is seen as innovation; and technology transfer is argued to require ongoing dedication to the process of change and the conscious management of mutual adaptation. Managers operate within parameters set by the technology's implementation characteristics: its transferability, organizational complexity, and divisibility. These implementation characteristics inherently constrain or influence the way managers operationalize their implementation strategies.

The introduction of IT in organizations has been conceptualized as comprising three broad subtasks: design of the IT system and the organization that will operate it, development of the enabling human resource policies, and management of the implementation process (McKersie and Walton, 1991). Effective IT implementation requires the alignment of the technology and the organization that operates it (Walton, 1989). In other words, the requirements of a given system (in terms of motivation, competence, and coordination), should be matched by the capabilities of the organization using it. Conversely, the requirements of the organization, e.g. decentralized decision making, continuous learning, challenging jobs, or attractive career paths, etc.- should be accommodated by the design of the IT system.

Alignment, in practice, can be achieved in several ways. First, some of the enabling conditions may already exist or can be developed in anticipation of the introduction of an IT system. Second, the organization can pull the technology into place. Organization and technology can be developed simultaneously, each adapted to the other. Third, management can focus exclusively on implementing the technology, letting the technology drive subsequent organizational adaptation (McKersie and Walton, 1991).

Organizational change is thus a necessary condition for the effective use of IT. McKersie and Walton (1991) identified three patterns of technological and organizational changes:

1. **Organizational change followed by IT implementation.** In a number of cases the conditions favorable to effective IT utilization

are already in place when an IT system is introduced. IT is pulled into place by the users rather than pushed into place by sponsors or superiors.

2. **Simultaneous implementation of organizational change and IT.** In some cases, management puzzles through the designs of the new technology and the operating organization at the same time. This pattern has the advantage of allowing mutual adaptation of the technical and social subsystems of an IT implementation.

3. **New technology as a catalyst for organizational change.** In this case, a firm moves ahead with the introduction of IT, leaving existing organizational arrangements in place, and subsequently attends to organizational changes on a responsive or adaptive basis. Practitioners argue that preemptive organizational change is not always feasible, nor desirable (e.g., "if we changed the organization first, there would be blood all over the floor," McKersie and Walton, 1991, p. 253). Conceptually, the introduction of the technology serves as a catalyst, an agent for unfreezing the status quo.

Organizational impacts of IT

The literature on the impacts of Information Technology (IT) on individuals has traditionally concentrated on jobs, skills and employment. (Osterman, 1991; Zuboff, 1988; Adler, 1986; Osterman, 1986; Salzman, 1985). Questions regarding the effect of IT on job content are often raised: Will work become more or less skilled? Broader or narrower? Will jobs be lost? How will the organizational transformations feed back upon careers?

A classic debate underlies these questions: Some believe that IT has negative consequences for individuals, in terms of employment loss, degradation of skill and job quality (Osterman, 1986). Others believe that creative use of IT can expand employment opportunities, increase the demand for workers with new and high skills (Spenner, 1987), and open up possibilities for creative expression and satisfaction at work (Salzman, 1985).

The rationale for increased skills levels implies moving away from manual skills and towards working with data. This aspect is at the heart of Zuboff's (1988) distinction between "automate" and "informate." She argues that IT will distance workers from the physical "feel" and will require instead that they learn the meaning of data generated by computers. In this context, training becomes key. An important element of new skills is responsibility. In most instances the system may perform well on its own, but the importance of spotting and understanding malfunctions increases sharply (Hirschhorn, 1984).

IT can reshape work beyond its impact on specific jobs. The way jobs link together may change and new forms of work organization may emerge. For example, work teams, in which job boundaries are diffuse, are more productive than traditional arrangements (Katz, Kochan, & Keefe, 1987; MacDuffie & Maccoby, 1986). IT-based networks affect the social relations of individuals within the firm, and the way their jobs interrelate, and by extension, they can lead to changes in organizational structure (e.g., flatter and more decentralized).

Different frameworks have been proposed to classify the wide range of (intra-) organizational impacts associated with IT adoption. The 7-S framework (Athos and Pascale, 1982) can be used to categorize the internal changes triggered by IT adoption. Cash and Konsynski (1985) used it to discuss the organizational changes caused by inter-organizational systems. Similarly, Holland, Lockett and Blackman (1992) describe the effects of EDI on organizational structure and processes, on individual roles and tasks, the size and productivity of the staff, and the creation of new organizational units. But perhaps one of the best known frameworks linking IT and organizational change is MIT's "Management in the 1990's" (Scott Morton, 1991). This framework categorizes IT impacts along the following inter-linked dimensions: management, technology, strategy, structure, individuals and roles.

Whatever the classification scheme, the different taxonomies suggest several interrelated levels of impact. On one hand, impacts can be assessed for different "unit of analysis," such as individuals, departments, or whole business units. On the other hand, the impacts can be experienced at different levels (e.g., changes in procedures, staff, or organizational structure). Figure 2 shows a diagram that incorporates these different dimensions of change.

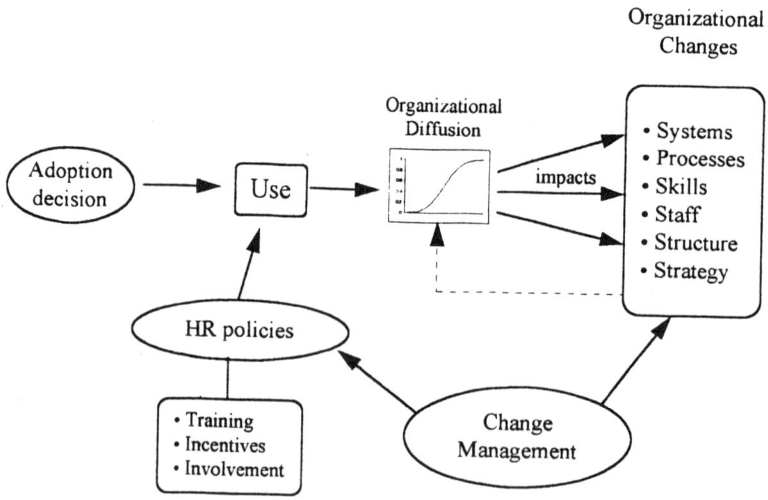

Figure 2: Technology Implementation

Implicit in this framework is a process view of technology implementation; i.e., a process that goes from the adoption decision, to actual technology use and its subsequent effects on the organization. More specifically, the process involves the following stages: (1) pre-existing organizational and environmental context, (2) technology assessment, (3) planning and adoption decision, (4) usage or internal diffusion, and (5) organizational changes.

"Context" refers to anything which is important to understand the setting in which technology is adopted, e.g., company characteristics (strategy, structure, processes), motivation to adopt the new technology, technology awareness and expertise, competitive situation, state of the technology diffusion in the industry, etc. The adoption decision is made right after the "Assessment" stage, when the potential technology applications are analyzed and prioritized (Technology Planning). Both the contextual factors and the assessment leading to the adoption decision have already been presented in figure 1. Figure 2 focuses on post-adoption effects, and should thus be considered as the continuation of Figure 1. In Figure 2, implementation is seen as the onset of organizational change, in a combination of technical, human, and organizational decisions intended to stimulate the proper use of IT by the organization (i.e., "intra-organizational diffusion").

An Integrated Framework for Diffusion of IT and Reengineering

The preceding discussion on technology implementation assumes that the organizational changes triggered by the technology are intended and, therefore, beneficial for the organization (i.e., they will contribute to increase the value of its products/ services, or reduce operating costs) and, ultimately, they should contribute to gain or to defend competitive advantage. From a management perspective, the question is thus how to translate the organizational changes associated with the adoption of new technology into performance improvements.

From organizational change to business performance: The missing link

The link between organization design and business performance has been debated extensively in the literature for a long time (e.g., Chandler, 1962; Thompson, 1967; Galbraith, 1977; Caves, 1980; Quinn, 1980; Porter, 1985). More recently, the role of IT as enabler of organizational design and organizational transformation has become a topic of interest both in the IT literature as well as in the general management literature (Hammer 1990, Scott-Morton 1991, Davenport 1993, Hammer and Champy 1993).

These developments should be welcome in the field of IT research, where for many years the explanatory link between IT adoption and competitive advantage was missing (Lucas, 1993). The interaction of IT and the adopter organization was treated like a black box. The adoption of IT was assumed to lead to competitive advantage, just because it seemed to have worked in a few organizations which had become popular success stories; e.g., American Hospital Supply, McKesson, American Airlines, etc. Banker et al. (1993) provide a comprehensive bibliography on the strategic and economic value of IT (pp. 607-657).

The recent emphasis on re-engineering, process-orientation, and networked organizations has led to the discussion of what is the role that IT plays in all these developments. Some argue that IT has a supporting role, others say IT is the enabler, and others posit IT is the key element. The most important by-product of this research stream is the realization that the difference between success and failure of strategic IT applications often has to do more with the post-adoption management of change, and less with the early stages of the process (i.e., technology assessment and the adoption decision). Markus and Soh (1993) call such process "IT conversion effectiveness." They provide a model which discusses factors leading to returns from IT spending, structural factors influencing firms' abilities to achieve returns from IT, and internal management processes that affect firms' conversion effectiveness.

The reengineering movement has brought the focus of attention to a more operational level (i.e. the business processes). It has opened the

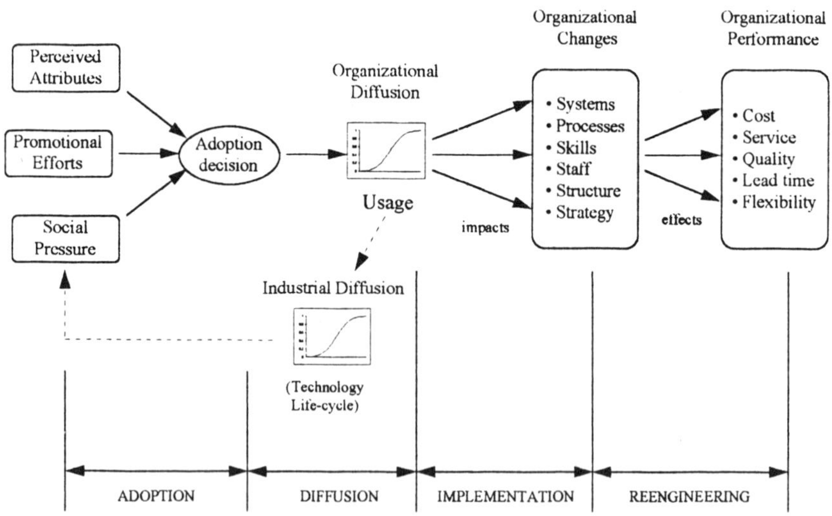

Figure 3: Integrating Framework

"black box" by focusing on the interaction of IT, people and business processes. By rethinking the way old business processes are performed, and using IT to enable new ones, many organizations are finally grasping the elusive link between IT adoption and competitive advantage and achieving significant performance improvements in key business drivers, such as cost, quality, service or lead times. Figure 3 shows an integrated framework which incorporates Figure 1 and Figure 2 (although somewhat simplified for the sake of clarity and parsimony) as well as this final link between organizational change and business performance. The framework brings together the four research domains that have been discussed above: Adoption, Diffusion, Implementation and Reengineering.

The case for Technology Pull

At first sight, the framework may seem to conform to the third implementation pattern identified by McKersie and Walton (1991) and described earlier (i.e., "new technology as a catalyst for organizational change"). This pattern implies a somewhat passive (or reactive) view of the links between technology adoption and organizational changes. From this perspective, the framework would be used to study the organizational impacts of some IT application which has been "pushed" into an existing organization, and where the prior nature and extent of these impacts are unknown or uncertain. A more interesting use of the framework, however, emerges when one takes a more proactive view of IT; i.e., as a tool for organization design and as enabler of organizational change (Keen, 1991; Venkatraman, 1991).

In this case, one should start with the outcome of the process first, and then work backwards (from right to left on the picture) just as the pundits of re-engineering would advocate. The process should start by defining the desired performance improvement targets, then (re)design the appropriate organization and, finally, make the appropriate technology choices. This approach can, in fact, be called "technology pull" because the chosen technology is "pulled" into the organization, as opposed to the previous situation in which the organizational impacts of the technology are uncertain, and arise as a result of a technology "push" (Zmud, 1984; King et al. 1994). The process would be similar to the first implementation pattern of McKersie and Walton (1991) which has been discussed above (i.e. "organizational change followed by IT implementation").

This other perspective suggests an IT planning approach that takes into account the proactive role of IT in organizational design and organizational transformation (e.g., Malone and Smith, 1984; Venkatraman, 1991). For research, the implication is the design of a research program that capitalizes on the effects of some technologies or applications (e.g., those that have been tested in a variety of organizational settings) and explores their proactive role, as tools to enable desired changes in organization and business strategy.

Applying the Framework: EDI As An Example

For many business organizations, Electronic Data Interchange (EDI) has been and still is a technological innovation. EDI can provide important business benefits to the organizations adopting them. At the same time, EDI can entail significant organizational changes. EDI seems thus an appropriate technology candidate to test the usefulness of the framework.

EDI adoption factors: Perceived Attributes

In light of the preceding theoretical discussion, it can hypothesized that the adoption of EDI can be explained by its perceived attributes, just as other technological innovation reported in the literature. For adopting organizations, "relative advantage" is the degree to which EDI transactions are perceived as being better than transacting by conventional methods. In short, "relative advantage" translates into customers' expected benefits. EDI linkages have three immediate effects on the quality of inter-organizational communications: 1) faster transmission, 2) greater accuracy, and 3) more complete information about the transactions (Stern and Kaufmann, 1985). A direct computer-to-computer or terminal-to-computer linkage eliminates the need for re-keying the document (e.g., purchase order) when it is received. By suppressing this manual step, clerical work diminishes, the errors associated with keying orders decrease, and lead times shorten. Companies are thus able to offer customers some value through greater productivity and/or better service (Dearing, 1990; O'Callaghan and Turner, 1993). In industries moving physical goods, shorter lead times allow customers to purchase more frequently, cheaper and in smaller lot sizes, thus reducing the need for inventory. In summary, two types of customer benefits can be expected from EDI adoption: service improvement, and cost reduction (efficiency gains).

The other two major innovation attributes that can be applied to EDI are Rogers' "complexity" and "compatibility." These attributes can be referred to as adoption costs. They are costs in a broad sense: as inhibitors of or barriers to adoption (O'Callaghan, Kaufmann, and Konsynski, 1992).

According to Rogers, complexity is the degree to which an innovation is perceived as relatively difficult to understand and use. In the case of EDI, the degree of difficulty is normally given by the kind of user-interface the system provides as well as other characteristics such as the way EDI is integrated with the back-office computer systems existing in the organization (e.g., whether modifications to existing systems are necessary). One can refer to all of these as system related costs, or simply system costs.

Compatibility is the degree to which an innovation is perceived as consistent with current practices and procedures. Like all technological innovations, EDI brings about change. In the adopting organization, EDI affects the tasks of the personnel involved: there are changes in procedures, disruptions due to start-up problems, and learning. These can be called organizational adaptation costs, or simply organizational costs.

External Factors

As with other innovations, the adoption of EDI may be influenced by factors other than the perceived attributes of the innovation itself. The promotional efforts by organizations that have launched EDI and try to enroll their trading partners constitute a clear example. Trade associations, on the other hand, have also been promoting the concept of EDI in an attempt to speed up diffusion within the industry (O'Callaghan et al., 1992).

The promotional efforts to spread the use of EDI are both direct and indirect. On one hand, there is a "push" by the developers of the systems who target their trading partners and offer them to link up. Solicitation sometimes turns into coercion, especially if there are significant differences in the relative bargaining power of the company with respect to its trading partner.

On the other hand, companies are often exposed to an on-going educational process through EDI courses, articles in trade publications, trade conference, etc. All these activities contribute to raise the level of awareness and, while not promoting any system in particular, they promote the generic concept of EDI as being beneficial for the industry and lowering knowledge barriers (O'Callaghan and Eistert, 1995). In conclusion, it can be hypothesized that companies that listen to the advice of these industry sources will be more likely to adopt EDI.

Finally, the "diffusion effect" is another factor influencing EDI adoption. The diffusion effect (Rogers, 1983) is the cumulatively increasing degree of influence upon a company to adopt or reject an EDI offer, resulting from the "activation of peer networks" in the industry. The activation of peer networks implies that a company listens to the advice of companies which have already adopted EDI. Some may behave as "imitators," and therefore be influenced in their adoption decision by the decisions and opinions of others that have already adopted. Of course the assumption here is that the experience with EDI is generally positive; and, therefore, this favorable experience will encourage non-adopters to adopt. The probability that an imitator will adopt is a function of the number of previous adopters and an "imitation index" (Bass, 1969) which somehow indicates the degree to which the imitator is influenced by other adopters (i.e., how much the "imitator" actually "imitates"). This imitation index can be interpreted as the extent to which a company seeks and follows the advice of other companies that have already adopted EDI. It can also be argued that as the number of adopters grows, user groups will start to emerge. These user groups constitute yet another mechanism for intra-industry information diffusion and, therefore, contribute to lower the knowledge barrier.

Organizational Impacts of EDI

Interorganizational systems, and EDI in particular, have been reported to have a wide range of impacts on organizations (Gupta and Neel, 1992; O'Callaghan et al., 1992; Holland et al., 1992; Emmelhainz, 1991; Dearing, 1990; Emmelhainz, 1987). At the lowest level, when an EDI transaction is introduced, the first impact is technical. It involves the integration into existing computer applications that must either generate or accept the data. At a higher level, EDI may affect operating procedures and even the definition of how the work is accomplished.

Changes in business processes are the most cited effects of EDI (Emmelhainz, 1987; Monczka & Carter, 1989; Gupta & Neel, 1992). Processes must change to conform with EDI messages and to take into account new procedures in communications patterns, report formats, and internal controls. Thus, EDI is often associated with business process redesign (BPR). In turn, the shift in the underlying business processes may bring changes in the skills and responsibilities of employees and, in some cases, even new employee categories (e.g., an EDI administrator). There may even be an impact on the basic attitudes that have existed internally, among business functions, which now must work closely together. When used in a key business function, EDI may imply changes not only in skills and organization structure but even in business strategy (Cash & Konsynski, 1985; Malone, Benjamin & Yates, 1987). The impact of EDI can thus ripple through the entire organization.

Improving Business Performance

Through changes to business processes, EDI allows companies to save both time and money because the data is transmitted rapidly, requires no data entry (no re-keying), and has been generated by a computer application. Additionally, it can greatly reduce the order cycle and/or the pay period, which in turn allows a company to reduce its inventory levels and improve cash flow (Dearing, 1990). EDI can thus be used to support business strategies that require flexibility and short lead times (O'Callaghan, 1995). For most companies, some of these potential benefits constitute the ultimate reason to adopt EDI. O'Callaghan and Turner (1995) provide a comprehensive review of these "normative expectations" associated with EDI adoption.

The link between EDI, organizational change and business performance has therefore been established. In addition, the proposed framework suggests that effective EDI planning is the planning of organizational change. For many companies, however, this is not easy. The organizational consequences of EDI are not always well understood. In these companies, changes are not planned, they just happen (i.e., the opposite of what the Reengineering pundits would advocate). The question thus is how to plan for EDI (Holland et al., 1992).

Planning for Organizational Change

In the above framework, the changes at the different levels can be analyzed in conjunction with the attitude of the firm towards EDI. The sequence of intra-organizational impacts (internal changes) can be hypothesized to vary depending on the pro-active / re-active posture of the firm; i.e., whether an organization is reacting to an EDI system launched by another company or whether the company itself is actually sponsoring the system (Cash and Konsynski, 1985). The logic is as follows: If a company joins in an EDI project proposed by another organization, general management is not likely to participate in the decisionmaking process, and will neither explicitly plan nor consider the implications of the system. Changes will occur first in business processes, then in skills and staff, and later, perhaps, in organizational structure. Eventually, the company may be forced to reassess its business strategy, but only on a reactive basis. On the other hand, if the organization is the initiator (has a proactive attitude), EDI will then be an enabling tool for changes in strategy. The system will be better planned and the order of impacts will be the reverse: first, strategy and organizational structure; then, selection and training of staff; and, finally, business processes and systems. In this case, EDI can be used as tool to enable or support re-engineering.

Conclusion

A general framework to study organizational adoption of technology has been developed. Based on literature from different fields, this framework constitutes an attempt to integrate several research domains: adoption, diffusion, planning, implementation and change management. Furthermore, the framework provides a link between classic diffusion theory and the more recent perspectives on BPR ad Re-engineering, Business Process Redesign (Short and Venkatraman,

1992; Davenport, 1993; Hammer and Champy, 1993).

The framework has been tested by applying it to EDI. The starting point has been a process view whereby adoption and implementation decisions take place in stages: context analysis, technology assessment, adoption decision, implementation, and organizational changes. The framework has been used to identify and classify different levels of organizational impact associated with the adoption of EDI as a technological innovation.

The proposed framework has also been used to highlight and discuss a new perspective on "technology pull." In this view, the adoption decision emerges as the result of a "proactive" planning process which considers IT as a key enabler of organizational transformation and which takes technology diffusion as an input. In other words, technology diffusion leads to increased learning and to lower knowledge barriers. From this perspective, the framework suggests that a "technology pull" strategy will make sense only when technologies have "matured." That is because their effects and the role they can play in organizational transformation are well known. Technology push, on the other hand, will make sense when technologies are relatively new, there is little collective knowledge, and organizations need to learn and understand their potential effects.

Future research might entail an empirical validation of the model on a larger scale (e.g., with several field studies and also applied to other technological innovations). Such effort would constitute the basis for a normative model linking IT diffusion, organizational change, organizational learning and proactive IT planing.

References

Adler, P. (1986). New technologies, new skills. *California Management Review*, 29 (1), 9-28.

Athos, A. & Pascale, R. (1982). The art of Japanese management. New York: Warner Books.

Attewell, P. (1992). Technology diffusion and organizational learning: The case of business computing. *Organization Science*, 3 (1), 1-19.

Banker, R. D., Kauffman, R. J., & Mahmood, M. A. (Eds.). (1993). *Strategic information technology management: Perspectives on organizational growth and competitive advantage.* Harrisburg, PA: Idea Group Publishing.

Bartlett, M. S. (1960). *Stochastic population models in Ecology and Epidemiology.* London: Methuen.

Bass, F. M. (1969). A new product growth model for consumer durables. *Management Science*, 15 (1), 215-227.

Bass, F. M. (1980). The relationship between diffusion rates, experience curves, and demand elasticities for consumer durable technological innovations. *Journal of Business*, 53, 851-867.

Beer, M. (1980). *Organizational development and change.* Goodyear Press.

Cash, J., & Konsynski, B. (1985). IS redraws competitive boundaries.

Harvard Business Review, 63 (2), 134-142

Caves, R. (1980, March). Industrial organization, corporate strategy and structure. *Journal of Economic Literature*, 18, 64-92.

Chandler, A. D. (1962). *Strategy and structure: Chapters in the history of the American Industrial Enterprise.* Cambridge, MA: The MIT Press.

Cron, W. J., & Sobol, M. G. (1983). The relationship between computerization and performance: A strategy for maximizing the economic benefits of computerization. *Information and Management*, 6, 171-181.

Crowston, K., & Treacy, M. E. (1986, December). Assessing the impact of information technology on enterprise level performance. *Proceedings of the Seventh International Conference on Information Systems*, San Diego, 299-310.

Curley, K. F. (1981). Word processing: First step to the office of the future? An examination of an evolving technology and its use in organizations. Unpublished doctoral thesis, Harvard Business School, Boston.

Davenport, T. H. (1993). *Process innovation: Reengineering work through information technology.* Boston: Harvard Business School Press.

Dearing, B. (1990). The strategic benefits of EDI. *The Journal of Business Strategy*, 11, January-February, 4-6.

Emmelhainz, M. A. (1987, February). The impact of electronic data interchange on the purchasing process (Working Paper No. 87-3). Dayton, Ohio: University of Dayton, School of Business Administration, The Center for Business and Economic Research.

Emmelhainz, M. A (1991). EDI: Does it change the purchasing process?. *EDI Forum* (journal published by The EDI Group, Ltd.), 4, 158-161.

Eveland, J. D., & Tornatzky, L. (1990). The deployment of technology. In L. Tornatzky & M. Fleischer (Eds.), *The process of technological innovation* (chap. 6). Lexington, MA: Lexington Books.

Galbraith, J.R. (1977). *Organizational design.* Reading, MA: Addison Wesley Publishing.

Gatignon, H., & Robertson T. S. (1989, January). Technology diffusion: An empirical test of competitive effects." *Journal of Marketing*, 53 (1), 35-49.

Gupta, Yash, P., & Neel, G. A. (1992, August). the origin of EDI and the changes associated with its implementation." *Industrial Engineering.*

Hammer, M. (1990). Reengineering work: Don't automate, obliterate. *Harvard Business Review*, 68 (4), 104-112.

Hammer, M., & Champy, J. (1993). Reengineering the corporation: A manifesto for business revolution. New York: Harper Business, HarperCollins Publishers, Inc.

Henderson, J. C., & Venkatraman, N. (1993). Strategic alignment: Leveraging information technology for transforming organizations. *IBM Systems Journal*, 32 (1), 4-16.

Hirschhorn, L. (1984). *Beyond mechanization: Work and technology in a postindustrial age.* Cambridge, MA: The MIT Press.

Holland, C., Lockett, G., & Blackman, I. Planning for Electronic Data Interchange. *Strategic Management Journal*, 13, 539-550.

Katz, H., Kochan, T., & Keefe, J. (1987). Industrial relations and productivity in the U.S. automobile industry. *Brookings Papers on Economic Activity* (Special Issue on Microeconomics), 3, 685-728.

Keen, P. (1977, July). *Implementation research in OR/MS and MIS: Description versus prescription.* (Research Paper No. 390). Stanford University, Graduate School of Business.

Keen, P. G. W. (1991). *Shaping the future: Business design through information technology.* Boston, MA: Harvard Business School Press.

King, C.W. (1966). *Adoption and diffusion research in marketing: An overview.* Fall Conference of the American Marketing Association, Chicago.

King, J. L., Gurbaxani, V., Kraemer, K. L., McFarlan, W., Raman, K. S., & Yap, C. S. (1994). Institutional factors in information technology innovation. *Information Systems Research,* 5 (2), 139-169.

Kotter, J., Schlesinger, L. A., & Gathe, V. (1979). *Managing the human organization.* Boston, Mass.: Harvard University Press.

Krcmar, H., Bjorn-Andersen, N., & O'Callaghan, R. (Eds.). (1995). *EDI in Europe: How it works in practice.* Chichester (UK): Wiley Series on Information Systems, John Wiley & Sons.

Leonard-Barton, D. (1987). Implementation structure software methodologies: A case of innovation in process technology. *Interfaces,* 17 (3), 6-17.

Leonard-Barton, D. (1989). Implementing new production technologies: Exercises in corporate learning. In M.A. Von Glinow & S. Morhman (Eds.), *Managing complexity in high technology industries: Systems and people.* Oxford Press.

Leonard-Barton, D., & Kraus, W. A. (1988). Implementation as mutual adaptation of technology and organization. *Research Policy,* 17 (5), 251-267.

Lewin, K. (1952). Group decision and social change. In Newcomb & Hartley (Eds.), *Readings in Social Psychology.* New York: Henry Holtt and Co.

Lucas, H. C. (1993). The business value of information technology: A historical perspective and thoughts for future research. In R. D. Banker, R. J. Kauffman & M. A. Mahmood (Eds.), *Strategic information technology management: Perspectives on organizational growth and competitive advantage* (pp. 359-374). Harrisburg, PA: Idea Group Publishing.

MacDuffie, J. P., & Maccoby, M. (1986, September). *The organizational implications of new technologies: remote work stations at AT&T Communications* (Discussion Paper No. 154D). Cambridge, MA: Harvard University, Kennedy School of Government.

Mahajan, V., & Peterson, R. A. (1985). *Models for innovation diffusion* (University Paper Series No. 48). Beverly Hills, CA: Sage Press.

Malone, T. W. (1985, August). *Organizational structure and information technology: Elements of a formal theory* (Working paper No. 130). Cambridge, MA: MIT, Center for Information Systems Research.

Malone, T. W., Benjamin, R. I., & Yates, J. (1987). Electronic markets and electronic hierarchies: Effects of information technology on market structure and corporate strategies. *Communications of the ACM,* 30 (6), 484-497.

Malone, T. W., & Smith, S. A. (1984). *Tradeoffs in designing organizations: Implications for new forms of human organizations and computer systems* (Working Paper No. 112). Cambridge, MA: MIT, Center for Information Systems Research.

Mansfield, E. (1961). Technical changes and the rate of imitation. *Econometrica,* 29.

Mansfield, E. (1968). *Industrial research and technological innovation: An econometric analysis.* New York: Norton.

Markus, L. M. (1983). Power, politics and MIS implementation. *Communications of the ACM,* 26 (6), 430-444.

Markus, L. M. (1984). *Systems in organizations: Bugs and features.* Marshfield, Mass.: Pitman Publishing, Inc.

Markus, L. M., & Soh, C. (1993). Banking on information technology: Converting IT spending into firm performance. In R. D. Banker, R. J. Kauffman & M. A. Mahmood (Eds.), *Strategic information technology management: Perspectives on organizational growth and competitive advantage* (pp. 375-403). Harrisburg, PA: Idea Group Publishing.

McKersie, R. B., & Walton, R. E. (1991). *Organizational Change.* In M. S. Scott Morton (Ed.), The corporation of the 1990's: Information technology and organizational transformation (pp. 245-277). New York: Oxford University Press.

Monczka, R. M., & Carter, J. R. (1989). Implementing Electronic Data Interchange. *Journal of Purchasing and Materials Management,* 25 (1), 26-33.

Nolan, R. (1973). Managing the computer resource: A stage hypothesis. *Communications of the ACM,* 16 (7), 399-405.

Nolan, R. (1979). Managing the crises in data processing. *Harvard Business Review,* 57 (2), 115-126.

Nolan, R., & Gibson, C. B. (1974). Managing the four stages of EDP growth. *Harvard Business Review,* 52 (1).

O'Callaghan, R. (1995). EDI in procurement and flexibility strategies: The case of Alcatel Bell Telephone in Belgium. In H. Krcmar, N. Bjorn-Andersen & R. O'Callaghan (Eds.), *EDI in Europe: How it works in practice* (pp. 21-46). Chichester, U.K.:: Wiley Series on Information Systems, John Wiley & Sons.

O'Callaghan, R., & Eistert, T. (1995). An association leadership for industry-wide EDI - The case of AECOC in Spain. In H. Krcmar, N. Bjorn-Andersen & R. O'Callaghan (Eds.), *EDI in Europe: How it works in practice (pp. 277-297).* Chichester, U.K: Wiley Series on Information Systems, John Wiley & Sons.

O'Callaghan, R., Kaufmann, P., & Konsynski, B. (1992, April). Adoption correlates and share effects of Electronic Data Interchange systems in marketing channels. *Journal of Marketing,* 56, 45-56.

O'Callaghan, R., & Turner, J. (1995). Electronic Data Interchange: Concepts and issues. In H. Krcmar, N. Bjorn-Andersen & R. O'Callaghan (Eds.), *EDI in Europe: How it works in practice* (pp. 1-19). Chichester U.K.: Wiley Series on Information Systems, John Wiley & Sons.

Osterman, P. (1986, January). The Impact of Computers upon the Employment of Clerks and Managers. *Industrial and Labor Relations.*

Osterman, P. (1991). Impact of IT on jobs and skills. In M. S. Scott Morton (Ed.), *The corporation of the 1990's: Information technology and organizational transformation* (pp. 221-243). New York: Oxford University Press.

Petrini, F. (1966). *The rate of adoption of selected agricultural innovations* (Report No. 53). Uppsala, Sweden: Agricultural College of Sweden.

Porter, M. (1985). *Competitive Advantage.* New York: The Free Press.

Premkumar, G., Ramamurthy, K., & Nilakanta, S. (1994). Implementation of EDI: An innovation diffusion perspective. *Journal of Management Information Systems,* 11 (2), 157-186.

Quinn, J. B. (1980). *Strategies for change: Logical incrementalism.* Homewood, Illinois: Irwin.

Radnor, M., Rubenstein, & Tansik. (1970, November). Implementation in Operations Research and R&D in Government and Business Organizations. *Operations Research.*

Rice, R., & Rogers, E. (1980). Reinvention in the Innovation Process. *Knowledge: Creation, Diffusion, Utilization,* 1 (4), 499-514.

Robertson, T. S., & Gatignon, H. (1986, July). Competitive effects on technology diffusion. *Journal of Marketing,* 50 (3), 1-12.

Rogers, E. M. (1983). *Diffusion of innovations* (3 rd. edition). New York: The Free Press.

Rogers, E. M., & Shoemaker, F. F. (1962). *Diffusion of innovations.* New York: The Free Press.

Rogers, E. M., & Shoemaker, F. F. (1971). *Communication of innovations: A cross-cultural approach.* New York: The Free Press.

Rosenberg, N. (1982). *Inside the black box: Technology and economics.* Cambridge University Press.

Salzman, H. (1985, May). *The new Merlins or Taylor's automatons? The impact of computer technologies on skill and workplace organization* (Working Paper No. 85-5). Boston University, Center for Applied Social Science.

Scott Morton, M. S. (Ed.). (1991). *The Corporation of the 1990's: Information Technology and Organizational Transformation.* New York: Oxford University Press.

Short, J., & Venkatraman, N. (1992). Beyond business process redesign: Redefining Baxter's business network. *Sloan Management Review,* 34 (1), 7-20.

Spenner, K. (1987, July). *Technological change, skills requirements, and education* (Mimeo). Duke University, Department of Sociology.

Srivastava, R. K., Mahajan, V., Ramaswami, S. N., & Cherian J. (1985). A multi-attribute diffusion model for forecasting the adoption of investment alternatives for consumers. *Technological Forecasting and Social Change,* 28.

Stern, L., & Kaufmann, P. (1985). Electronic Data Interchange in selected consumer goods industries: An interorganizational perspective. In R. Buzzel (Ed.), *Marketing in an electronic age* (pp. 52-73). Boston: Harvard Business School Press.

Thompson, J. D. (1967). *Organizations in action.* New York: McGraw-Hill.

Tornatzky, L. G., & Klein, R. J. (1982). Innovation characteristics and innovation adoption-implementation: A meta-analysis of findings. *IEEE Transactions on Engineering Management,* EM-29.

Venkatraman, N. (1991). IT-induced business reconfiguration. In M. S. Scott Morton (Ed.), *The corporation of the 1990's: Information technology and organizational transformation* (pp. 122-158). New York: Oxford University Press.

Walton, R. E. (1989). *Up and running: Integrating information technology and the organization.* Boston: Harvard Business School Press.

Zmud, R. W. (1984). An examination of 'push-pull' theory applied to process innovation in knowledge work." *Management Science,* 30 (6), 727-738.

Zuboff, S. (1988). *In the age of the smart machine: The future of work and power.* New York: Basic Books.

CHAPTER
17

Information Systems Innovation: A Framework for Research and Practice

Tor J. Larsen
The Norwegian School of Management

The present (and quite likely the future) business environment requires more innovation occurring in a more distributed and organic pattern with less time available to make decisions. Information systems (IS) innovation plays an increasingly important role in facilitating and supporting business innovation. This chapter argues that a holistic understanding of IS innovation is needed and proposes a process based on the systems thinking "human activity system" view. This holistic view is developed through paying attention to key innovation issues and key issue structure, these being the basis for the development of an IS innovation framework. The IS innovation framework is used to develop models for single as well as multiple IS innovation processes. These are exciting research questions that need our attention.

The number of issues that may be studied within the field of management information systems (MIS) is vast and the interaction among elements virtually infinite. A frequent tactic employed to manage the complexity of the field has been to concentrate on specific domains, for example, strategic use of information systems (IS), the role of change champions, the diffusion of IS, systems development, end-user computing, and, less often, maintenance.

By focusing on one domain, researchers risk leaving out elements that are essential to organizational success. That is, too much focus on detailed studies increases the danger of reductionism (Lakatos, 1970). Commonly used methods may direct developers attention to particular, perhaps more often than not technological, IS development aspects. Since IS development requires the integrated solution of a wide range of phenomena, the danger is that the end IS result does not satisfy business needs and human requirements. Support for the argument that organizations quite frequently do not master their IS innovation processes can be found in surveys of user opinions, indicating that as much as 80 percent of IS development efforts are judged as either direct or partial failures (Mowshowitz, 1976; Vowler, 1991). Although the IS failure rate may not be this high, these reports tell us that we need to increase our understanding of how IS innovations unfold. After all, from the viewpoint of the organization, the total benefit of IS deployment is the combined utility of all its various uses.

Elements of IS innovations include technical issues, human concerns, managerial actions and knowledge, interactions among line employees and information technology (IT) experts, strategic, tactical and operational requirements, organizational elements, and vision. A holistic approach to IS development and use is needed to counteract the dangers of reductionism, that is, paying too much attention to a subset of these elements while glossing over others. This chapter offers innovation theory as the basis for the development of an integrated understanding of IS development and use.

In the next section, the background of the IS innovation view adopted here is presented, followed by the development of an IS innovation framework. The framework is used to suggest future trends within organizational IS innovation research and practice.

Background

The first challenge in developing a holistic IS innovation view is the identification of the basic positioning that will determine the subsequent content. The chosen positioning must be followed by the identification of key issues embedded within IS innovation. Actors will interpret key issues in specific ways. The basic positioning, the key issues, and key issue interpretation serve as the basis for the development of an IS innovation framework.

The Basic Positioning

The basic question here is whether IT itself is the prime mover of innovation or whether IS innovation must be understood in a human actor context. This is the question of technology push versus pull (Zmud, 1984).

Obviously, firms in the IT industry sector have a market potential in mind in their IS/IT product development efforts. Because of the immense vendor and software houses IS/IT product innovations, the possible areas of IS/IT use today is radically larger than what was the case in, for example, the early 1970s. The development of efficient mass data storage devices, multi-media processing, PCs, powerful microcomputers, enhanced telecommunication services, and flexible software applications have vastly increased the range of the areas where IS/IT can be utilized, or existing IS solutions adapted to address new problems. In this regard the innovation process is technology driven.

Diffusion of IT innovation models may help us understand how IS/IT products are disseminated from vendors and software houses into user organizations. When they have the resources, user organizations may choose to combine IS/IT products and internally developed applications to derive a benefit. The unique application of IS/IT to the development of organization-specific applications is the organizational IS innovation.

It seems appropriate to observe that the quality of the IS/IT product is a necessary but not sufficient prerequisite for IS innovation success. The *people* within the organizations determine the outcome. Consider three points relevant to the coupling between technology issues and the human element.

First, of importance to researchers, and managers of organizational IS innovation are the relative contributions of IS experts versus employees (users and managers) within the line organizations. IS experts are frequently, albeit often indirectly, blamed for IS innovation failures because the innovation process is driven by IT rather than business considerations (Morton, 1991). Second, the critical factor in ensuring appropriate utilization of IS is the line-managers' knowledge of IS business potentials and IT platforms. Line managers who take responsibility for coupling business needs and IS solutions and support avoid the negative impacts that can come from focus on technical, rather than business considerations (Keen, 1991). Third, middle managers engaged in business innovation and active in discussing IS-related issues are responsible for implemented IS innovations. Usage may not play a decisive role (Larsen, 1993a).

The critical role of the human element in the innovation process is clearly articulated in the following definition of innovation:

> "An innovation is the development and implementation of new ideas by people who over time engage in transactions with others within an institutional context." (Van de Ven, Andrew, "Central Problems in the Management of Innovation," *Management Science*, Vol. 32 (5), May 1986, p.604.)

The view that people are key to understanding an innovation process is also parallelled in systems thinking with the notion that only people, through a ***human activity system***, can create artifacts (Checkland, 1981; Checkland and Scholes, 1990). The term "human activity system" denotes *how* people, with a purpose in mind, execute their efforts. An artifact is an outcome that is other than a natural made phenomenon, that is, a human-made physical or abstract product.

These observations lead to the conclusion that a holistic understanding of IS innovation must take as its foundation the people within an institutional setting. An IS innovation is an artifact and can only be explained as a result of human activity. The position taken here does not mean that studies of IS/IT diffusion are irrelevant or of less value. Diffusion studies incorporate people issues. However, people issues are attached to the IS/IT phenomenon defined as the anchor element studied. The human actor as the basis unit for investigation has simply not gotten the level of attention it needs.

The IS Innovation Framework

The IS innovation framework is developed in two steps. First, key issues that a framework might contain are discussed. Next, key issue interpretations are presented. These are the basis for the IS innovation framework.

Key Issues

Leavitt (1965) was among the first to present a coherent framework for issues that may regulate organizational processes: task, structure, technology and people. His model has been further developed to include factors, such as organizational culture (Davis and Olson, 1985) or strategy, management processes and roles (Morton, 1991). Leavitt's model has been extensively used and cited, but Larsen (1993b) has argued that the model allows direct interaction among items, for example, between task and technology, without human interference, a highly unlikely occurrence.

The recent approach to IS innovation—business process reengineering—advocates a conceptual process framework consisting of five phases (Davenport, 1993). The phases are: identifying process for innovation, identifying change levers, developing process vision, understanding existing processes, and designing and prototyping the new process. It has strong similarities with phase models commonly found within the IS strategy literature (for example, Wetherbe, 1993) and systems development literature (for example, Olle, Hagelstein, Macdonald, Rolland, Sol, Van Assche, and Verrijn-Stuart, 1988). While the concept of phases may help us develop methods and understand the uses of IS, the innovation process may not be linear. It is more likely that human beliefs, attitudes, interests and behaviors determine how the process unfolds and influence process content (Van de Ven, 1986, 1992; Van de Ven and Poole, 1995). Human intervention may result in continuous execution of processes other than those initially planned and planned processes may not be carried out.

This author suggest from the discussion conducted so far, we can draw upon five key issues to explore IS innovation. First, human activity unfolds within an *organizational* setting. Second, as Keen has pointed out, *knowledge* plays a vital part. Third, drawing upon systems thinking principles, the objective of an IS innovation process is the creation of an *artifact*, that is, an information system that satisfies certain needs. Fourth, since IS innovations occur on strategic, tactical, and operational levels, the *time horizon* for the innovation process is an element. Fifth, the creation of an IS artifact needs an *innovation process*.

Key Issue Interpretation

Each of the five key issues consist of components and these components may be put into a specific order. The combination of components and order are labelled "structure" and compromises a chosen way to understand and interpret a key issue. The chosen structure may not represent the subject matter true components and component order but it is what an actor uses to understand it. A model for key issues and structure is presented in Figure 1 (see p. 419).

Time Horizon Structure

The most common view is to differentiate among strategic, tactical and operational issues. Although an IS innovation may not be clearly identified with one of these, it is generally true that: (1) some innovations have a long-term horizon and strategic nature, (2) some systems have an intermediate-term horizon (for example, tactical planning, and control,) and (3) some systems have a short-term focus (for example, transaction systems or systems that support workers and office personnel on a daily basis).

Note, however, that a transaction system (for example, an order processing system in an international business setting) may also be a strategic solution. Additionally, transaction systems may also provide valuable information to top and middle managers and therefore be of tactical importance. Clearly, one IS may be placed within each of the three levels of strategic, tactical and operational.

Knowledge Structure

Among others, Keen (1991) and Swanson (1994) have convincingly shown that IS knowledge among line managers and business knowledge among IS professionals play a vital role in IS success. In the present model knowledge is divided into three levels; (1) in-depth expertise - the knowledge actors should posses within their own job domain, (2) principal understanding - the knowledge actors should have of critical issues outside their own job domain, and (3) rudimentary insight - the shallow knowledge actors may have of issues outside their own job domain.

Organizational Structure

Organizational structure creates both opportunity and limitation for innovation. Some organizations are more innovative than others, a theme that has been studied in its own right (Henderson, 1979; Peters and Waterman, 1982; Kanter, 1983; Lawrence and Dyer, 1983). The minimum differentiation of structure needed to account for IS innovation are the levels of the organization, group and individual. The organization level includes traditional arrangements of units (divisions, departments and offices) and designed networks binding these units together. Because many organizations purposefully choose to cooperate through legally binding contracts over considerable time, networks may include units in more than one organization.

The group level should be divided into two types: formal and informal. Formal groups exist at all levels of the organization, making decisions on issues in regularly scheduled meetings. Among the formal groups of the organization, there is ample evidence that the power elite (top managers within the line-organization and IT department) is a critical body (Thompson, 1967; Hage and Dewar, 1973). The power elite, because they have decision mandate and resources available, initiate more change than managers lower in the organizational hierarchy. To a large extent, the power elite use innovation to promote their own organizational views and interests (Kling, 1980). They may suffer from self-induced protective behavior to further their own interests and protect themselves against competing views (Janis, 1971). It can take considerable pressure to induce them to initiate radical change (Pettigrew, 1985).

Informal groups emerge through individual action to introduce change. They may work as much through informal contacts as through established managerial command chains in the organizational hierarchy (Schein, 1980; Kanter, 1983, Albrecht and Ropp, 1984). Informal groups may be quite stable over time.

Individual employees, and specifically middle managers in the core organization, also take action to initiate change (Kanter, 1983; Beath, 1991; Larsen, 1993a). Often, individuals employ newly introduced IS before the organization has developed active strategies for its use (Brancheau and Wetherbe, 1990).

Artifact Structure

Usually, the introduction of an IS/IT component/tool is regarded as the IS innovation. Most diffusion of innovation studies use a clearly defined IS/IT component as their study object, for example, spreadsheets (Brancheau and Wetherbe, 1990) or a work-station (Moore and

Benbasat, 1991.) It can be inferred that Swanson's (1994) IS innovation topology had concrete IS and IT components/tools as his anchor (electronic data interchange systems as an example of type III innovations and executive information systems as an example of type II innovations).

The basic assumption the present definition of the artifact structure rests on is that the implemented IS or IT component/tool is a concrete and visible component of a wider phenomenon. The artifact is a mix of people's views of information needs, what an IS is and its purpose, and the IS/IT as it physically exists. The German philosopher Emanuel Kant defined the differentiation made here as "how people interpret a phenomenon" (Das Ding an Mich) and "what the phenomenon is in the real world" (Das Ding an Sich).

Consequently, the definition of the artifact must span from people's views to the concrete and installed IS/IT solutions. It is suggested it consists of five levels: the core business level, the information needs and requirements level, the IS/IT expert level, the IS level and the database level.

The starting point adopted here is that actors in the line organization have ideas - some nebulous, others concrete - about the core business processes in their organization (this is the core business level). These ideas provide the basis for what the actors think are their information needs and requirements (this is the information needs and requirement level). In most cases IS/IT experts assist actors in the line organization in formalizing their business views and information needs and translate the formalized descriptions into concrete solutions (this is the IS/IT expert level).

The core business level, the information needs and requirements level and the IS/IT expert level are the basis for the physical part of the artifact structure. The physical part contains two elements. First, at the IS level organizations create principal IS/IT solutions that subsequently guide the development and use of IS. It includes: 1) the IT architecture, a foundation for an organization's IS portfolio (which includes methods and tools for the development and maintenance of the IS portfolio), necessary rules and guidelines, 2) IS applications, including in-house-developed IS and packaged systems, and 3) IT components, including hardware, telecommunication equipment and operating systems. The database level includes all the data the organization and its actors store: personal files, transaction databases, document archives and data warehouse.

Innovation Process Structure

It is quite common to advocate that an innovation develops over time in phases, but a development process may contain many elements not included in a phase model (Van de Ven, 1992). The process may reiterate phases more often than thought and each reiteration may contain different elements and include actors not previously involved in the process.

Many IS innovations are large, complex undertakings that require formal project organization to develop. However, a project organization is not the first step in the innovation process. A project is established because actors in the organization decided there was a need for it at an earlier point in time. Nor is a project organization an end. It is a vehicle for creating an IS solution. The demise of a project organization occurs when its product has been transferred to the users.

In the actor oriented model, then, we see that most IS innovations develop through three distinct phases. In the idea phase, actors within the organization give birth to an idea and process it until a decision is reached to create a project organization. In the creation phase the project organization tests the soundness of the initial idea and conducts the necessary activities needed to create the new IS solution. In the usage phase the new IS solution has been handed over to the core organization and taken into daily use.

The idea phase, the creation phase and the usage phase are the absolute minimum number of phases needed to understand most IS innovation processes, from birth to death. The division of the total process into only three phases makes the model robust. The names for phases describe the principal process within each. For example, the traditional label for the creation phase is systems development. However, systems development has strong technological overtone and because of IS failures, many actors within the core organization react negatively to the term.

The three phases put forward here are fewer than those Cooper and Zmud (1990) suggest in their technological diffusion model. Their stages are initiation, adoption, adaption, acceptance, routinization and infusion. Since an external IS/IT innovation is their anchor, these phases mirror technology penetration into user organizations. The basic premise made is that the IS/IT solution to a well defined problem already exists. This is a good illustration of the subtle difference between the present framework and diffusion of innovations. The principal element in the IS innovation framework is the human actors within the organization. The anchor of the innovation process is the development of peoples' ideas over time and how these ideas attach IS/IT as part of the solution.

The nature of the three phases requires elaboration.

The Idea Phase. Idea generation is the source of organizational renewal. Organizations must encourage individual creativity as well as provide a structure to guide and encourage the development of novel business approaches.

Ideas may come from many sources: strategic processes, individuals within the line organization, or IS/IT experts. Regardless of source, an idea needs to gain organizational support for its development (Kanter, 1983; Van de Ven, 1986; Beath and Ives, 1989; Beath, 1991). Both informal groups and the power elite of the organization must be convinced of its benefits and allocate resources for its development. The idea phase may consist of two sub-phases; idea percolation and idea molding. Innovation is fostered or hindered in both.

The Idea Percolation Sub-Phase. The freedom to generate ideas—idea percolation—may vary substantially among organizations. Some researchers argue that nearness to demanding customers is the most effective way to secure innovative behavior (Kohli and Jaworski, 1990; Narver and Slater, 1990). Others point to an organizational climate favorable to change. In a positive climate ideas are welcomed and are not interpreted as a danger to the existing order. Ideas are tested objectively and rewards for idea generation and implementation are appropriate. People are not incorrectly punished for failures (Feldman, 1986; Frost, Moore, Louis, Lundberg, and Martin, 1991; Linstead and Grafton-Small, 1992). According to Kanter, mechanisms the organization may use to foster innovation are inter departmental job rotation and reward in the form of recognition rather than monetary premium or promotion.

Strategic planning may lead to the identification of areas for IS innovation (Clemons and Harris, 1991; Morton, 1991; Clemons and Row, 1992). Since the power elite (e.g. top managers) are more active in initiating business change (Hage and Dewar, 1973; Ettlie, 1983), one might think that the power elite also initiate IS innovation as well. However, this is not necessarily so. Larsen (1993a) found no relationship between implemented IS innovation and the managerial hierarchy. Some evidence suggests that in cases where top management initiate IS innovation, organizations experience more problems than if IS management starts the process (Lederer and Sethi, 1988).

Individuals within the line organization are important sources of ideas (Beath, 1991) and middle managers may play a particularly critical role (Kanter, 1983; Larsen, 1993a). Younger and better educated employees are more likely to try newly developed information technology (Brancheau and Wetherbe, 1990). IS/IT experts also forward ideas (Allen and Morton, 1994). All these categories of people display a high degree of innovativeness.

The Idea Molding Sub-Phase. Kanter (1983) found that innovative organizations strengthen coalition building through the dispersal of resources — monetary and human. These resources should be spread throughout the organization to ensure that innovators have to negotiate with people outside their own department to secure funding and support. In the process, innovative managers use the informal network to engage the support of their peers and the formal network to get resources allocated from their superiors.

Rogers (1993) argued that the probability of innovation acceptance may depend on the innovation's fundamental characteristics. Relative advantage, result demonstrability and visibility were found to be the best predictors for adoption versus non-adoption (Moore and Benbasat, 1991). (These constructs were tested on the adoption of personal work stations. They may not apply in the situation where only conceptual ideas about IS/IT exists.)

At the organizational level very little is said in the literature about how the idea molding process should be carried out. Within the IS strategy literature, the focus is on phases and their contents. The human actor is not treated as the major focus (Clemons and Harris, 1991; Morton, 1991; Clemons & Row, 1992). Material related to customers, vendors, competitors, the threat of new entrants, and the probability of substitute products is also commonly found (Porter, 1985; Porter and Millar, 1985).

The Creation Phase

At some time during the molding sub-phase an idea will have gained enough support for a decision to be made about its development into an IS. This is the creation phase. The MIS literature presents a variety of approaches for this phase: systems development, the waterfall model, the socio-technical approach, heuristic development, prototyping or object oriented analysis and design. Each has differences and similarities.

In principle, the creation phase consists of two sub-phases. First, a formal testing of the soundness of the underlying idea must be carried out (this is the change process definition sub-phase). Second, if the solution is found to be both systemically desirable and culturally feasible (Checkland and Scholes, 1990), the new IS is created and implemented (this is the change creation sub-phase).

The Change Process Definition Sub-Phase. The traditional phrase for this phase is feasibility study but I propose we substitute the term "change process definition" for two reasons. The high failure rate of IS innovation justifies shifting the testing of an idea from the systems development process and formalizing it as a preconditional activity. Second, the term "change process definition" more clearly signals that this sub-phase is the bridge that takes the innovation process from the molding sub-phase to the creation sub-phase.

The scope of change process definition is wider than that of traditional feasibility studies. It incorporates three issues. First, an idea, its promoters and opponents already exist. These elements must be consciously incorporated as the basis for the work carried out in the change process definition sub-phase. Kanter (1983) found that innovative middle managers active in idea generation do not take on the role of project managers, or take part in carrying out the detailed work of the project but rather act as brokers between the project team and the rest of the organization. Their contributions are their ability to anticipate problems and to take an active role in solving emerging problems and political issues. They maintain project momentum and keep the effort going. Kanter called this "boundary walking."

Second, the idea, alternative technical solutions, and effects of these solutions on the innovation process participants (organization, power elite, IT, IS experts, informal networks and individuals) must be documented. Research indicate that projects have a higher probability of success when a line manager, rather than an IS/IT expert takes on the role of project manager (Johnston and Carrico, 1988).

Third, as advocates of total quality management argue, the requirements guiding the detailed IS development that will take place in the following sub-phase and later, during maintenance, must be determined (Gillies, 1992). This would include selection of quality standards, methods, information technology support (CASE), budgeting, scheduling, and milestones (Swanson, McComb, Smith, and McCubbrey, 1991).

The Change Creation Sub-Phase. The change creation sub-phase includes traditional systems development activities: detailed information requirements development, detailed IS evaluation documentation, IT procurement installation, program development and testing, and implementation (see, for example, De Marco, 1978; Mumford, 1983; Olle et. al., 1988; Swanson, 1988; Westerman and Donoghue, 1989; Snow and Couger, 1991). This illustrates the complexity of the area.

The quality of the change creation execution is crucial to the IS innovation success. Many of these research issues are examined in Cotterman and Senn (1992). Systems development may be said to be one of the most researched areas within MIS. Despite these efforts many issues seem unresolved, for example: securing positive and limiting negative IS effects, the role of user participation, the line management versus IS management role or implementation strategies. The purpose here is not to argue the pros and cons of various approaches to change creation but to define where this sub-phase fits into the IS innovation context.

The Usage Phase

The usage phase in a product market life-cycle includes at minimum: initiation, maturation and decline (Kotler, 1991). With an IS innovation, parallel phases include change anchoring, change refinement and change termination.

The Change Anchoring Sub-Phase. A new information system is a complex and novel experience for users and IS/IT experts. Some users may be pessimistic, others overly optimistic. They need time to adapt to the new IS. The time required for this phase may vary but cannot be shorter than the time it takes people to adapt to a new situation cognitively and emotionally. If the IS contains elements that are used seasonally, a year may be an appropriate time frame.

People's reactions to an IS innovation may depend upon whether that person initiated the change effort or whether the innovation is the result of organizational activity, and the characteristics of the particular hardware and software being introduced. Preparing the ground for successful use of an IS innovation starts in the idea phase and should be systematically implemented in the creation phase (Swanson, 1988). The preparation includes idea generation participation and responsibility. As the innovation process progresses, informing others about the idea and IS effects, securing user participation and establishing educational programs.

The Change Refinement Sub-Phase. Following completion of change anchoring, the information system enters its longest period. From an innovation standpoint, the critical aspect of this phase is the continuous changes made to the information system. The term maintenance has been used to describe this process, however, maintenance implies fixing something that is wrong. A more appropriate view, this author suggests, would be to consider the difference between the concepts of product versus process innovation (Abernathy and Utterback, 1982). The product innovation is the radical IS renewal undertaken in the creation phase. However, the new IS includes principal design features that limit the degree of change that can be made in the usage phase. Consequently, the change potential in the usage phase is incremental relative to the change potential in the creation phase. Therefore, the change potential in the usage phase can be characterized as process innovations. They are *refinements* within the established IS innovation design features.

Although the change refinement sub-phase lasts longer than any other and, quite probably, is the most important period in an IS's life-cycle, relatively little research has been reported. I see it as symptomatic that comprehensive IS textbooks do not use the word "maintenance," or its equivalent, in the table of contents or in the index, include it under subheadings, or give it less than one page of space (for example: Cash, McFarlan, and McKenney, 1988; Wysocki and Young, 1990, Laudon and Laudon, 1996). Journal articles often have a technical flavor, focusing on professional problems or purely technology considerations (Edwards, 1984; Kim and Westin, 1988). The exception to the rule is the work of Swanson and Beath (1989). Their book gives a comprehensive view of the many issues of maintenance and how it could be better managed.

The focus of literature on the change refinement phase has mainly been on how the IS operation should be organized and run (Dearden, 1987; La Belle and Nyce, 1987; Wysocki and Young, 1990). The most important line management concern seems to be calculating the right sum of money that should be paid for IS services (Hufnagel and Birnberg, 1989).

Regarding end-user computing systems, change refinement is

viewed as problematic - seemingly because people in the line organization do not have the discipline needed to analyze their own information requirements, or to test and document their programs properly (Davis, 1989). Managerial challenges have been more thoroughly investigated in the area of end-user computing (Nelson, 1989).

Frameworks for the evolution of the management of end-user computing over time have also been presented (Henderson and Treacy, 1986). Recent publications have initiated a discussion of the overall responsibilities of line managers (Keen, 1991; Boynton, Jacobs, and Zmud, 1992). In conclusion, so far change refinement has been investigated sparsely and fragmentally. Generally, it has not been addressed as a critical IS innovation opportunity.

The Change Termination Sub-Phase. Because of the inherent limitations in any IS, and because of IS/IT developments, the gap between business needs and deliverables from an IS will increase over years of use. Therefore, every IS will enter a period where the question of its demise and substitution with a new system surfaces. This is the "change termination phase".

To the knowledge of this author there is no published work that specifically addresses this phase. Rather, this dimension of a system's life is subsumed under other fields in a structural and technical manner. For instance, when strategic planning or an application project effort define an opportunity or a problem, it is automatically assumed that a new system will render an old one obsolete. During implementation the emphasis is on how an old system's data will be converted into the new IS and how functional transition from the old system to the new may be achieved.

Probably, change termination and idea percolation are two sides of the coin. If an IS can be said to have progressed into the change termination sub-phase, the organization must have come to the conclusion that this IS can no longer be acceptably enhanced through change refinements. This state of affairs must fertilize thoughts about IS potentials and opportunities, the heart of idea percolation.

The problem is that many organizations do not have a conscious understanding of how far in the usage "life-cycle" their ISs have progressed. Because they do not practice a systematic and strategic evaluation of their IS portfolio, they do not evaluate IS quality in an appropriate manner. This scenario leads to a very jumpy and abrupt transition from the change termination phase to a new idea percolation phase.

The IS Innovation Framework Defined

The framework presented in Figure 1 shows the key issues and key issues structure.

IS innovation processes may vary substantially in regard to the

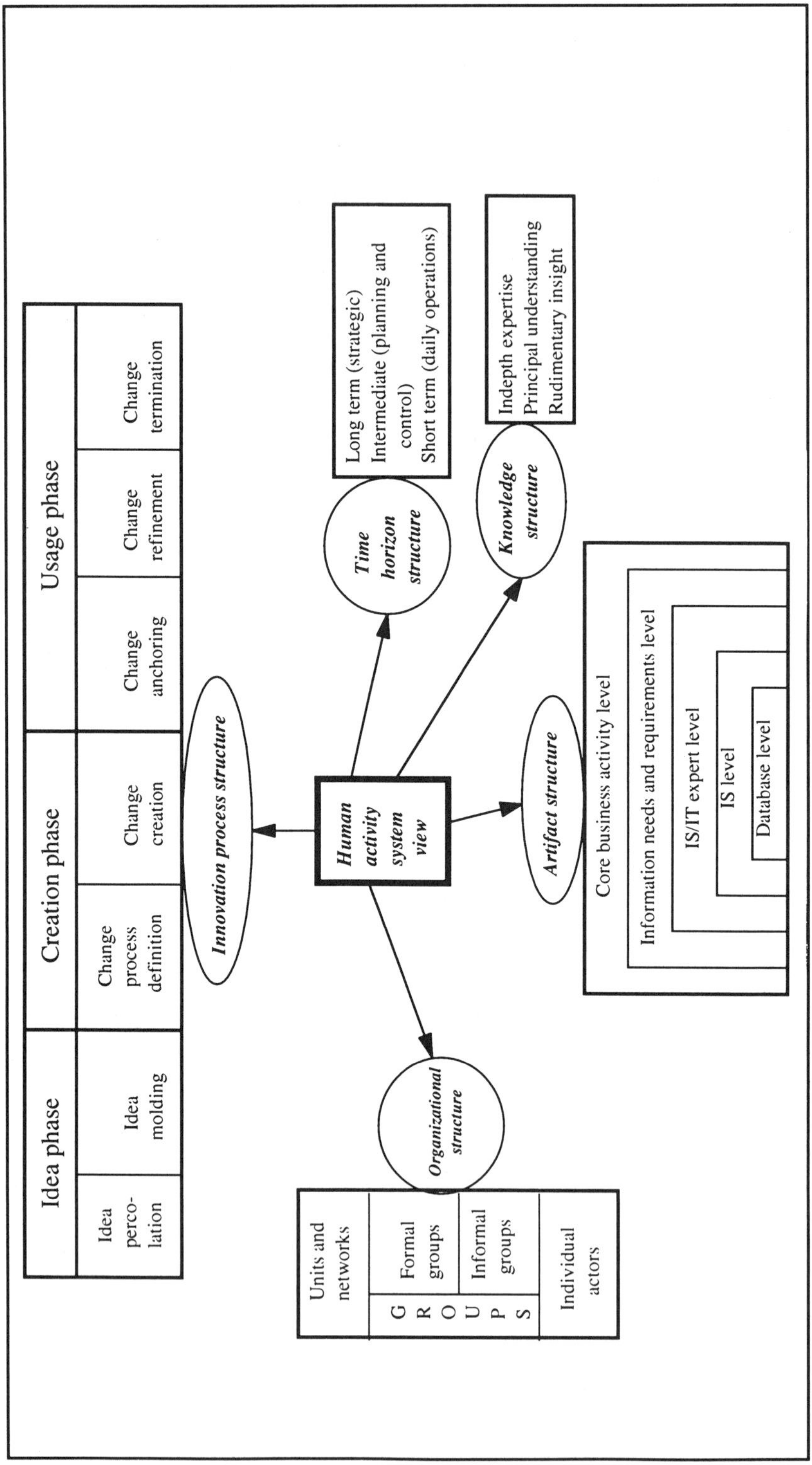

Figure 1: The IS Innovation Framework: Key Issues and Key Issues Structure

particular set of key issues and structural parts involved. For example, a brainstorming activity in an IS strategy project is the idea percolation in a formalized, top-down IS innovation process, the time horizon may be a mix of long term, intermediate and short term issues, within the knowledge structure it may combine indepth expertise as well as principal knowledge, within the organizational structure the participants form a contemporary formal group, and among the elements in the artifact structure core business concerns and information needs and requirements may be the main focus. Because research findings directed at particular interactions often are overlapping or inconclusive, the framework does not include lines among key issues and their structure.

The framework can be used as a checklist to map the elements actors include in their innovation undertakings and increase the awareness of IS innovation aspects not yet considered. It can be used as a vehicle for identifying actors' perceptions of key issues and key issues structure.

Suggested Future Trends Within IS Innovation

The framework in Figure 1 can also serve as a vehicle to develop integrated models for a single IS innovation process and multiple IS innovations.

Single IS Innovation Process

As a starting point, let us consider a particular application (for example, a payroll system) and its innovation process. Every one of the five key issues are relevant in this context. However, since an integrated model presented on paper only can represent three key issues and their structure, a choice has to be made. This choice depends on what the actors involved regard as the most important key issues. My own understanding and experience with systems development efforts, leads me to suggest that most actors would select the innovation process structure, the organizational structure and the artifact structure. Parallelling the partial derivation technique in mathematics, the key issues of time horizon and knowledge are treated as constants, see Figure 2.

This model is similar to traditional "systems development" models in respect to including phases. However, while most systems development models contain the detailed phases (for example, information analysis, technical analysis and implementation) within the creation sub-phase, Figure 2 includes macro oriented phases from application birth (idea percolation) to termination. The innovation model may help actors understand the application in its complete IS innovation

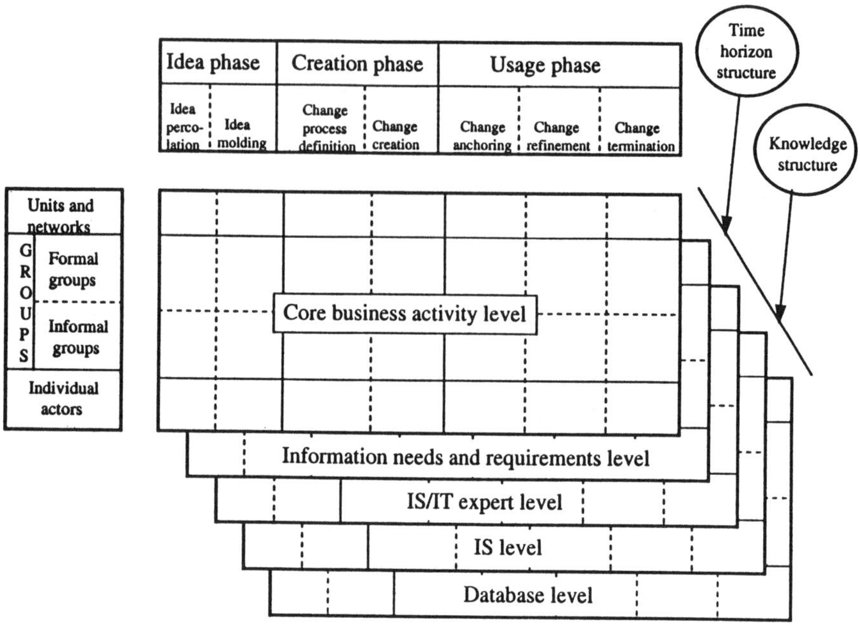

Figure 2: Model for a Single IS Innovation Process

life-cycle.

Most systems development models describe in depth the tasks that should be carried out within each detailed phase. The innovation model presented here does not contain such tasks but shows the interactions among innovation process structure, organizational structure and artifact structure. Each cell in the model represents one specific combination of the three key issues. Traditional systems development tasks can be mapped into the innovation model, showing a way to carry out the analysis for a specific cell or how integration among cells can be achieved. The innovation model can be used to check the completeness of a specific systems development approach and point to areas it does not cover.

Multiple IS Innovations

The distinction made between the creation phase and change refinement in the usage phase may not be as distinct as portrayed; that is, an innovation process may not be linear. Additionally, multiple IS innovation processes may unfold simultaneously over time.

The Validity of Linear IS Innovation Models

In most cases, systems development projects develop through a highly iterative and recurring process (Weinberg, 1992). More often than not, planned sub-phases are never executed and other sub-phases critical to successful project completion emerge because of unforseen needs or personal preference. The emerging sub-phases may be intuitively recognized and carried out without ever becoming a formally recognized part of project planning and control.

The proposition that radical IS innovation occurs in the creation phase and that only incremental IS innovations are possible in the usage phase may be questionable. Through increased quality management and IT tool flexibility, the ability to make radical changes to an IS currently in use may be possible to the degree that it would be difficult to differentiate between the degree of change introduced through the initial creation of an IS and the degree of enhancements added to the IS system in the usage phase. In principal terms, the challenge is to build systems that can suggest new directions for action based on detection of shifts in the change rate (Walls, Widmeyer, and El Sawy, 1992). A signal that this may be the case is the increased focus on concept and tool selection in the creation phase, rather than perceiving information requirements elicitation and programming as the most important systems development issues (Dekleva, 1992).

On a strategic level, dividing the IS innovation process into a creation phase and a usage phase might be even more problematic. Van de Ven and Poole (1995) show that a business strategy may develop in four distinct ways (life-cycle, teleological, dialectical and evolutionary) or in any combination of these four. Among the alternatives, only the life-cycle approach is a more traditional task based phase-linear model. A recent *Business Week* report on business strategy planning among major US Fortune 500 companies stated that the business strategy process is no longer viewed as a top-down logical planning and control activity (Byrne, 1996). CEOs understand that multiple business strategies emerge within a system where power, responsibility and knowledge are distributed throughout the corporation.

If IS strategy is an integral part of the business strategy (as suggested by Morton, 1991; Banker et al., 1993; Davenport, 1993), then we would expect that an IS strategy develops like a business strategy. Apparently, this is not the case. IS professionals talk about IS and business strategy integration but forward phase-based linear models for IS strategy development that are not a plausible extension of business strategy evolution. Ciborra (1993) found that most IS strategies were extensions from major applications, an observation that indicates that IS strategies are not developed in accordance with the dominant linear IS strategy thinking.

The IS innovation processes unfolding in the change refinement sub-phase within the usage phase may be more complex and more important than previously thought. The model presented in Figure 3 reflects this on the strategic as well as application level.

Simultaneous IS Innovation Processes

In regard to the link between IS strategy and application development, the common view is that the outputs from the strategy process (for example, the IS/IT infrastructure, development policies, project prioritation, and educational strategy) direct how applications should be developed and used. Although highly desired, a smooth integration among the strategic, tactical and operational levels has always been problematic (Keen, 1991).

From the beginning of IS strategy completion until the completion of application development, business needs and strategy may change. A company may have hired new people with more knowledge and different agendas, and key people may have left. IT may have improved, creating new possibilities. Although members of an organization may understand that substantial changes have taken place, they may not see the need or have the energy to initiate another round with IS strategic thinking.

These considerations are incorporated in Figure 3.

The stippled line from the strategic level change creation sub-phase to the IS level and database level (see Figure 1 - artifact structure) indicates that this connection may not be as dominant as traditional IS strategy models portray it. Additionally, the two-sided arrow lines from the strategic level change refinement sub-phase and the application level change creation sub-phase to the IS level and database level, denotes that the coupling between the strategic level and the application level is complex and may be unidirectional. The line pointing from the IS level and database level to the application level change refinement sub-phase indicates that the principal IS/IT design features determine the degree of change that can be made.

We must expect that the more organizations adapt organic organizational principles, the more difficult it becomes to develop and maintain appropriate couplings among these elements. We critically need more research into strategic and application IS innovations and how these levels interact.

Conclusions

The objective to develop a holistic view of IS innovation may be an undertaking that even Sisyphus would recognize. However, the increased technological and business pressures to create integrated IS within less time, require conceptual IS innovation understanding to guide research and practice.

Based on the systems thinking "human activity system" view, the first challenge was seen as identifying key issues. It is obvious that the chosen five key issues are exactly that - and five is not a magical number. Although they were derived from a literature study, it cannot be said that they emerged from a complete and rigorous investigation.

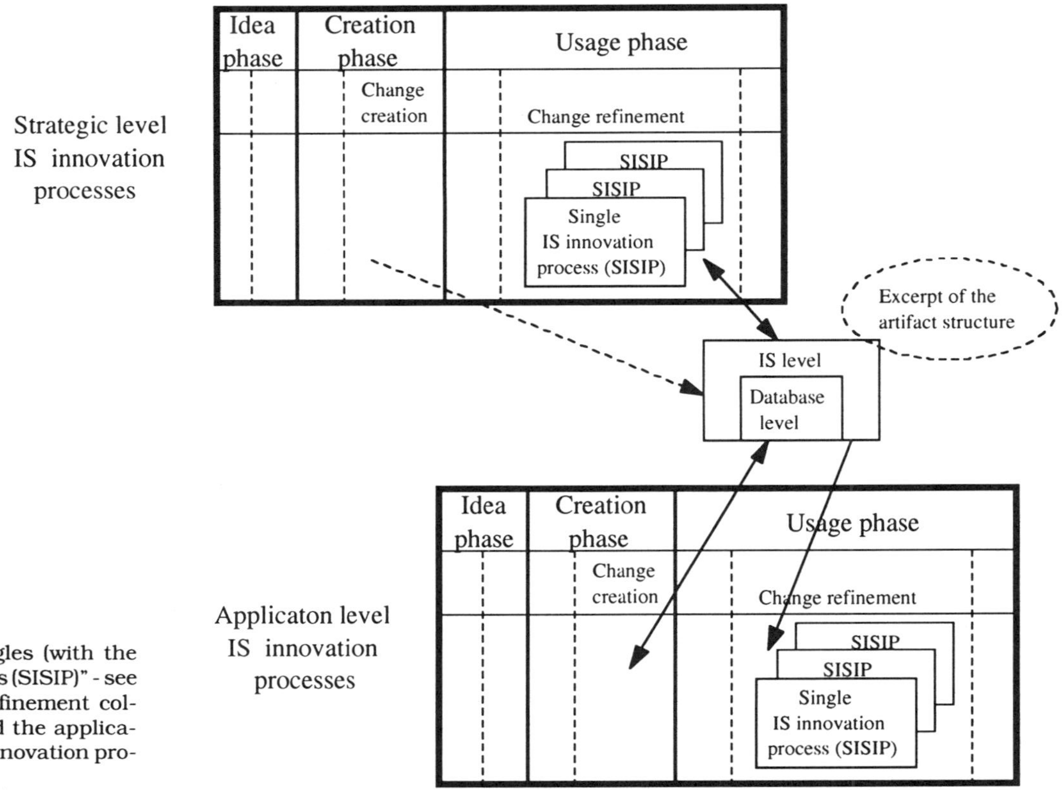

Notes: The inclusion of rectangles (with the text "Single IS innovation process (SISIP)" - see Figure 2) within the change refinement columns on the strategic level and the application level denote that multiple innovation processes can occur.

Figure 3: Multiple IS Innovation Processes

They mainly serve as an illustration of the topic areas a set of key issues might cover.

The second point made was that people interpret key issues in specific ways. The key issues structures presented in this chapter are examples of how key issues might be understood. Despite the fact that our interpretations determine later action, these structures are frequently not made explicit. The effect of unclarity at this level is that the critical thinking behind an IS innovation process cannot be properly communicated and shared. Consequently, development efforts run the risk of uncoordinated or unfocused intuitive actions. It is a sobering fact that people will do as best they can when necessary.

The interactions among key issues and their structural elements are too numerous to be represent in one model. Therefore, the framework in Figure 1 was used to develop a partial innovation model for three of the five key issues (see Figure 2). The model's strength is that it is logically derived from the framework. Rather than including detailed phases and tasks, as is the case in most traditional IS application development models, the IS innovation model shows the possible interactions among the IS process structure, the organizational structure and the artifact structure. The innovation model is robust because it does not depend on differences in methodological scope or technological specialties.

An obvious limitation is that the innovation phase structure displays its phases as a linear sequence (from idea percolation to change termination). As discussed, there is ample evidence that many IS innovation processes do not progress in this manner. The models presented here are macro level analyses. Taking theater plays as an analogy, the present material addresses the thinking behind the determination of the play plot and staging. The micro level, that is, the detailed script and actor interactions are not discussed.

Since the innovation model incorporates the innovation process from birth to termination, and not just the creation phase as in most systems development models, it provides an opportunity to discuss the more complex phenomenon of the combination of innovation in the creation phase and innovation in the change refinement sub-phase, as shown in Figure 3. In a strategic IS innovation setting, versus an application development effort, the relative importance of innovation in the creation phase and innovation in the change refinement sub-phase may vary. The question was also raised whether stand-alone IS strategy processes play the vital role many authors claims. The forces and sources determining the IS strategy that organizations should follow may be more varied and complex than traditionally thought.

This chapter has suggested that more attention should be paid to key issues, key issues structure and the IS innovation process view that logically can be derived from these. The IS innovation view must have power to portray single as well as multiple IS innovations. These issues are exciting research questions that need our attention.

References

Abernathy, W. J. and Utterback, J. M. (1982). "Patterns of Industrial Innovation," in Tushman, Michael L. and Moore, William L.(Eds), *Readings in the Management of Innovation.* Boston, MA: Pitman.

Albrecht, T. L. and Ropp, V. A. (1984). "Communicating about Innovation in Networks of Three U.S. Organizations," *Journal of Communication*, Summer, pp. 78-91.

Allen, T. J. and Morton, M. S. S. (1994). *Information Technology and the Corporation of the 1990s.* New York, NY: Oxford University Press.

Banker, R. D., Kauffman, R. J. and Mahmood, M. A. (Eds) (1993). *Strategic Information Technology Management: Perspectives on Organizational Growth and Competitive Advantage.* Harrisburg, PA: Idea Group Publishing.

Beath, C. M. (1991). "Supporting the Information Technology Champion," *MIS Quarterly*, 15(3), September, pp. 355-372.

Beath, C. M. and Ives, B. (1989). "The Information Technology Champion: Aiding and Abetting, Care and Feeding," in Gray, P., King, W. R., McLean, E., and Watson, H. (Eds), *The Management of Information Systems.* Chicago, IL: The Dryden Press.

Boynton, A. C., Jacobs, G. C. and Zmud, R. W. (1992). "Whose Responsibility is IS Management?," *Sloan Management Review*, Summer, pp. 32-38.

Brancheau, J. C. and Wetherbe, J. C. (1990). "The Adoption of Spreadsheet Software: Testing Innovation Diffusion Theory in the Context of End-User Computing," *Information Systems Research*, 1(2), pp. 115-143.

Byrne, J. A. (1996). "Strategic Planning," *Business Week*, August 26, pp. 46-52.

Cash, J. I. Jr., McFarlan, F. W. and McKenney, J. L. (1988). *Corporate Information Systems Management: The Issues Facing Senior Executives.* Homewood, IL: Irwin.

Checkland, P. (1981). *Systems Thinking, Systems Practice.* Chichester, England: John Wiley & Sons.

Checkland, P. and Scholes, J. (1990). *Soft Systems Methodology in Action.* Chichester, England: John Wiley & Sons.

Ciborra, C. U. (1993). *Teams, Markets and Systems: Business Innovation and Information Technology.* Cambridge, UK: University Press.

Clemons, E. K. and Harris, S. E. (Eds) (1991). "The Strategic Use of Information Systems (special issue)," *MIS Quarterly*, 15(3), September.

Clemons, E. K. and Row, M. C. (Eds) (1992). "Strategic and Competitive Information Systems (special issue)," *Journal of Management Information Systems*, 9(2), Fall.

Cooper, R. B. and Zmud, R. W. (1990). "Information Technology Implementation Research: A Technological Diffusion Approach," *Management Science*, 36(2), February, pp. 123-139.

Cotterman, W. W. and Senn, J. A. (Eds). (1992). *Challenges and Strategies for Research in Systems Development.* Chichester: John Wiley & Sons.

Davenport, T. H. (1993). *Process Innovation: Reengineering Work through*

Information Technology. Boston. MA: Harvard Business School Press.

Davis, G. B. (1989). "Caution: User-Developed Systems Can Be Dangerous to Your Organization," in Nelson, R. Ryan (Ed), *End-User Computing: Concepts, Issues and Applications.* New York, NY: John Wiley & Sons.

Davis, G. B. and Olson, M. H. (1985). *Management Information Systems: Conceptual Foundations, Structure, and Development*, New York, NY: McGraw-Hill Book Company.

De Marco, T. (1978). *Structured Analysis and System Specification.* New York, NY: Yourdon Inc.

Dearden, J. (1987). "The Withering Away of the IS Organization," *Sloan Management Review,* Summer, pp. 87-91.

Dekleva, S. M. (1992). The Influence of the Information Systems Development Approach on Maintenance," *MIS Quarterly*, 16(3), September, pp.355-372.

Edwards, C. (1984). "Information Systems Maintenance: An Integrated Perspective," *MIS Quarterly*, 8(4), December, pp. 237-256.

Ettlie, J. E. (1983). "A Note on the Relationship Between Managerial Change Values, Innovative Intentions, and Innovative Outcomes in Food Sector Firms," *R & D Management*, 13(4), pp. 231-244.

Feldman, S. P. (1986). "Management in Context: An Essay on the Relevance of Culture to the Understanding of Organizational Change," *Journal of Management Studies*, 23(6), November, pp. 587-607.

Frost, P. J., Moore, L. F., Louis, M. R., Lundberg, C. C. and Martin, J. (Eds) (1991). *Reframing Organizational Culture.* Newbury Park, CA: Sage Publications.

Gillies, A. C. (1992). *Software Quality: Theory and Management.* London, England: Chapman & Hall.

Hage, J. and Dewar, R. (1973). "Elite Values Versus Organizational Structure in Predicting Innovation," *Administrative Science Quarterly*, September, pp. 279-290.

Henderson, B. (1979). *Henderson on Corporate Strategy.* Cambridge, MA: Abt Books.

Henderson, J. C. and Treacy, M. E. (1986). "Managing End-User Computing for Competitive Advantage," *Sloan Management Review*, Winter, pp. 3-14.

Hufnagel, E. and Birnberg, J., G. (1989). "Perceived Chargeback System Fairness in Decentralized Organizations: An Examination of the Issues," *MIS Quarterly*, 13(4), December, pp. 415-429.

Janis, I. L. (1971). "Groupthink," *Psychology Today*, November, pp. 378-384.

Johnston, H. R. and Carrico, S. R. (1988). "Developing Capabilities to Use Information Strategically," *MIS Quarterly*, 12(1), March, pp. 37-47.

Kanter, R. M. (1983). *The Change Masters: Innovation & Entrepreneurship in the American Corporation.* New York, NY: Touchstone.

Keen, P. G. W. (1991). *Shaping the Future.* Cambridge, MA: Harvard Business School Press.

Kim, C. and Westin, S. (1988). "Software Maintainability: Perceptions of EDP Professionals," *MIS Quarterly*, 12(2), June, pp. 167-185.

Kling, R. (1980). "Social Analyses of Computing: Theoretical Perspectives of Recent Empirical Research," *ACM Computing Surveys*, 12(1), March, pp. 61-110.

Kohli, A. K. and Jaworski, B. J. (1990). "Market Orientation: The Construct, Research Propositions, and Managerial Implications," *Journal of Marketing*, 54(April), pp. 1-18.

Kotler, P. (1991). *Marketing Management: Analysis, Planning, Implementation, and Control.* Englewood Cliffs, NJ: Prentice-Hall International, Inc.

La Belle, A. and Nyce, H. E. (1987). "Wither the IT Organization," *Sloan Management Review*, Summer, pp. 75-85.

Lakatos, I. (1970). "Falsification and the Methodology of Scientific Research Programs," in Lakatos, Imre and Musgrave, Allen (Eds), *Criticism and the Growth of Knowledge.* Cambridge, England: Cambridge University Press, pp. 91-196.

Larsen, T. J. (1993a). "Middle Managers' Contribution to Implemented Information Technology Innovation," *Journal of Management Information Systems*, 10(2), Fall, pp. 155-176.

Larsen, T. J. (1993b). "Globalization of Information Technology: Here to Stay or a Passing Fad?" in Khosrowpour, Mehdi and Loch, Karen (Eds), *Global Information Technology Education: Issues and Trends.* Harrisburg, PA: Idea Group Publishing, pp. 122-155.

Laudon, K. C. and Laudon, J. P. (1996). *Management Information Systems: Organization and Technology.* Upper Saddle River, NJ: Prentice Hall International, Inc., Fourth Edition.

Lawrence, P. R. and Dyer, D. (1983). *Renewing American Industry: Organizing for Efficiency and Innovation.* New York, NY: The Free Press.

Leavitt, H. J. (1965). "Applying Organizational Change in Industry: Structural, Technological and Humanistic Approaches," in March, J. G. (Ed), *Handbook of Organizations.* Illinois: Rand McNally.

Lederer, A. L. and Sethi, V. (1988). "The Implementation of Strategic Information Systems Planning Methodologies," *MIS Quarterly*, 12(3), September, pp. 445-461.

Linstead, S. and Grafton-Small, R. (1992). "On Reading Organizational Culture," *Organization Studies*, 13(3), pp. 331-355.

Moore, G. C. and Benbasat, I. (1991). "Development of an Instrument to Measure Perceptions of Adopting an Information Technology Innovation," Information Systems Research, 2(3), September, pp. 192-222.

Morton, M. S. S. (Ed) (1991). *The Corporation of the 1990s: Information Technology and Organizational Transformation.* New York, NY: Oxford University Press.

Mowshowitz, A. (1976). *The Conquest of Will: Information Processing.* Reading, MA: Addison-Wesley.

Mumford, E. (1983). *Designing Human Systems For New Technology.* Manchester, England: Manchester Business School, Printed by Wright's (Sandbach) Ltd.

Narver, J. C. and Slater, S. F. (1990). "The Effect of a Market Orientation on Business Profitability," *Journal of Marketing*, 54(October), pp. 20-35.

Nelson, R. R. (Ed) (1989). *End-User Computing: Concepts, Issues, and Applications.* New York, NY: John Wiley & Sons.

Olle, T. W., Hagelstein, J., Macdonald, I. G., Rolland, C., Sol, H. G., Van Assche, F. J. M. and Verrijn-Sturart, A. A. (1988). *Information Systems Methodologies: A Framework for Understanding.* Wokingham, England: Addison-Wesley Publishing Company.

Pettigrew, A. (1985). *The Awakening Giant: Continuity and Change in ICI.* Oxford, England: Basil Blackwell.

Peters, T. J. and Waterman, R. H., Jr. (1982). *In Search of Excellence: Lessons from America's Best-Run Companies.* New York, NY: Warner Books.

Porter, M. E. (1985). *Competitive Advantage: Creating and Sustaining Superior Performance*, New York, NY: The Free Press.

Porter, M. E. and Millar, V. E. (1985). "How Information Gives You Competitive Advantage," *Harvard Business Review*, July/August, pp. 149-160.

Rogers, E. M. (1983). *Diffusion of Innovations.* New York, NY: The Free Press.

Schein, E. H. (1980). *Organizational Psychology.* Englewood Cliffs, NJ: Prentice-Hall, Inc.

Snow, T. A. and Couger, J. D. (1991). "Creativity Improvement Intervention in a Systems Development Work Unit," *Proceedings of the Twenty-Fourth Annual Hawaii International Conference on Systems Sciences*, IV, pp. 412-418.

Swanson, E. B. (1994). "Information Systems Innovation Among Organizations," Management Science, 40(9), September, pp. 1069-1092.

Swanson, E. B. (1988). *Information Systems Implementation: Bridging the Gap Between Design and Utilization.* Homewood, IL: Richard Irwin, Inc.

Swanson, E. B. and Beath, C. M. (1989). *Maintaining Information Systems in Organizations.* Chichester, England: John Wiley & Sons.

Swanson, K., McComb, D., Smith, J. and McCubbrey, D. (1991). "The Application Software Factory: Applying Total Quality Techniques to Systems Development," *MIS Quarterly*, 15(4), December, pp. 567-579.

Thompson, J. D. (1967). *Organizations in Action.* New York, NY: Mc Graw-Hill.

Van de Ven, A. H. (1992). "Suggestions for Studying Strategy Process: A Research Note," *Strategic Management Journal*, Special Issue on "Strategy Process Research," 13(Summer), pp. 169-188.

Van de Ven, A. H. (1986). "Central Problems in the Management of Innovation," *Management Science*, 32(5), pp. 590-607.

Van de Ven, A. H. and Poole, Marshall Scott (1995). "Explaining Development and Change in Organizations." *Academy of Management Review*, 20(3), July, pp. 510-540.

Vowler, J. (1991). "A Risky Investment that Business Has To Make," *Computer Weekly*, 6 (November).

Walls, J. G., Widmeyer, G. R. and El Sawy, O. (1992). "Building an Information System Design Theory for Vigilant EIS," *Information Systems Research*, 3(1), March, pp. 36-59.

Weinberg, G. M. (1992). *Quality Software Management: Volume 1: Systems Thinking.* New York, NY: Dorset House Publishing.

Westerman, J. and Donoghue, P. (1989). *Managing the Human Resource.* New York, NY: Prentice Hall.

Wetherbe, J. C. (1993). "Four-Stage Model for MIS Planning Concept, Techniques, and Implementation," in Banker, Ravjiv D., Kauffman, Robert J., and Mahmood, Mo Adam (Eds), *Strategic Information Technology Management: Perspectives on Organizational Growth and Competitive Advantage.* Harrisburg, PA: Idea Group Publishing, pp. 83-117.

Wysocki, R. K. and Young, J. (1990). *Information Systems: Management Principles in Action.* New York, NY: John Wiley & Sons.

Zmud, R. W. (1984). "An Examination of 'Push-Pull' Theory Applied to Process Innovation in Knowledge Work," *Management Science*, 30(6), pp. 727-738.

CHAPTER
18

Complexity, Chaos and Software Development: Issues in Innovation and Diffusion

Eugene G. McGuire
American University

The growing global economy for software products and services has driven many organizations to redefine the very paradigms by which they operate. A larger and more demanding customer base, the necessity for reduced cycle time, the growing complexity of software systems, and the increased mission-critical status of these software systems are only some of the factors that are pressuring organizations to deliver higher quality, more complex software products in the face of an ever turbulent and competitive global environment.

In contrast to the activities associated with most functional areas of organizations, successful information systems (IS) development does not depend on achieving a stable equilibrium between all the various task, process, and infrastructure requirements. Instead, rapid technological change demands managing within an environment best characterized as *bounded instability*.

According to Waldrop (1992), many organizations are operating at or approaching the "edge of chaos" as the complexity of their environment increases and as the metabolic rate of their functions and processes escalates to keep pace. "The edge of chaos is where new ideas and innovative genotypes are forever nibbling away at the edges of the status quo, and where even the most entrenched old guard will eventually be overthrown....The edge of chaos is the constantly shifting battle zone between stagnation and anarchy, the one place where a complex system can be spontaneous, adaptive, and alive." (Waldrop 1992, p. 12). Successfully operating at the "edge of chaos" requires first recognizing the environmental, organizational, structural, and process elements that lead to this state and then identifying appropriate management practices to balance these elements.

Chaos is not predictable. It depends on or arises from feedback. Small initial differences lead to immense later differences. One of the big lessons of chaos is that a slight change in circumstances can lead to a massive change in the orderliness and predictability of the results.

Chaos is a mathematical concept that can be described as deterministic randomness. The title of a paper by mathematicians Yorke and Li (1975) is the source of current usage of the term. Pool (1989) explains that chaos appears in an otherwise well-behaved physical system when it is "pushed"—when a parameter is increased so high that irregular behavior sets in. At some point the system behavior becomes so complicated that it is unpredictable, but not completely random. Alternately, this behavior can be described as sensitivity to initial conditions. Chaotic systems do obey certain simple equations, but the overall behavior of the system is so complicated that it appears random. In fact, long term behavior is impossible to predict because any uncertainty in initial conditions increases exponentially over time.

Michaels and Marshall (1992) describe chaos as the science of complex systems, i.e., those that exhibit apparent randomness and non-linear behavior. The new paradigm for complex systems suggests that they survive only when they are permitted to be in a state of constant change. This is in direct contrast to the old systems development thinking of seeking balance, harmony, equilibrium, and steady state after a period of turbulence.

McMaster (1996) defines complexity as being concerned with the phenomena that occur when systems are beyond linear understanding, beyond simple cause and effect, and beyond particular analysis, but have not yet spilled over into the area of chaos, i.e., these phenomena are occurring at the "edge of chaos" where patterns can be seen and even understood but the interplay of elements cannot be reduced to the individual elements.

Seeing phenomena through the lens of complexity continuously brings more of the unknown and unmanageable into the realm of adaptation and innovation. Complexity is a way of seeing things that transforms the chaotic and complicated into something simple and amenable to understanding.

An organization is a complex adaptive system. It is capable of learning and adapting to its environment. Of course, not all organizations do this equally well. It is especially difficult for organizations in rapidly changing environments, e.g., software development organizations, to learn, adapt, innovate and diffuse the learning and innovations quickly enough to remain continuously successful and competitive.

This chapter suggests that traditional ways of looking at technology innovation and diffusion need to be replaced by new approaches that strongly and directly address issues of turbulence, rapid change, complexity, contradiction, and chaos. These new approaches will be particularly useful when applied to organizations heavily dependent on or influenced by information systems technology and even more particularly, software development organizations.

Organizational Complexity

McMaster (1996) notes that between 1970 and 2000 there will be an approximately 500 million-fold increase in the ability to access, move, create, and process information. This shift is made possible by computers and includes increasingly extensive and effective use of information technology, but the shift is not really about computers. It is about information itself and includes communication technology, organizing work practices, organizing information, accessing information and generating information.

Compared to the increase in the capacity to move, process and store information, methods of organizing and working together have not been adequately transformed to meet that increase. This era requires new capacities for transforming chaotic data into useful information and new levels of innovation capable of designing practical applications for that information.

Tushman and O'Reilly (1997) state that an organization's long-term success in today's environment depends on the ability to manage different kinds of innovation over time, to lead a range of innovations that must be simultaneously developed within the firm. They identify three fundamentally different kinds of innovation that need to co-exist: incremental, architectural (innovation that reconfigures existing technology), and discontinuous. They note that while incremental innovation can be managed within the existing organization, architectural and discontinuous innovation must be coupled with more radical organizational change.

These three types of innovations have contradictory requirements and the challenge for organizations is to succeed at all three types of innovations. Complexity abounds in such an approach. This perspective redirects organizational attention away from a particular orientation (e.g., market push versus technology pull) and towards a series of contrasting innovations. The contradictions inherent in the multiple types of innovation create conflict and dissent between organizational units as each mode of innovation requires a different type of support structure and fosters a different organizational culture. The challenge is now to create co-existing highly differentiated and highly integrated organizations that exist in complex, "edge of chaos" environments.

Chakravarthy (1997) asserts that in industries related to information and communication, traditional industry boundaries have disappeared and that for firms competing in these industries, the number of competitive forces they face has expanded and technological innovations have accelerated the rate of change. He proposes a new strategy framework for coping with this turbulence which he claims is characterized by multiple, unpredictable equilibrium points. Frequent technological innovation quickly makes competitive advantages obsolete.

Tapscott and Caston (1993) are among the many researchers who contend that a paradigm shift is underway that will revolutionize the management of IS. These paradigmatic shifts occur in organizations as they begin treating information as an enterprise resource and initiating systems development and integration as part of their strategic planning. Although easy to articulate, the practical issues involved with linking information systems development to organizational strategic change are immense. Cause and effect linkages are influenced by a large number of internal and external variables and linear information and feedback models quickly become limited in usefulness.

A new conceptual model is needed that facilitates understanding of how small, loosely structured, and conflicting changes have long-term, widespread consequences that can neither be known nor intended at the start of a systems development effort. The emerging science associated with chaos theory has served well in other disciplines for this purpose. Chaotic systems are highly sensitive to small changes and amplify tiny fluctuations. It is important to remember that chaos in the scientific sense is not utter confusion or disorder. Rather, it describes specific kinds of instability within recognizable patterns. This appears to be an accurate description of many software development projects today and has implications for how innovations are adopted and diffused throughout the software development environment.

Chaos Theory and Information Systems

Ramackers and Verrijin-Stuart (1991) suggested that information systems were subject to two types of changes. First order dynamics—well represented by conventional models that provide functional specifications of time-dependent data and second order dynamics—changes in the organizational environment that lead to new information requirements. Until now, developers have relied on the tools of system maintenance in attempts to address second order dynamics. This distinction is generally not recognized and neither is the reality that these second order changes further necessitate new first order descriptions. Their results support the call for a new perspective that allows IS resources to function as second-order, dynamic architectural networks rather than first order, engineered hierarchical structures.

There exists an abundance of literature that attempts to predict the future of the information systems discipline (Constnatine, 1993; 1995; Glass, 1995; Humphrey, 1997; Rasch and Tosi, 1992). Common themes include: systems developers rapidly approaching their limits in terms of handling the associated complexity, managers increasingly disappointed by the lack of contribution of information systems to profits, the chaos of personal computing, and the overall failure of existing approaches for developing successful systems. Traditional information systems development methodologies seek balance, harmony, equilibrium, and steady state in response to periods of turbulence. Newer object-oriented approaches presume that the way to understand information systems to begin by keep isolating the basic data ingredients.

DeMarco and Lister (1987) contend that it is basic human nature to replace chaos with order, but advocate constructive reintroduction of small amounts of disorder to instill energy into information systems project work. An even more direct link between chaos theory and information can be established using Shannon and Weaver's (1963) classic text, which introduced information theory as a way of measuring the substance transmitted over communication lines and radio waves. Information theory enabled an understanding of how noise in the form of random errors interfered with the flow of binary digits. The relevant concepts included redundancy (a predictable departure from randomness) and entropy (from thermodynamics—any system tends to slide toward a state of increasing disorder). Thus, chaos theory is directly associated with the creation and distribution of information.

Pool (1989) contends that the appealing aspect of chaos is that it offers a way to understand complicated behavior as something that is purposeful and structured rather than extrinsic and accidental.

Innovation Diffusion in Software Development

A recent study by Zelkowitz (1996) examines how software engineering differs from other engineering disciplines in following technology transfusion models. In studying NASA software development projects, Zelkowitz concluded that:

- Infusion mechanisms do not address software engineering technologies well because software engineering is more process-driven and less product-oriented and more dependent on programmer expertise than upon implemented technology;
- Quantitative data is crucial for understanding software development

processes but few organizations collect effective data on their development practices because measurement and experimentation are not part of the software development culture;
- technology infusion is not free and in software development projects these costs usually have to be borne by project development budgets which greatly increases the risk of a cost overrun and thereby severely inhibits using new innovations.

Zelkowitz also noted some similarities between software engineering and other domains concerning technology infusion:

- Most software professionals are resistant to change, thereby decreasing the likelihood of rapid adoption of new technology;
- Technology transfer is more than simply understanding the new technology and may involve multiple instances of training and pilot projects;
- Technology infusion is part of the total environment of the consumer organization — it does not occur in a vacuum;
- The government can have an impact on technology infusion, although more by utilizing advocates in key positions than by imposing rules mandating certain technologies;
- People contact is the main transfer agent of change and finding the appropriate advocate to act as gatekeeper or guru for the technology is a crucial component of any technology transfer mechanism;
- Timing is a critical decision and may be either a positive or negative influence on the overall technology infusion process.

He concludes that "Technology infusion today is generally ignored and left up to the individual engineer to discover what is needed and available. With today's shrinking budgets and the need to work 'better, faster, cheaper,' management needs to address this issue and help in the search for new effective technology to use."

Nonlinearity in Software Development Organizations

Nonaka and Takeuchi (1995) address the successful management of the new product development process that is characteristic of software organizations. They state that organizations must maintain a highly adaptive and flexible approach to new product development because it rarely proceeds in a linear or static manner and instead involves an iterative, dynamic, and continuous process of trial and error. They cite the software industry as best embodying this spiral approach and reference a recent study by Cusumano and Selby (1995) showing that software developers move around in phases, iteratively designing, coding, and testing as the project progresses.

In addition, Nonaka and Takeuchi believe that self-organizing project teams should oversee new-product development. Organizations should be willing to give a high degree of autonomy to such teams especially as they cope with ambiguity, fluctuation, and creative chaos in the startup phases of new-product development where little prior knowledge exists and taking initiatives and risks are necessary.

Nadler, et al (1992) views the future organization in much the same way. One of the characteristics they see becoming prevalent for organizations is that organic, adaptable teams will be the norm at all organizational levels. Not only will this provide the capability for faster response time but these teams will be relatively autonomous in comparison to previous organizational eras. Teams will have far greater latitude to be innovative and solve problems than before and management will assist rather than direct this process.

In addition, Nadler et al see an emphasis on system-level learning to be one of the cornerstones of the new organizations. This will enable organizations to greatly increase its capacity to innovate and learn because they will require members to seek out new ideas and adapt them to internal use. Increased innovation and experimentation will result in successes and failures that will be analyzed to extract knowledge that can be applied to new situations.

Leonard-Barton (1995) sees much the same thing happening to organizations. Certain types of tool implementation such as intensive codevelopment with user groups leads to the opportunities for *mutual adaptation* of both technology and user work environment. Mutual adaptation is the reinvention of the technology to conform to the work environment and the simultaneous adaptation of the organization to use the new technical system. Two major aspects of mutual adaptation are: it occurs in small and large recursive spirals of change; and, it often requires attention to all four of what Leonard-Barton identifies as the organization's core capabilities — employee skills and knowledge, physical technical systems, managerial systems, and values and norms. She integrates these two aspects of mutual adaptation into a vision of the new organization addressing all of its activities in an "iterative, return-loop nature" that is characterized by prototypes of all different kinds and a "try-it-and-learn" attitude. Nonlinearity and higher-order learning (learning beyond the current operating level and problems of the organization) become organizational norms.

The concept of organizational nonlinearity is also explored by Goldstein (1994) who describes one of the advantages of a nonlinear system as its potential for self-organization since change does not need to be imposed but just released under appropriate conditions. Linear systems, in contrast, do not have this potential. He notes that the radical organizational change that strategies such as reengineering, cycle time management, concurrent engineering, TQM, and continuous improvement call for are more congruent with the spontaneous system reorganization of nonlinear systems than with the more stable nature of linear models. He reasons that if organizations are nonlinear, then only a nonlinear strategy takes advantage of organizational nonlinearity with its evolutionary potential.

Nonlinearity is also seen as the norm in software development by Olson (1993) who argues that "highly detailed linear development processes" do not match the way real software development is done but plans and schedules are nevertheless constructed according to this model. He states that the failures in software development are the result of applying valid and significant disciplines (such as linear process models) in areas where the base assumptions of the disciplines do not match the problems that are to be solved or the organizations that are involved.

Olson believes that software development is characterized by a chaotic approach where "there's always another way. You should achieve a fit between the discipline, the problem, the people, and the organization." (p. 54). He claims that the fundamental stages of software development such as problem solving, requirements gathering, system design, user interface design, coding and data structures are all chaotic in nature as they involve multiple iterations and feedback loops that result in a maturing sequence of mental models. The progression from model to model, he states, is not linear but instead proceeds in fits and starts as software developers study and worry about problems, find solutions, and then discover new problems. These mental models, explains Olson, behave like drops of water, with change over time, internal dynamics, and feedback between cycles. Chaotic behavior, therefore, should be expected.

Managing Complex Change

The above discussion strongly indicates that traditional models of managing innovation and change within organization are quickly becoming outmoded. The literature shows a strong call for new models of change management that are more appropriate for organizations operating in complex, chaotic environments.

Orlikowski and Hofman (1997) note that traditional models of change as characterized by Lewin's three-stage model of "unfreezing, change, and refreezing" are appropriate only for organizations that are relatively stable and bounded and whose functionality is fixed. These models, however, are not appropriate in the more turbulent, flexible, and uncertain organizational and environmental conditions that often exist today.

Orlikowski and Hofman suggest that a more appropriate way of thinking about change would be an approach that reflects the "unprecedented, uncertain, open-ended, complex, and flexible nature of the technologies and organizational initiatives involved." Such a model, they note, would enable organizations to systematically absorb, respond to, and even leverage unexpected events, evolving technological capabilities, emerging practices, and unanticipated outcomes. This type of approach would accommodate and encourage ongoing and iterative experimentation, use and learning. The proposed model relies on two assumptions: first, the changes associated with technology implementations are an ongoing process; and second, not all of the technological and organizational changes made during an ongoing process can be anticipated ahead of time.

New change models suggest the need for new roles for change agents. Markus and Benjamin (1996) have recently addressed the need for an expanded definition of the traditional model of IS change agentry. They conclude that the traditional IS model wherein the IS change agent views the technology as the agent of change is rapidly becoming unviable. They further state that IS professionals should probably begin integrating aspects of the "facilitator" model of change agentry wherein the facilitator views people as the agent of change but views himself/herself as the expert in the process of change and the "advocate" model of change agentry wherein the advocate views his/her role as one of influencing others' behaviors in certain directions into their traditional IS role. They note that IS professionals need to become better organizational change agents "because change agentry will most likely become the largest and most important part of interorganizational IS work in the future" as more information technology is implemented into organizations.

Conclusion

Much has been recently written about the growing complexity and turbulence of the global business environment. Nowhere is this environment more unpredictable and ever changing than in the area of information technology. Organizations that are dependent on or influenced by information technology, particularly software development organizations, are faced with unprecedented complexity and unknowns as they operate at the "edge of chaos." New models of technology innovation and diffusion that strongly and directly address this environment would be highly appropriate and useful for this environment. In addition, new models of change agentry that are able to successfully operate in these conditions are also needed. The original, creative, and integrative thinking and work currently underway in these areas needs to be supported and expanded.

References

Chakravarthy, B. (1997). "A New Strategy Framework for Coping with Turbulence," *Sloan Management Review*, Winter 1997, pp. 69-82.

Constantine, L.L. (1993). "Work organization: Paradigms for project management and organization," *Communications of the ACM*, Vol. 36, No. 10,

October 1993, pp. 35-43.

Constantine, L.L. (1995). *Constantine on peopleware.* Yourdon Press, Englewood Cliffs, NJ.

Cusumaud, M. & Selby, W. (1995). *Microsoft Secrets.* new York: The Free Press.

DeMarco, T. & Lister, T. (1987) *Peopleware: Productive Projects and Teams.*

Glass, Robert L. (1995). *Software creativity.* Prentice-Hall, Englewood Cliffs, NJ.

Goldstein, J. (1994). *The Unshackled Organization,* Portland, OR: Productivity Press.

Humphrey, W. (1997). *Managing Technical People: Innovation, Teamwork, and the Software Process,* Reading, MA: Addison-Wesley Publishing.

Leonard-Barton, D. (1995). *Wellsprings of Knowledge: Building and Sustaining the Sources of Innovation,* Boston: Harvard Business School Press.

Markus, M.L. & Benjamin, R.I. (1996). "Change Agentry — the Next IS Frontier," *MIS Quarterly,* December 1996, pp. 385-407.

McMaster, M.D. (1996). *The Intelligence Advantage: Organizing for Complexity,* Boston: Butterworth-Heinemann.

Michaels, M. & Marshall, L. (1992). "The Chaos Paradigm."

Nadler, D.A., Gerstein, M.S, & Shaw, R.B. (1992). *Organizational Architecture: Designs for Changing Organizations,* San Francisco: Jossey-Bass Publishers.

Nonaka, I. & Takeuchi, H. (1995). *The Knowledge-Creating Company,* New York: Oxford University Press.

Olson, D. (1993). Exploiting Chaos: Cashing in on the Realities of Software Development, New York: Van Nostrand Reinhold.

Orlikowski, W.J. & Hofman, J.D. (1997). "An Improvisational Model for Change Management: The Case of Groupware Technologies," *Sloan Management Review,* Winter 1997, pp. 11-21.

Pool, R. (1989) "Is It Chaos, or Is It Just Noise." *Science.* 243 (January) : 25-28.

Ramackers, G. & Verrijin-Stuart, A. (1991). First and Second Order Dynamics in Information Systems. In *Dynamic Modeling of Information Systems,* ed. Henk G. Sol and K.M. Van Hee. Amsterdam: North-Holland.

Rasch, R.A. & Tosi, H.L. (1992). "Factors affecting software developers' performance: An integrated approach," *MIS Quarterly,* September 1992, pp. 395-409.

Shannon, C. & Weaver, W. (1963). *The Mathematical Theory of Communication.* Urbana: University of Illinois.

Tapscott, D. & Caston, A. (1993). *Paradigm Shift: The New Promise of Information Technology.* New York: McGraw-Hill, Inc.

Tushman, M.L. & O'Reilly, C.A. III (1997). *Winning Through Innovation,* Harvard Business School Press, Boston, MA.

Waldrop, M. M. (1992). *Complexity: The Emerging Science at the Edge of Order and Chaos.* New York:Touchstone.

Yorke, J. & Li, T. (1975). "Period Three Implies Chaos." *American Mathematical Monthly* 82: 985-992.1

Zelkowitz, M. V. (1996). "Software Engineering Technology Infusion Within NASA", *IEEE Transactions on Engineering Management,* Vol. 43, No. 3, August 1996.

AUTHORS

Book Editors

Tor J. Larsen received his Ph.D. in Management Information Systems from the Carlson School of Management, University of Minnesota, in 1989. At present, he is Associate Professor at the Norwegian School of Management, Oslo, Norway. Dr. Larsen's main research interest areas are innovation processes, managers' use of information, and effects from the deployment of information systems. He has published in international journals, conferences, and books. He serves as associate editor for *Management Information Systems Quarterly* and *Journal of Global Information Management.*

e-mail: Tor.J.Larsen@BI.NO

Eugene McGuire is an Associate Professor in the Computer Science and Information Systems Department at the American University in Washington, D.C. His primary research interests are in the areas of software process improvement, software team development and organizational change. He is on the editorial board of the *Journal of End User Computing* and has served as guest editor for special issues of that journal as well as *Journal of Global Information Management.* He is currently working on another book *Software Process Improvement: Concepts and Practices* scheduled for publication by Idea Group Publishing in Winter 1998.

e-mail: mcguire@american.edu

Chapter Contributors

Jonathan P. Allen is an assistant lecturer in the Department of Engineering and the Judge Institute of Management Studies, University of Cambridge. Jonathan studies the implementation and use of information systems in organizations, and the interplay between information technology and social change. He completed a Ph.D. in Computers, ORganizations, Policy, and Society (CORPS) at the University of California, Irvine.

Brian Butler holds a B.S. in Computer Science and a M.S. in Information Systems from Carnegie Mellon University. He is a doctoral candidate in the information systems group in the Graduate School of Industrial Administration at Carnegie Mellon University. His research develops communication-based perspectives on the use and impact of information technology in complex social systems. His work examines the role of social context in application of information technologies in organizations, the impact of the developing telecommunication infrastructure on commercial activities, and the use of new communication technologies for supporting education.

e-mail: bb26@andrew.cmu.edu

Colin Charlton is a Senior Lecturer in the Department of Computer Science at the University of Liverpool. He has over 25 years experience in system software and his early work focused on operating systems, distributed systems and system development tools. In 1992, Colin set up the Hewlett-Packard Worldwide Public Domain Archive, followed in 1994 by the establishment of the Connect Internet Centre. Colin is Director of both projects.

J. Roberto Evaristo is an Assistant Professor in MIS at the Daniels College of Business at University of Denver. He holds a bachelor's degree in Civil Engineering from Universidade Federal do Rio Grande do Sul, Brazil, a Master's degree in Transportation Engineering from Universidade Federal do Rio de Janeiro, Brazil, and a Ph.D. in Management Information Systems from the University of Minnesota.

Current research interests include the impact of information technology on information load (his doctoral dissertation) and cross-cultural research in information systems. Roberto has published in *Database, International Information Systems,* ICIS Proceedings, and *Competitive Intelligence Review.* He was the guest co-editor of a special issue of *Information Technology and People* on cross-cultural research in Information Systems.

Deborah Gibbons is an assistant professor of management at Georgia State University. She obtained her bachelor's degree in Psychology at the University of Washington. She earned her M.S. and Ph.D. in Organizational Behavior and Theory at Carnegie Mellon University's Graduate School of Industrial Administration. Her research program examines social relations within and between organizations, particularly where social structures overlap with knowledge transfer, technology diffusion, interpersonal influence, and control.

e-mail: mgtdeg@langate.gsu.edu

Rick Gibson is an Assistant Professor at the American University in the Department of Computer Science and Information Systems (CSIS). He completed his Ph.D. in Information Systems at the University of Maryland in 1992. His research has been focused on international computing with a particular focus on software development by offshore companies that are now competing heavily with U.S. based firms. In his consulting work during the past year, he has been helping a major telecommunications company client by evaluating the software process maturity of their software vendors. Dr. Gibson has over 15 years of software engineering experience, with a background that includes system development using both structured and object-oriented software development methodologies. He is certified by the Software Engineering Institute to conduct software capability evaluations.

e-mail: rgibson@american.edu

Chris Gittings has worked in the Department of Computer Science at the University of Liverpool since 1993. His research in parallel logic simulation and distributed systems has developed into current interests in many Internet-related issues, including novel distributed parallel languages and security. He has considerable experience in various research and development projects within Connect.

Georg Herzwurm holds a Master's degree and a Ph. D. in Business Administration from the University of Cologne, researcher and lecturer at the chair of Business Computing, University of Cologne, EOQ Quality Auditor, member of the QFD Institute, spokesman of the German non-profit organization QFD Institut Deutschland (QFD-ID), and project manager of projects in cooperation with industry about computer-aided software engineering and software quality management. His fields of research: customer-oriented software development, particularly Software Quality Function Deployment.

e-mail: herzwurm@informatik.uni-koeln.de

Arild Jansen is a Senior Lecturer at Finnmark College, where he has been teaching in informatics and doing research on the diffusion and use of IT in rural areas. He earned a Master's degree in informatics (computer science) in 1974, and is currently working on his Ph.D. Previously, he worked 14 years at the University of Oslo in system development, including the planning and construction of the Norwegian and Nordic academic networks. In addition, he also spent 6 years as head of planning studies on

IT policy matters in the Ministry of Public Administration, including running the secretariate for the Norwegian IT policy board from 1985-1990.
e-mail: arildj@ifi.uio.no

Thomas Joseph received his Ph.D. degree in Information Systems from the Anderson School at UCLA in 1996. His dissertation research focuses on spatial and linguistic modes of interaction with relational databases. His research interests include IT infrastructure, innovations, implementation, and human-computer interaction. While a doctoral student, T.J. also worked for UCLA's Administrative Information Systems division, where he coordinated the design and implementation of a campus-wide data warehouse for reporting and decision support. He holds a Master's degree in Industrial & Systems Engineering and a Bachelor's degree in Mechanical Engineering.

Elena Karahanna is an Assistant Professor in MIS at the Department of Public and Business Administration at the University of Cyprus. She holds a bachelor degree in Computer Science and a Master in Business Administration from Lehigh University, and a Ph.D. in Management Information Systems from the University of Minnesota. Current research interests include the adoption, implementation, and use of information technology innovations and the effect of media choice and use on individuals and organizations. Her work has been published in *Management Science*, *Organization Science*, and the *Journal of Management Studies*.

Paul Leng graduated in Mathematics from London University and has an MSc in Computing from Liverpool University. After working for a time in the Computing Industry, he joined the staff of Liverpool University in 1971, and he is now a Senior Lecturer in, and Head of Department of Computer Science. His research interests and publications have embraced Computer Architecture, software tools for computer and digital systems design, logic simulation, program execution analysis, and parallel and distributed computing. His work with Colin Charlton in the latter areas led to the creation of the Connect Centre in 1994, and he continues to participate strongly in the planning and work of the centre. He is a member of ACM and the British Computer Society.

Janet Little is a lecturer in the Department of Computer Science at the University of Liverpool. She is Director of the Software Technology Programme which retrains unemployed computing professionals. Her current interests are focused on establishing the Information Society on Merseyside and include activities such as Internet awareness-raising and community-based Internet skills training.

Riccardo D. Macri graduated from the University of South Australia in 1988 with a Bachelor of Applied Science in Computer Studies. That same year he began development with Doug Seeley on the *Sim*View graphically programmed simulation tool using the (then new) C++ programming language. Currently he is continuing to grow and enhance the product for InterDynamics Pty. Ltd. He also works at the Institute for Telecommunications Research, at the University of South Australia, where he provides software development support in satellite channel simulation, user interface development and codec simulation on PC systems. His interests include graphics & animation, discrete event simulation, object oriented software, audio/video/multimedia, communications/signal processing and digital audio recording.

Gregory R. Madey earned his M.S. and Ph.D. degrees in Operations Research from Case Western Reserve University. He also earned a B.S. and M.S. in Mathematics at Cleveland State University. He worked in aerospace engineering and management for over 12 years with Gould Ocean Systems (now Westinghouse), Goodyear Aerospace, and the Loral Corporation. His last position at Loral was Director of Advanced Programs and

Business Analysis. He is currently Associate Professor in the Administrative Sciences Department at Kent State University. His research interests include the Internet, expert systems, hypermedia & multimedia, chaos theory and neural computing. He has published in various journals including the Communications of the ACM, IEEE Transactions on Engineering Management, Decision Sciences, The European Journal of OR, Omega, The Journal of Intelligent Manufacturing, Computers in Personnel, The Journal of Systems Management, and The Journal of Expert Systems.
e-mail: gmadey@synapse.kent.edu

Matthew K. McGowan earned his Ph.D. in MIS and an M.B.A. from Kent State University. He also earned a B.S. in Mathematics from Bucknell University, and an M.S. in Computer Science from Bowling Green State University. He worked in industry for 12 years as a systems programmer, systems analyst, project manager, and applications development supervisor. He is currently Assistant Professor of Business Computer Systems in the Foster College of Business Administration at Bradley University in Peoria, Illinois. He teaches courses in managerial applications of personal computers, management of information technology, electronic commerce, and database management. His research interests include the diffusion of innovation theory, electronic data interchange (EDI) and electronic commerce, ethics related to information technology, and pedagogical issues of information technology. He has presented papers at several regional and national conferences.
e-mail: mmcgowan@bradley.edu

Werner Mellis, Ph. D. in Mathematics in 1980. His previous work includes four years as a research assistant at the University of Cologne; five years at Nixdorf Computer AG leading the department knowledge engineering tools, project manager of national and international research projects about knowledge-based systems, machine learning and computer integrated manufacturing, He spent four years with Daimler-Benz AG: head of the software technology department, since 1993: Professor, Chair of Business Computing at the University of Cologne. His research interests include software quality management, and software process improvement.
e-mail: mellis@informatik.uni-koeln.de

Irene Neilson is the most recent member of the Department of Computer Science, Liverpool University having joined as a lecturer in November 1995. Prior to her appointment, she worked as a member of the research staff at Strathclyde, Edinburgh and Heriot Watt Universities, Scotland in the areas of computer-based learning, natural language and graphical interfaces to knowledge bases, computer-based architectural design and dialogue control. Her current research interest centers on the effective exploitation of the infrastructure provided by the WWW in the service of collaborative learning.

Sue Newell is a lecturer at the Warwick Business School, University of Warwick, in the UK. She has a B.Sc. and Ph.D. from University of Wales, Cardiff. Her central research interest relates to understanding the diffusion and adoption of new technologies, focusing on the role of inter-organizational networks in this process. The current work is based on research being funded by the ESRC which involves a comparison of diffusion processes in four European countries. Other research interests include evaluating the effectiveness of management development programs and looking at the changing role of women in work.
email: irobsn@wbs.warwick.ac.uk

Fred Niederman currently serves as Associate Professor at the Merrick School of Business, University of Baltimore. He earned his Ph.D. in MIS at the Univserity of Minnesota in 1990. He has published in the *MIS Quarterly, Decision Science, Group Decision and Negotiation, Journal of Strategic Information Systems, Decision Support Systems, Small Group Research, The Journal of Business Communication*, and *Computer Personnel*. His research interests are in problem formulation, meeting support systems,

decision support systems, IS personnel, and global/international information systems.
e-mail:fniederman@ubmail.ubalt.edu

Ramon O'Callaghan holds the degrees of D.B.A., Harvard; M.B.A., IESE; and Electrical Engineer (Telecommunications), Polytechnic University of Barcelona. His education included further graduate courses at MIT. He worked several years at Texas Instruments as systems specialist and project manager. As researcher at the Harvard Business School, he became interested in the use of IT for competitive advantage. His doctoral thesis studied the adoption of EDI in the insurance industry in the US. He joined IESE, the business school in Barcelona, as Assistant Professor of Control and IS. Some years later, he moved to the Netherlands as Associate Professor of IS and Director of the MBA Program at Nijenrode University. Recently, he has been appointed Fortis Professor of Information Systems at Tilburg University in the Netherlands. He has worked in consulting and executive education for many European organizations. His current interests focus on the use of IT in marketing and distribution, and, more generally, on the role of IT in organizational design and organizational transformation.
e-mail: ocallaghan@nijenrode.nl

Mark Paulk is a Senior Member of the Technical Staff at the Software Engineering Institute. He has been with the SEI since 1987, initially working with the Software Capability Evaluation project. He has worked with the Capability Maturity Model project since its inception and was the project leader during the development of Version 1.1 of the CMM. He is the product manager for version 2 of the CMM and a contributor to ISO's SPICE (Software Process Improvement and Capability dEtermination) project, which is developing a suite of international standards for software process assessment. He is a senior member of the IEEE and a senior member of the ASQC.
e-mail: mcp@sei.cmu.edu

Jonathan Palmer is Assistant Professor of Management and MIS at the University of Oklahoma. Prior to joining the faculty at OU in 1995, he was director of corporate relations at The Peter F. Drucker Graduate Management Center at the Claremont Graduate School in California. His previous academic experience includes administrative positions at The Fletcher School of Law and Diplomacy and The Harvard Business School. He teaches MBA and undergraduate courses in MIS, databases, business strategy and electronic commerce. His research interests include the strategic use of information, electronic commerce, the use of information technology in retailing, and digital documents. He has been actively involved with establishing the Internet Coalition, a student, faculty and staff group exploring the use of the Internet.
e-mail: jpalmer@ou.edu

Philip Powell is Director of the Information Systems Research Unit, Warwick Business School and an ICAEW Academic Fellow. He has authored books on information systems and financial modelling, and has published in *Omega, JORS, Accounting and Business Research, Information Systems Journal, British Journal of Management, Organizational Computing, Journal of Strategic Information Systems, Journal of Management Systems*, and *Decision Support Systems* amongst others. He is associate editor of the Information Systems Journal, the Journal of Strategic Information Systems and OR Insight. His main interests are the organisational impacts of IS and IT, especially decision support systems and expert systems, and the ways such systems might be evaluated throughout the project cycle.

Maxine Robertson has a B.A. and an M.A. in Industrial Relations from University of Warwick. She was a systems analyst in industry for eight years. Formerly a Research Fellow at Warwick Business School working on the ESRC project mentioned above, she has recently taken up an appointment as Senior Lecturer at Coventry Business School. Her research interests include the role of inter-organizational networks and the process of management consultancy in innovation.

Douglas Seeley is a founding Director of InterDynamics Pty. Ltd. and the author of *SimView*. He is the designer of InterDynamics' Whole Systems methodology, and its Dynamic Decision Support and Trusted Measurements approaches. He is currently at Simon Fraser University in British Columbia, having recently left the University of South Australia where he was a Senior Lecturer. He was an Assistant Professor in Computer Science at the University of British Columbia where, in the early 1970's, he developed with his graduate students some of the first Visual Interactive Modelling systems for both discrete and continuous systems. He obtained his MASc (operations research) and PhD (cybernetics) from University of Toronto. His research interests include object-oriented software engineering, human-computer interaction, complex systems and the evolution of random digraphs. He is currently extending *SimView* as a visual modelling tool for complex systems (including their financials) and embedding this capability in tailored decision support instruments.

Dr. Dirk Stelzer holds a Masters degree and a Ph.D. in Business Administration from the University of Cologne, researcher and lecturer at the chair of Business Computing, University of Cologne, participated in several international research and consulting projects, chairman of a software process improvement network of German software suppliers, current fields of research: software quality management, software process improvement, organizational learning.
e-mail: stelzer@informatik.uni-koeln.de

Jacky Swan is a lecturer in Organizational Behavior at Warwick Business School, University of Warwick. She has a BSc. in Psychology from Sussex University and a PhD from the University of Wales, Cardiff. Her research interests and publications focus on inter-organisational networks and cognitive and political processes as they relate to innovation diffusion and appropriation. She is currently Principal Researcher on an ESRC-funded research project investigating the roles of European professional associations and IT suppliers in technology diffusion. Other research interests include understanding the nature of managerial cognitions and occupational stress.
e-mail: irobjs@wbs.warwick.ac.uk

E. Burton Swanson is Professor of Information Systems at the Anderson School at UCLA, where he also presently serves as Academic Unit Chair and Co-Director of the Information Systems Research Program. He has been at UCLA since 1974. He received his Ph.D. degree in Business Administration from the University of California, Berkeley. He has held visiting appointments in Germany, Belgium, and England. He was the founding Editor-in-Chief of the journal, Information Systems Research, 1987-92, and was earlier a co-founder of the International Conference on Information Systems (ICIS). Burt is an active member of the Southern California chapter of the Society for Information Management (SIM). His research interests are in information systems innovation, implementation, utilization, and maintenance.
e-mail: burt.swanson@ anderson.ucla.edu.

Vasilis Theoharakis is a doctoral candidate in Marketing and Strategic Management at Warwick Business School. He received an M.B.A. (with distinction) in Marketing and International Business from the Stern School of Business, New York University, an M.Sc. from Polytechnic University and a B.Eng. (cum laude) both in Electrical Engineering from the State University of New York at Stony Brook. Vasilis is presently the Product Line Manager at 3Com Corporation in Santa Clara, California responsible for the server products business line. Prior to this position, he spent over five years conducting research at the IBM T.J. Watson Research Center in the area of high speed communications. His current research interests include diffusion processes for innovative products and market modelling.
e-mail: phdvt@razor.wbs.warwick.ac.uk

Johannes N. van Dijk and **Marcel J. Jobing** are recent graduates in Applied Computer Science from the Department of Information Systems of Delft University of Technology, The Netherlands. Their primary interest is the use of simulation modelling tools for decision support. The work presented herein incorporates elements of their thesis research conducted at the University of South Australia.

James Warren is Senior Lecturer in Computer and Information Science at the University of South Australia, where he leads the Intelligent Systems research group. He received a Ph.D. in Operations Analysis from the University of Maryland. His primary research interests are the use of simulation as a tool for understanding business processes, and the application of advanced information processing techniques to the clinical patient record.
email: warren@cis.unisa.edu.au.

Veronica Wong is a Senior Lecturer in Marketing and Strategic Management at Warwick Business School. Previously she worked for Ciba-Geigy and has taught at Loughborough University Business School. She has a B.Sc. in Pharmacology and an M.B.A. from Bradford University. Her Ph.D. focused on the area of innovation management. Dr. Wong has researched and taught extensively in the areas of product innovation management, marketing strategy, planning and implementation. She has conducted extensive surveys into the marketing strategies of Japanese and Western multinationals including Sony, Hitachi, Toshiba, Matushita, Ford, General Motors, General Electric, Vickers, IBM and Hewlett-Packard. She has also worked on a consultative basis on marketing and international strategy issues with medium-sized and large corporations including Rover, Kodak, Celltech, BT, BAA and Honeywell. To date, she has over 40 publications in top research and management journals of national and international standing. At Warwick Business school, Dr. Wong is associate director of the Marketing Strategy Research Unit and associate staff member of the Technological Innovation Research Unit (TIRU). She also manages their successful distance learning MBA marketing courses and is an active contributor to their Executive Short Course programs.
e-mail: msmvw@wbs.warwick.ac.uk

INDEX